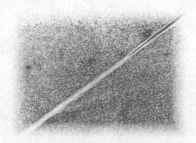

INTRODUCTION TO
ADIABATIC SHEAR
LOCALIZATION
Revised Edition

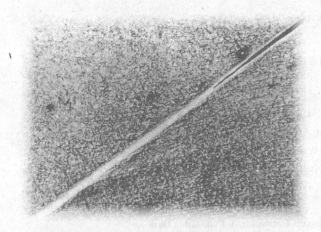

INTRODUCTION TO
ADIABATIC SHEAR
LOCALIZATION

Revised Edition

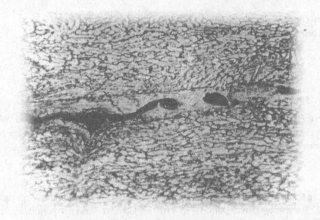

Bradley Dodd · Yilong Bai

Imperial College Press

ICP

Published by

Imperial College Press
57 Shelton Street
Covent Garden
London WC2H 9HE

Distributed by

World Scientific Publishing Co. Pte. Ltd.

5 Toh Tuck Link, Singapore 596224

USA office: 27 Warren Street, Suite 401-402, Hackensack, NJ 07601

UK office: 57 Shelton Street, Covent Garden, London WC2H 9HE

Library of Congress Cataloging-in-Publication Data
Dodd, Bradley.
    Introduction to adiabatic shear localization / Bradley Dodd, (Imperial College London, UK),
Yilong Bai, (Chinese Academy of Sciences, China). -- Revised edition.
    pages cm
    Includes bibliographical references and indexes.
    ISBN 978-1-78326-432-2 (hardcover : alk. paper) -- ISBN 978-1-78326-433-9 (pbk. : alk. paper)
    1. Shear (Mechanics) 2. Plasticity. I. Bai, Yilong. II. Title.
    TA417.6.D65 2014
    620.1'123--dc23

                                                    2014022818

British Library Cataloguing-in-Publication Data
A catalogue record for this book is available from the British Library.

Typeset by Stallion Press
Email: enquiries@stallionpress.com

Printed in Singapore

Preface

Localized deformation in the form of narrow shear bands are often observed to develop after large plastic deformations in metals, polymers and powders. Shear bands, being a form of large plastic deformation, are usually the precursors of ductile fracture. Therefore, an improved knowledge of localized deformation, including instability, shear bands, damage and fracture, is particularly significant for a wide variety of engineering topics. One example is material processing. Since the 1970s shear banding has been extensively studied by mechanical and metallurgical engineers. There is a pressing requirement in physics and engineering to summarize the knowledge gained and to assist students and researchers to apply this knowledge in their respective areas of technology.

The formation and evolution of these so-called adiabatic shear bands are a typical form of localized deformation. The importance of these tiny shear bands in materials failure in impact dynamics and metalworking has focused the interest of engineers on the phenomenon. The research that has been carried out in this area has produced a number of very interesting results. These may be helpful in further studies and applications.

We intend the book to be an explicit reference source on the topic of adiabatic shear localization. We hope that it provides a systematic description of various aspects of adiabatic shear banding, and that there is sufficient data available and case studies described to show how to apply the knowledge of adiabatic shear localization to various applications. Readers should easily be able to follow the different approaches and transfer the concepts and techniques to help solve the problems they encounter in their own areas. At the end of each chapter, where necessary, there is a section called Further Reading, which includes some more advanced references.

Y. Bai and B. Dodd

March 2014

Acknowledgements

We are highly indebted to the Chinese Academy of Sciences and the Royal Society for their support of our joint programme, which has guaranteed the completion of the book. Y. Bai also greatly appreciates the support of the National Natural Science Foundation of China and the Chinese Academy of Sciences under special grant No. 87-52.

The discussions, comments and contributions made by our friends and counterparts around the world have been invaluable in the preparation of this book. Our great thanks go to all of them.

We are very grateful to Richard Dormeval (upper photograph) and Xian Mengmai (lower photograph) for allowing us to reproduce their photographs on the front cover.

Finally, we would like to thank our wives, Fujiu Ke and Ruth, for their patience, understanding and support.

Contents

Chapter 1

Introduction

1.1. What is an Adiabatic Shear Band?

On cross-sections and surfaces of loaded materials, narrow bands of intense plastic shear strain are sometimes observed; these zones of intense shear strain are called shear bands. In dynamic processes a type of shear banding forms due to rapid local heating, resulting from the intense plastic shear deformation. In some cases the remains of phase transformations produced by the local high temperatures and rapid quenching by the cool surroundings are observed in these shear bands. In steels the shear bands often appear as distinctive white etching bands. These shear bands, formed in dynamic processes and related to rapid and local heating effects, sometimes during phase transformations, are called adiabatic shear bands (see, for example, Fig. 1.1).

The term adiabatic shear is usually attributed to the original report of Zener and Hollomon (1944). They not only called the bands adiabatic shear bands but also postulated the mechanisms that predominate in the formation of the bands and is generally accepted as the explanation today.

Their postulation is straightforward and is as follows. When plastic deformation occurs in a material, a large proportion of the work done is converted into heat. If the strain rate in a region is high, then there may not be enough time for the heat to diffuse away from the deforming zone. This causes a local thermal softening effect. If the strength loss due to thermal softening becomes greater than the increase in strength due to strain or strain-rate hardening, the plastic

1

(a)

(b)

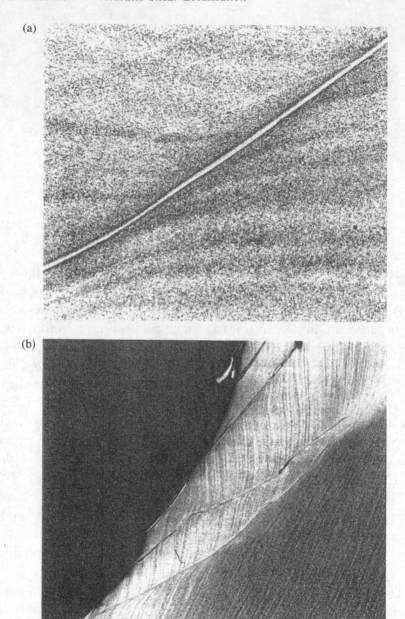

Fig. 1.1.   (a) Transformed shear bands in a steel target impacted at a velocity of 3000 m/sec by a steel projectile, magnification ×252. (b) General view of the target in (a) showing the network of shear bands. Courtesy A.J. Stilp.

deformation will become unstable. Therefore, homogeneous plastic deformation will give way to a localized band-like deformation mode. This is an adiabatic shear band. A prerequisite for adiabatic shear banding is a sufficient accumulation of localized plastic work in a time that is shorter than that required for heat diffusion away from the plastic zone.

Tresca's early work on heat lines and thermal crosses has been discussed in Bell's brilliant exposition on solid mechanics (1973). Also, Johnson (1987) reaffirms that the phenomenon of adiabatic shear banding was reported firstly by Tresca in English in 1878 and later by Massey (1921). Of course, Tresca may have been one of the first researchers to report the appearance of heat lines, but any blacksmith, ancient or modern, must have realized the effects of plastic work on localized heating through the change in colour of the steel.

Tresca's meticulous experimental work led him to report:

> that when a bar of platinum, at the moment of forging, has just cooled down to a temperature below that of red heat,... the blow of the steamhammer, which at the same time made a local depression in the bar and lengthened it, also reheated it in the direction of two lines inclined to each other, forming on the sides of the piece the two diagonals of the depressed part; and so great was this reheating that the metal was along these lines fully restored to a red heat.... These lines of increased heat even remained luminous for some seconds, and presented the appearance of the two limbs of the letter X. They were the lines of greatest sliding and also therefore the zones of greatest development of heat — a perfectly definite manifestation of the principles of thermodynamics.

Tresca carried out further experiments in which he treated a metal bar with wax or tallow, so that when the bar was compressed the wax melted and assumed in certain cases the form of the letter X. From a series of calculations of the amount of plastic energy expended, Tresca concluded that in excess of 70% of the plastic work was always converted into heat. This was the first attempt to apply the thermodynamic principles of Joule (1859). The accuracy of Tresca's calculations was not improved upon until the work of Taylor (1937).

Massey (1921) called these bands heat lines. Adiabatic shear bands have also been referred to as: catastrophic thermoplastic shear bands (Recht, 1964) thermal crosses or heat lines (Massey, 1921; Johnson, 1972) white bands of streaks (Bedford *et al.*, 1974).

No matter what name is used for this phenomenon, the essence of it has been recognized from Tresca's work to that of Zener and Hollomon to the most recent investigations. The essential feature is as Tresca described it: 'the zones of greatest development of heat — a perfectly definite manifestation of the principles of thermodynamics'.

Adiabatic shear bands may be clearly observed in many steels because of their distinctive white band appearance. The material constituting the white band has been generally identified as martensite. Metallurgically, this implies fast heating, causing a high temperature within the band. This temperature is in excess of 1200 K, which leads to transformation of the body-centred cubic ferrite to the high-temperature face-centred cubic austenite. The austenite is quenched very quickly by the surrounding cooler material to form thin bands of martensite. These bands, which etch up white with nital etchant, are evidence of fast or 'adiabatic' heating to high temperatures. This heating is one of the most conspicuous features of adiabatic shear bands and sets them apart from other types of shear band. Because rapid heating is involved in the formation of adiabatic shear bands, this may be the reason that researchers still prefer to use this term, although in many cases the evidence for rapid heating is not as apparent to the naked eye as the white bands that form in steels.

It has been emphasized so far that the term adiabatic shear band is an oversimplified abbreviation of a complicated mechanical and thermodynamic phenomenon. It is possible that use of such a term for such a phenomenon may lead to some misunderstandings. However, use of the adjective 'adiabatic' pinpoints as well as exaggerates one of the most distinguishable aspects of adiabatic shear bands. When an adiabatic shear band is formed, the heat generated is significantly greater than the heat lost. The abrupt gradient in temperature across and normal to a band can be very important in determining many features of the process, such as the evolution and structure of the shear band, the quenching of the heated material after loading as well as the metallurgical microstructures. Therefore, it is wise to bear all these aspects in mind and apply the concept of adiabatic shear banding properly when examining phenomena relevant to so-called adiabatic shear bands.

Thermodynamically, an adiabatic process is defined as one in which there is no heat lost or gained to or from the surroundings by the system. Clearly, from this strict thermodynamic definition, no heat can be conducted away from the band during its formation for this to be regarded as a truly adiabatic process. Thus the use of the term more widely, for situations in which some heat may be conducted away, is an approximation, which most researchers understand and accept.

Is an adiabatic shear band a type of crack? Generally speaking, an adiabatic shear band is not a type of crack. However, they may be frequently and closely related to cracking. In Fig. 1.1(a) the adiabatic shear band appears to coincide with the separation of two deformation zones, but there is no decohesion and no surfaces of separation in the shear band. So the deformed body remains integral and continuous with additional thin layers or bands, each with large gradients of shear strain associated with them. Adiabatic shear bands also play an important role in the formation of the solid state bond formed in the explosive welding of two pieces of metal (see Chapter 9). More often, though, a shear band serves as a precursor to ductile and brittle fracture (see Chapter 3).

Are adiabatic shear bands slip-lines? This is not exactly so. Adiabatic shear bands do usually appear on planes of maximum shear stress. Hence, in the region of inhomogeneous plastic deformation, adiabatic shear bands generally follow the trajectories of slip-lines. However, the theory of slip-line fields is based on rigid perfectly plastic material behaviour, while on the other hand adiabatic shear bands result from thermal softening outweighing deformation hardening. The physics of the two phenomena are different and therefore it is not strange that differences appear. Figure 1.2 compares a slip-line field in a plate undergoing quasi-static flat punch indentation with adiabatic shear bands formed by plugging of a plate by a blunt projectile. More details concerning the differences and relations between the two classes of physical phenomena can be found in Section 2.1.

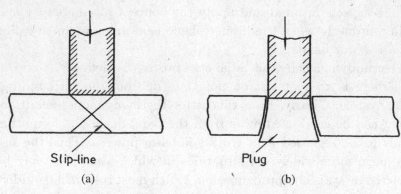

Fig. 1.2.   Comparison between (a) a simple slip-line field for quasi-static plane strain indentation by a flat punch into a plate and (b) plugging of a plate impacted by a blunt-nosed projectile.

Adiabatic shear banding is a unique mode of deformation in materials. Its morphology, mechanisms, metallurgical structures and effects on various materials and structures make the phenomenon unique and in need of special study.

Several review papers have appeared in the last 40 years. These describe advances made in the understanding and other features of adiabatic shear banding. The following reviews are recommended for first reading:

Bedford and co-workers (1974)
Rogers (1974, 1979, 1983)
Kou (1979)
Stelly and Dormeval (1986)
Dormeval (1987)
Shockey (1986)
Dodd and Bai (1987)
Bai (1989, 1990).

Original papers on specified features of adiabatic shear bands can be found in the references given in the following chapters. A comprehensive bibliography of pre-1974 references is given by Hargreaves and Weiner (1974).

1.2. The Importance of Adiabatic Shear Bands

Adiabatic shear bands are fine bands with widths between a few and hundreds of microns in components that are at least centimetres in size. Why is it that researchers in engineering, physics and materials science are so interested in these tiny bands?

Severely localized shear deformation can lead to premature failure in structures. Figure 1.3 shows two modes of failure, one with large plastic bulging and the other with plugging. In fact, less impact energy is needed to make the plate fail due to plugging. The following simple analysis provides a rough estimate of the difference in energies between adiabatic shear banding and other modes of failure (see Bai and Johnson, 1981, 1982). Consider a flat-ended cylinder of diameter D indented into a plastically deforming body. The required kinetic energy can be estimated from the work done on the indented body:

$$\frac{mv^2}{2} = Fd = \sigma_f \pi \frac{D^2}{4} d \tag{1.1}$$

where F and σ_f are the resistance force and the corresponding stress, respectively, and d is the depth of indentation. Now if a plate of the same material with a thickness b is perforated by the same cylinder, but failed in plugging, the required limiting kinetic energy will be:

$$\frac{mv^2}{2} = F_p d_p = \tau_f \pi D b \, d_p \tag{1.2}$$

where F_p and τ_f are the shear force resistance and the shear stress, respectively, and d_p is the critical displacement for plugging. d_p is given by:

$$d_p \approx \gamma_f w \tag{1.3}$$

where γ_f is the ultimate shear strain to failure and w is the width of the adiabatic shear band. Provided that the input kinetic energy remains the same, let us examine the difference between the indentation depth, d, and the thickness of the perforated plate, b. After

(a)

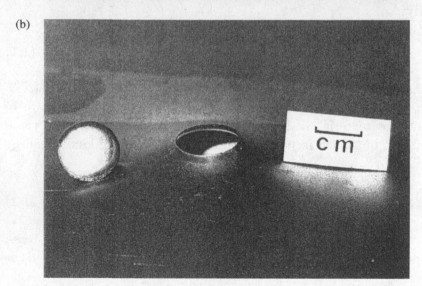

(b)

Fig. 1.3.   Transverse impact of a metal plate. (a) Formation of a plastic dome with high energy absorption. (b) Plugging failure mode with low energy absorption. From Harding (1987), courtesy Elsevier Applied Science Publishers Ltd.

substituting Eq. (1.3) into Eq. (1.2), we obtain:

$$\frac{d}{b} \approx \frac{4\tau_f}{\sigma_f} \frac{\gamma_f w}{D}. \tag{1.4}$$

For the purposes of estimation, we can assume $\tau_f/\sigma_f \approx 1/2$ and $\gamma_f \approx 5$ for steels. Then expression (1.4) becomes:

$$\frac{d}{b} \approx \frac{10w}{D} \approx 0(10^{-2}\text{--}10^{-1}). \tag{1.5}$$

The width of adiabatic shear bands in steels ranges from tens to hundreds of microns; namely, two orders of magnitude less than the projectile diameter of about a centimetre. Very approximately, we can say that the kinetic energy required to perforate a centimetre-thick plate by plugging corresponds to that to make an indentation of less than one millimetre. It is clear that adiabatic shear banding is a subtle failure mode, which can never be ignored in practice.

The local rapid and intense plastic deformation in an adiabatic shear band induces a high temperature locally. Let us now estimate how high this temperature rise can be. From the adiabatic assumption, the temperature rise, $\delta\theta$, resulting from the plastic deformation work is:

$$\delta\theta = \int_0^{\gamma_f} \frac{\tau \, d\gamma}{\rho c} \approx \frac{\tau_{\text{flow}}\gamma_f}{\rho c} \tag{1.6}$$

where τ_{flow} is the flow stress, which is about 400 MPa in steels, ρ is the density and c the specific heat. In most steels, $\rho \approx 7800 \, \text{kg/m}$, $c \approx 450 \, \text{J/kg K}$. γ_f we will still take to be about 5. On substitution, Eq. (1.6) gives a local temperature rise of about 600 K. This indicates a local temperature as high as 900 K. If γ_f exceeds 8, which is commonly reported in the literature, the local temperature can be 1200 K above the critical temperature for the $\alpha \rightleftharpoons \gamma$ phase transformation in steel. In energetic materials, such as high explosives, such high temperatures may form so-called hot spots, which are generally accepted to be the driving force for the initiation of detonation under impact. Therefore, unusually high temperatures have very special implications in certain areas.

Not only is the temperature within an adiabatic shear band very high owing to the plastic work done on it, once shear loading ceases the cooling of the hot shear band is very rapid. Since the hot shear band remains in intimate contact with the highly heat-conductive and comparatively cool metal body, the quenching rates are substantially higher than for conventional quenching processes. The quenching rate can be estimated from Fourier's law for heat conduction:

$$\rho c \dot{\theta} = \frac{d}{dy}(\lambda \, d\theta/dy) \tag{1.7}$$

where θ is the temperature, λ is the heat conduction and y is the axis perpendicular to the shear band. For simplicity, λ is taken to be a material constant, which for steels ~80 W/m K. Therefore the heat diffusion $\kappa = \lambda/\rho c$ is also a material constant and has a value of about $0.25 \times 10^{-4} \, \text{m}^2/\text{sec}$ in steels. The cooling rate can be expressed as:

$$(-\dot{\theta}) = -\kappa \frac{d^2\theta}{dy^2}. \tag{1.8}$$

The differential $d^2\theta/dy^2$ within the shear band can be estimated from:

$$\frac{d^2\theta}{dy^2} \approx -\frac{\Delta\theta}{\delta^2} \tag{1.9}$$

where $\Delta\theta$ is the temperature difference between the centre of the shear band and the surroundings, δ is the half-width of the band and $d\theta/dy \approx 0$ at the centre of the shear band because of symmetry. Therefore, the cooling rate can be estimated simply to be:

$$(-\dot{\theta}) \approx \frac{\kappa\Delta\theta}{\delta^2}. \tag{1.10}$$

In steels $(-\dot{\theta}) \approx 0.24 \times 10^{-4} \times 10^3/(10^{-4})^2 \approx 10^6 \, \text{K/sec}$, where we have assumed $\Delta\theta \approx 10^3 \, \text{K}$ and $\delta \approx 10^{-4} \, \text{m}$. This is an extremely high quenching rate, bearing in mind that conventional workshop quenching rates are a maximum of about $10^2 \, \text{K/sec}$. This extremely fast quenching rate makes the material within the shear band harder. The martensite formed in steels is significantly harder than that formed in workshops. Often brittle cracks are observed to form and grow within or across the shear band. Nevertheless, there are observations of strings of voids formed within shear bands, which indicates that the shear band material is not always brittle. The very high local temperatures and subsequent rapid cooling rates of adiabatic shear bands often make them serve as precursors to failure, whether brittle or ductile, during the impact process itself or in subsequent processing.

Adiabatic shear bands present the most unusual and unique features in mechanical, thermodynamic and metallurgical behaviour. It is impossible from conventional material data, such as strength, fracture toughness, fracture strain, etc., to obtain an indication of an ordering or a susceptibility of materials to adiabatic shear banding. Well-known examples are titanium and its alloys. From the mechanical point of view — strength, toughness and weight — these alloys are excellent. However, these materials suffer from a serious weakness: their susceptibility to adiabatic shear banding. Because of this it is very important to understand the significant variables that promote adiabatic shear banding so that it will eventually be possible to rank them in accordance with their susceptibility to adiabatic shearing. It is obviously very important when designing with a high-performance material to be sure that it is not prone to adiabatic shearing.

It now seems that adiabatic shear bands can be the precursors to catastrophic failures that affect the whole body, such as plugging failure, brittle and ductile fracture, impact ignition of explosives, etc. All of these catastrophic changes are due to the extraordinary features of very fine adiabatic shear bands.

Recently there has arisen an increased interest in adiabatic shear banding in academic areas such as physics and mathematics. In fact, an adiabatic shear band appears to be a good representation of mechanical instability in a material. Also, it is an example of a coupled thermal and mechanical deformation mode and is a type of dissipative structure within the purview of irreversible thermodynamics (see Hutchinson, 1984; Balakrishnan and Bottani, 1986). Adiabatic shear bands in mechanics provide an instructive example of a discontinuity and a related narrow structure dominated mainly by coupled dissipative work and heat conduction. All of these aspects merit a profound fundamental study of adiabatic shear bands. Phenomenological observations through to basic concepts of the process are all needed. Also, the mechanisms of evolution of bands and the factors governing this evolution should be a major focus of study in the future.

1.3. Where Adiabatic Shear Bands Occur

When a metal is plastically deformed most of the plastic work expended is dissipated as heat. After the pioneering work of Tresca and that of Taylor and Quinney (1934), it is generally accepted that the amount of plastic energy dissipated as heat can be in excess of 90%. The remaining 10% of the plastic work is called the latent energy of cold work and is stored in the specimen in the form of the new defect structure.

We know that the flow stress of most materials decreases with increasing temperature. Therefore any high strain-rate process can produce significant localization of plastic flow. This flow localization often leads to shear banding. Once shear banding begins, this can in turn lead to sudden catastrophic fracture.

Adiabatic shear banding has been observed in many different processes. For example, armour penetration, bomb fragmentation, dynamic blanking and cropping, high-speed machining and forming (such as impact extrusion), cryogenic plastic deformation, impact erosion and surface rubbing.

Zener and Hollomon (1944) observed the effects of localized shear deformation in punching. When a slow pressure was applied to the punch, the plastic deformation was not localized. However, when the pressure was applied dynamically, the plastic deformation was highly localized. Zener and Hollomon estimated that the thin band of material had undergone a shear strain of about 100. Further, they showed that a shear strain of 5 would produce a temperature rise in excess of 1000°C. In projectile impact studies, temperatures of this order can be produced in a matter of microseconds and therefore projectile impact and contained explosive experiments are used to study adiabatic shear bands.

In perforation studies, Zener (1948) assumed that the shear bands would coincide with the maximum shear stress planes. In high-speed machining Recht (1964) calculated a shear zone temperature of about 650°C for a cutting speed of 43 m/min. Dao and Shockey (1979) measured temperatures in orthogonal machining of steel and 2016-T6 aluminium using an infrared microscope. Also, they measured the width of shear bands produced and showed them to be periodic in the steel chips. Metal-forming processes can constrain metal to flow through complex paths. In compressive forming processes, such as upset forging with high tool–workpiece interfacial friction, the magnitude of the shear deformation can be large, often giving high local strain rates and the possibility of lines or zones of thermal discontinuity (Johnson, 1972).

Because strain rates can be so high in explosive loading, such as of the order of $\sim 10^7$/sec, adiabatic shear bands are common in these circumstances. In explosive welding, shear bands have also been observed. Two types of band have been observed. One is located in the bonded area and the other is inclined to the zone.

Adiabatic shear has been proposed as the cause of jerky flow in cryogenic tensile testing. The temperature rise due to a given amount of plastic work is dependent on the specific heat of the material. As the specific heats of metals decrease with decreasing temperature, a temperature increase produced at cryogenic temperatures will be appreciably higher than that produced at room temperature. This temperature rise can make the flow stress significantly lower than otherwise (see Basinski, 1957; Chin *et al.*, 1964a,b).

1.4. Historical Aspects of Shear Bands

Adiabatic shear banding is an important mechanism of plastic flow localization. It has been observed in processes in which materials are subjected to large strains at high strain rates (or equivalently for short times). Adiabatic shear bands have been observed in processes such as armour penetration, blanking, cropping, impact erosion, frictional contact, forming and cutting.

We have already outlined the valuable contribution made by Tresca to the study of localized heating due to plastic deformation. Bell (1973) and Johnson (1987) both describe Tresca's meticulous experiments, in which he observed deformation heating during plastic compression. Tresca (1878) carried out an extensive series of experiments on gridded lead, iron and copper bars and slabs. The results

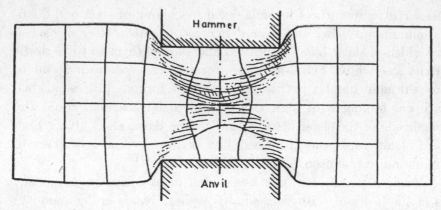

Fig. 1.4.   The plastic distortion of lead stamped by a hammer. After Tresca (1878).

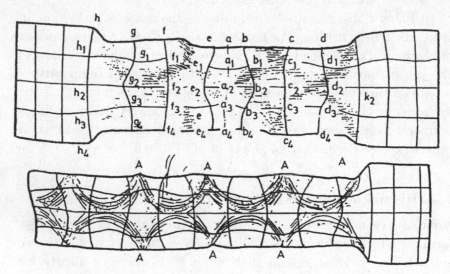

Fig. 1.5.   As Fig. 1.4, but successively stamped after transverse movement a second and third time. After Tresca (1878).

of some of these experiments are shown in Figs 1.4 and 1.5. In these experiments Tresca used a steam hammer and described how, in certain cases, these lines become luminous and form a letter X.

Tresca's work with wax and tallow coated bars has already been briefly described. When the steam hammer impacted the bar, the wax or tallow melted along an X shape. Tresca calculated the efficiency

Table 1.1. The area of melted wax is that included between the two strong dotted lines AA and BB in Figs 32 to 34, and between the four dotted lines AA and BB in Fig. 35 in Tresca's paper.

No. of figure		32	33	34	35
Metal		Iron	Iron	Iron	Copper
Form of impression		Rectangular	Rectangular	Curved	Rectangular
Area of wax melted	in^2	0.22	0.23	0.34	0.27
Thickness of forging	in	0.98	0.98	0.98	0.78
Volume of corresponding prism	in^3	0.22	0.23	0.34	0.21
Heat-units absorbed in raising prism to 122°F	units	0.5944	0.6138	0.9003	0.5273
Equivalent work, taking 1 heat unit = 772 ft-lb	ft-lb	460	475	696	408
Actual work in fall of hammer	ft-lb	579	651	796	434
Percentage of efficiency		79.6	73.1	87.7	94.2

Thus in the last experiment, taking as melted the area of wax included between the hammer and the crosses, a calculated useful effect of 94% is obtained.

of the process by calculating the ratio of the plastic work done to the kinetic energy of the striking hammer. From Table 1.1, which is taken from Tresca's paper, we can see that efficiencies ranging from 73% to 94% were obtained. These were remarkable results obtained at such an early date. The amount of heat produced by plastic work was next calculated by Taylor and Quinney (1934), over 50 years later.

After the presentation of all this experimental data by Tresca at the Institution of Mechanical Engineers, Tresca was asked by Hopkinson to distinguish between the flow of a specimen subjected to a blow either by a light hammer moving at high velocity or a heavy hammer moving at low velocity, both doing the same work. According to Bell's account, Hopkinson stated that he did not think that any difference could depend entirely on the inertia of the material. It appears then, at that time, that Tresca was more aware of the differences between quasi-static and dynamic loading than Hopkinson was.

Massey (1921) describes how it is possible to observe heat lines when hot forging at a relatively low temperature. Massey observed, on occasion, the same X-type patterns, but was unaware of Tresca's original paper. Very recently, a Russian research report has been discovered written by N. Dawidenkow and I. Mirolubov published in 1935 titled *A Particular Kind of Upset-forging of Steel (The Krawl–Tarnawskij Effect)*. This work shows that some researchers in the USSR were very active in the study of adiabatic shear bands. There are many photographs of shear bands in compressed specimens of steel. The paper appears in *Technical Physics of the USSR*, **2(1)**, 281–291. We feel sure that other references will appear from the Russian Federation soon, particularly because of the reference to the 'Krawl–Tarnawskij Effect'. The paper was discovered by Stephen Walley at the Cavendish Laboratory in the Philosophy Library.

Slater (1965) attempted to account for the temperature rises obtained by Massey using the upper bound technique. In this technique the workpiece is divided into a number of elements by lines or surfaces of velocity discontinuity. Plastic work is carried out on any element of the workpiece that traverses a velocity discontinuity. Since most of the plastic work expended is dissipated as heat, it is possible to estimate the temperature rise for a given velocity discontinuity configuration. These lines of thermal discontinuity were also discussed by Tanner and Johnson (1960), Johnson and co-workers (1964) and in the book by Slater (1977).

Trent (1941) mentioned that parts of the surfaces of wire ropes could become very hard, consistent with a martensitic surface. This indicates that the surface of the rope has been heated to about 900°C by friction and then quenched to room temperature by the surrounding material and air. Indeed, in the *National Coal Board's Ropeman's Handbook* (1982), martensitic fractures in wire ropes are discussed at some length. When a rope rubs or grinds against a metal obstruction, the friction can cause, eventually, a martensitic layer to a depth of 0.02 mm. So-called martensitic embrittlement is recognized as being very dangerous. The first time that such a rope is bent, the outer strands can fracture.

High-speed blanking, shearing and cropping are all cutting processes in which the applied impact velocity of the tools with the workpiece and the resulting temperature influence the process. High-speed blanking has been studied, for example, by Johnson and Travis (1968) and the fundamental nature of guillotining and related processes has been reported by Atkins (1981). A description of the mechanics of blanking with reference to the punch force–punch displacement curve has been given by Johnson and Slater (1967). The common feature of all these processes is that plastic deformation is constrained to be intense shear along narrow zones and therefore adiabatic shear bands are often reported in these processes.

Temperatures can also be very high in metal-cutting. Reichembach (1958) and Boothroyd (1981) measured chip temperatures experimentally and Wright (1971) measured tool temperatures by a quick stop and metallurgical etching technique. In orthogonal machining the temperatures reached in the primary and secondary shear zones of the chip can be extremely high (see Childs and Rowe, 1973). This topic will be returned to in Chapter 10.

Metals subjected to impact by projectiles are an obvious potential source of adiabatic shear bands and catastrophic fracture (see, for example, Rinehart and Pearson, 1965, which includes excellent photographs of shear bands and spalling in metal plates that have been subjected to explosive loading). A vast amount of literature has built up in this area. Of particular interest is the review by Backman and Goldsmith (1978) and the paper by Woodward (1984). Woodward classified the failure modes for pointed and flat projectiles as shown in Fig. 1.6. He observes that the major difference between the two forms of projectile can be seen in ductile low-strength targets. With pointed projectiles, a large degree of plastic flow away from the projectile occurs, but for a blunt projectile the material is constrained to move ahead of the punch, forming a plug.

1.5. Adiabatic Shear Bands and Fracture Maps

There are numerous mechanisms and modes of fracture, but most fractures are classified either as brittle or ductile. Brittle fractures require little or no plasticity to initiate and propagate, but ductile fractures require extensive plasticity to develop.

Ashby (1981) and Frost and Ashby (1982) describe what they call fracture mechanism maps. These are maps that indicate the different modes of fracture with axes of stress divided by Young's modulus

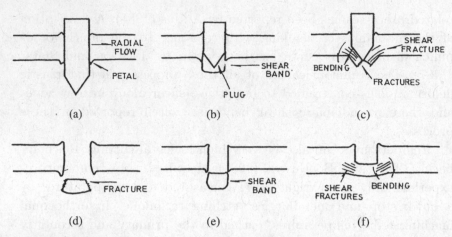

Fig. 1.6.   Classification of failure modes for pointed projectiles: (a) ductile hole formation, (b) adiabatic shear plugging and (c) discing; and for blunt projectiles; (d) ductile plugging; (e) adiabatic shear plugging and (f) discing. After Woodward (1984).

and homologous temperature, θ/θ_M, where θ_M is the melting point in degrees K.

In uniaxial tension, there are three basic classes of fracture: cleavage, void formation and rupture. Ashby (1981) subdivides brittle cleavage into cleavage I, II and III. Cleavage I requires pre-existing cracks that are large enough to propagate at a stress lower than the yield stress. Cleavage II fractures are initiated by the nucleation of cleavage nuclei at slip bands or twins. This normally involves a little plastic deformation. Cleavage III requires more extensive plastic deformation than that required for cleavages I and II. In the case of cleavage III, cracks can nucleate at grain boundaries, for example, and extend stably to some critical size before unstable fracture occurs. Like ductile fracture, cleavage is transgranular. However, with brittle grain boundary precipitates or imperfections it is possible for brittle fracture to occur by intergranular fracture (BIF), with three similar subgroups to cleavage being possible.

Rupture and void nucleation and growth are both modes of ductile failure. Rupture is equivalent to a 100% reduction in area in a tensile test and is more probable under conditions that promote dynamic recrystallization. Void nucleation is influenced by the nature of the second phase-particles and non-metallic inclusions in the microstructure. Thus control of cleanliness and second-phase particle shape can influence material ductility (see Dodd and Bai, 1987).

At higher temperatures, typically about 0.3 to 0.4 θ_M, deformation occurs through creep mechanisms. A number of modes of creep are common. These are: power law creep controlled by dislocation glide and climb and at sufficiently low stresses Harper–Dorn linear viscous creep. At higher stresses the simple power law creep breaks down. At high temperatures lattice diffusion controls the overall process and this is called Herring–Nabarro creep. At lower temperatures, Coble grain boundary creep occurs.

Figure 1.7 shows a typical fracture mechanism map for tungsten tested in tension and Fig. 1.8 shows fracture mechanism maps for the

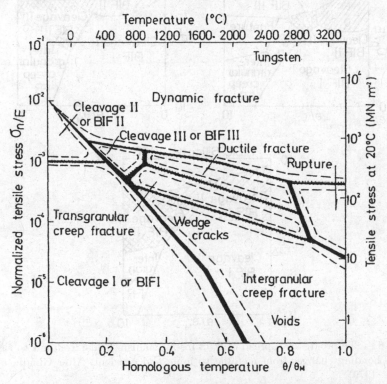

Fig. 1.7. The progressive spread of fracture-mechanism fields, from left to right as the bonding changes from metallic to ionic and covalent. After Gandhi and Ashby (1979).

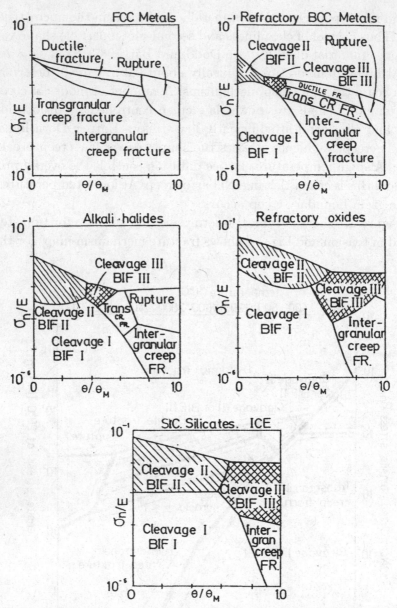

Fig. 1.8.   The progressive spread of fracture-mechanism fields, from left to right as the bonding changes from metallic to ionic and covalent. After Gandhi and Ashby (1979).

five groups of solids: face-centred cubic metals, body-centred cubic metals, alkali halides, refractory oxides and covalently bonded solids, according to Gandhi and Ashby (1979). From the general trends shown here, it is clear that the fracture behaviour of solids depends on a large number of microscopic and submicroscopic phenomena. Important variables are the nature of bonding, crystalline structure, nature of grain boundaries and the presence of pre-existing flaws, cracks and inclusions.

When loading is carried out at high rates, early cracking often results because of the increase in yield stress and the shrinking of the plastic deformation range. The micromechanisms of plastic deformation are also strain-rate dependent. At low loading rates, the velocity of dislocations is limited by the periodic lattice resistance or Peierls force. At higher strain rates, phonon and electron drag can influence dislocation velocity. At very high loading rates, relativistic effects become important.

Kocks and co-workers (1975) defined a dislocation drag coefficient, B, which can be subdivided into electron and phonon drag. It has been observed that the total drag coefficient varies approximately linearly with temperature:

$$B = B_e + B_p\theta/300 \qquad (1.11)$$

where B_e and B_p are the electron and phonon drag coefficients at 300 K. Electron drag is normally only significant at low temperatures. The relativistic effects show themselves when the dislocation velocity approaches the velocity of sound in the material. A constriction of the strain field causes the elastic energy to rise, giving the dislocation a limiting velocity.

Figures 1.9 and 1.10 show strain rate versus homologous temperature maps for titanium and zirconium, after Sargent and Ashby (1983). These maps show well the dislocation mechanisms operative over a range of strain rates. Drag-controlled flow is extensive at high strain rates and elevated temperatures. The adiabatic shear zones for both metals are similar in shape, but unlike the various modes of microscopic flow, adiabatic shear is unique because it is dependent on the strain rate and strain.

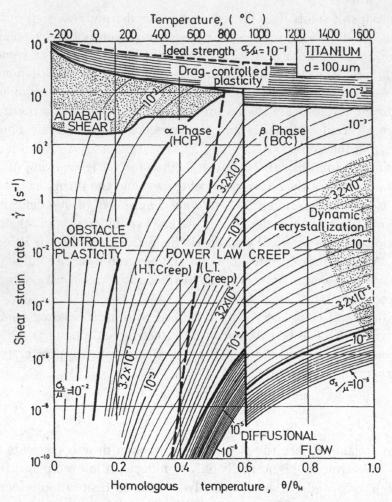

Fig. 1.9.   Strain rate/homologous temperature deformation map for titanium. After Sargent and Ashby (1983).

Sargent and Ashby (1983) included the effects of a nucleation strain γ_i for adiabatic shear. They assumed that two conditions must be met simultaneously for adiabatic shear to occur. These are a critical strain and a critical strain rate.

Figure 1.11 shows part of the deformation map with contours of critical strain for the onset of adiabatic shear. From Fig. 1.11 it is observed that larger strains are required for the nucleation of adiabatic shear the lower the strain rate.

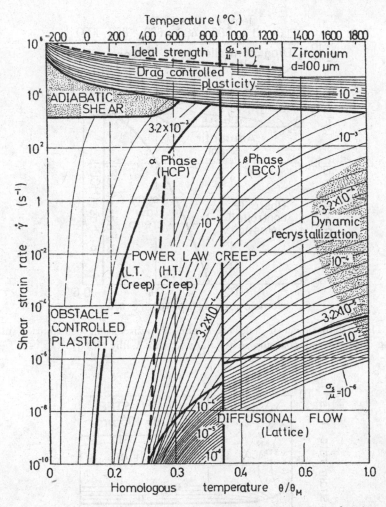

Fig. 1.10. Strain rate/homologous temperature deformation map for zirconium. After Sargent and Ashby (1983).

Figures 1.12, 1.13 and 1.14 show the adiabatic shear zones for mild steel, titanium–6% aluminium–4% vanadium and pure aluminium. Figure 1.14 shows the effect of strain rate on the critical strain for the onset of adiabatic shear. The partial map for pure aluminium does not exhibit adiabatic shear below about 0.06 θ_M because of its high thermal conductivity. The effect of small amounts of alloying elements on the shape of the adiabatic shear zone is

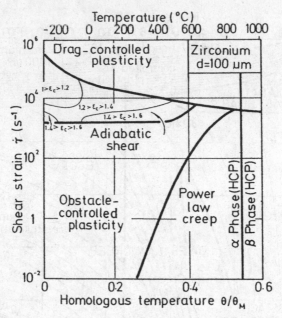

Fig. 1.11. Adiabatic shear field on a partial α-zirconium deformation map showing critical strain contours. After Sargent and Ashby (1983).

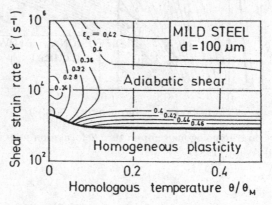

Fig. 1.12. Adiabatic shear field on a partial mild steel deformation map showing critical strain contours. After Sargent and Ashby (1983).

shown in Fig. 1.15. Here there is no limiting homologous temperature below which adiabatic shear is impossible. Also, there is no high temperature cut-off, as for pure aluminium, caused by homogeneous heating.

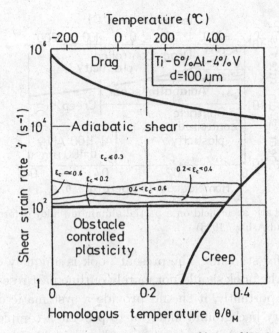

Fig. 1.13. Adiabatic shear field on a partial Ti–6% Al–4%V deformation map. After Sargent and Ashby (1983).

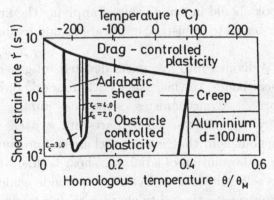

Fig. 1.14. Adiabatic shear field on a partial aluminium deformation map showing critical strain contours. After Sargent and Ashby (1983).

1.6. Scope of the Book

Although a number of review papers published in journals and conference proceedings provide good introductions to various aspects of adiabatic shear bands, it was felt that a reference book on the topic

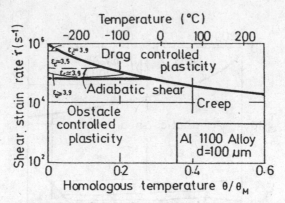

Fig. 1.15.   Adiabatic shear field on a partial aluminium alloy deformation map. After Sargent and Ashby (1983).

was required for students. The present book is uniquely devoted to this subject. The book should not merely outline past research in the area. More importantly, it should provide a systematic description of the necessary background material to provide a complete picture of the phenomenon. Also, as a reference book, it should provide the necessary data on materials that are relevant to adiabatic shear banding and which are not readily available in conventional handbooks. Finally, the book should indicate ways of applying the knowledge of adiabatic shear bands to industrial applications and related unsolved problems.

The book is divided into five parts devoted to: general features of adiabatic shear bands, the necessary background knowledge, basic concepts, theories and mechanisms, some case studies of applications and several appendices. These parts are written around the core of the phenomenon, that is the mechanisms predominating in the occurrence and evolution of adiabatic shear bands. Nevertheless, each chapter is self-contained and can be read independently of the others, with reference to the other chapters being made where necessary. For example, readers who are working in the field of impact dynamics and are familiar with some of the phenomena of shear banding can start reading at Chapter 6. Alternatively, readers who require a quick grasp of the key points of the subject can read the appendices directly.

The first three chapters provide a general picture of adiabatic shear bands. In particular, they cover characteristic features of the phenomenon, their relationships with and differences from other localized deformation modes, the position of the adiabatic shear zone in fracture maps and connections with brittle and ductile fractures as well as some general concepts involved in the phenomenon. All of these features are necessary for the establishment of a correct physical approach to this complicated coupled thermomechanical process.

Chapters 4 and 5 are concerned with experimental techniques and constitutive examples for the description of the mechanical properties of materials at high rates of strain. Normally, adiabatic shear bands are observed and studied in mechanical engineering with the help of metallurgical investigations. The description of the mechanical behaviour of a material is through the constitutive equation. We need constitutive equations to help in the study of adiabatic shear bands, although often the theoretical background to a constitutive equation is limited. More often than not a straightforward curve-fitting exercise is carried out on experimental results to obtain an equation that describes the strain, strain rate and temperature dependence of the flow stress.

The main body of the book can be regarded as Chapters 6, 7 and 8. These chapters are devoted to the mechanisms and the formulations of the whole process of adiabatic shear banding. Starting from a homogeneous deformation mode, due to an added disturbance, instability occurs. Beyond the point of instability, inhomogeneous plastic deformation develops and adiabatic shear bands may occur. The complete formation of adiabatic shear bands is usually accompanied by an abrupt increase in the local strain rate and temperature as well as a rapid decrease in flow stress. Eventually, the quasi-steady late-stage morphology of adiabatic shear bands develops. It is clear that these aspects consist of multiple stages and various mechanisms. The dominant parameters in each of these stages and mechanisms are discussed.

Chapters 9 and 10 contain some selected topics in impact dynamics and metalworking. The different sections in each chapter are case studies, which illustrate how a knowledge of adiabatic shear bands can be applied to industrial processes. It is shown that in some cases the application of the theories of adiabatic shear banding leads to satisfactory quantitative predictions, and therefore they are quite helpful. However, in most cases the applications are still limited to observations. These areas have not been studied widely and are worthy of further exploration.

The appendices are an important part of the book. They consist of a quick reference to adiabatic shear banding, lists of materials in which adiabatic shear has been investigated and data sheets on the thermomechanical properties of materials. It is hoped that the subject matter contained within the appendices will be helpful even after the reader has finished the book.

Chapter 2

Characteristic Aspects of Adiabatic Shear Bands

2.1. General Features

It is clear that the presence or absence of adiabatic shear bands is very sensitive to the stress state. In the exploding cylinder test, shear bands almost always occur on or near the inner wall where the detonation of the explosive creates extreme pressures. However, shear bands rarely appear on the outer surface of the same cylinder where the controlling factor is the tensile hoop stress. If we consider as another example oblique penetration, shear bands are only found on the side of the projectile that is subjected to compression. In addition, there are very few reports of observations of adiabatic shear bands formed in tensile stress states, even at high strain rates (see Rogers, 1974, 1979). The suppression of adiabatic shear bands in tensile stress states was interpreted by Rogers (1979) as being the early intervention of ductile fracture. This is quite probable because of the sensitivity of crack propagation to tensile stresses. This form of localization is very different from other types of localized behaviour such as local necking.

Generally, the outstanding feature of an adiabatic shear band is the high local value of the shear strain, shear strain rate, temperature and distinct mechanical behaviour. The local shear strains usually range between 5 and 100 (Rogers, 1979; Timothy, 1987). The local shear strain rate was estimated to be as high as 10^7/sec when a local strain of 100 was attained within about $10\,\mu$sec. At least we can say that the local strain rates reached 10^4 to 10^6/sec within the shear band. The temperature rises of several hundred degrees locally are experienced commonly and can be measured directly (Costin *et al.*, 1979; Hartley, 1986; Hartley *et al.*, 1987; Marchand and Duffy, 1988). At the centre of some shear bands thin solidified melt layers have even been observed in the tested materials (see Hartmann *et al.*, 1981). The width of the shear bands range approximately from $10^1 - 10^2\,\mu$m. The harder the material of the shear band is, the narrower is the shear band, for both deformed and transformed shear bands. Additionally, transformed shear bands are usually narrower and harder than deformed bands in the same material. Shear bands are often observed to span several grains in width. However, in recent years there have been a number of observations of fine intergranular adiabatic shear bands with widths of typically 0.1–0.3 μm formed under shock wave loading conditions. To conclude, then, the general features of an adiabatic shear band are a very thin band with extremely high strain rate, strain and temperature.

Usually, adiabatic shear bands are non-crystallographic and transgranular. A shear band can simply be visualized as consisting of severely distorted and elongated grains. However, both in transformed bands and at the centre of deformed bands, fine equiaxed grains have been observed (Timothy, 1987). Perhaps these fine grains and the distorted structures are responsible for the higher hardness of adiabatic shear bands.

It is necessary to clarify the differences between adiabatic shear bands and a number of other deformation bands, although we have already outlined the salient features of shear bands.

Slip bands occur along active crystallographic planes and are therefore almost always confined to a single grain. Even a glide packet has a thickness of the order of 10^{-6} m. Compared to slip bands, adiabatic shear bands are usually non-crystallographic and transgranular and normally cover tens or hundreds of grains and hence have a width of 10^{-4} m. In this sense, adiabatic shear bands are macroscopic in nature. Some observations have shown that grains within adiabatic shear bands are distorted along some active slip plane, but as a whole, adiabatic shear bands do not show any preferred crystallographic orientation.

A band of local necking occurs in a sheet subjected to tensile loading. At the onset of a local neck, all plastic deformation is limited to a narrow band. Within the band the material is subjected to zero extension along the band. Certainly, macroscopic plastic shear deformation develops within the band and finally a local neck forms. Hence this band is formed due to geometrical softening and is not connected in any way with thermal softening.

Lüders bands are another form of localized deformation within which material is plastically deformed, with the outer material remaining elastic. The occurrence of Lüders bands is closely related to discontinuous yielding, i.e. they appear at the upper yield point and then diffuse on either side of the band, finally disappearing when the specimen is fully plastic. Clearly, Lüders bands have no connection with the adiabatic shear bands discussed here.

For slip-lines there are two families of characteristics along which inextensional shear occurs. This has been most clearly delineated and formulated in rigid perfectly plastic material subjected to a plane strain stress state. In this case, the governing equations are hyperbolic; therefore, there must be two families of characteristics intersecting each other orthogonally on the physical plane. Strictly speaking, there are two families of characteristic orientations on the plane. Along these characteristics a tangential velocity discontinuity, that is an infinite shear strain rate, is permitted. This implies that intense shear within an infinitesimal width can occur along these characteristics. Certainly, this picture of zero-width shear bands is attributed to the rate-independent rigid perfectly plastic assumption. According to this approximate model of plasticity, these characteristics, more commonly called slip-lines, provide the potential orientations for possible intense localized shear deformation. The true appearance of some kind of band along a certain characteristic that eventually becomes a visible shear band depends very much on the detailed loading conditions. Importantly, adiabatic shear bands do not form under tension.

For a general isothermal rate-independent model of plastic materials, the onset of shear bands may coincide with the occurrence of a bifurcation from elliptical to hyperbolic governing equations (see Hill and Hutchinson, 1975).

In these isothermal shear bands, the key mechanism responsible for the occurrence of the shear bands is a yield surface vertex, which is governed by discrete crystallographic slip. This feature of plasticity is the reason that shear bands occur in strain hardening materials (Rice, 1977; Dodd and Bai, 1987).

In isothermal rate-dependent deformation of materials, shear bands may occur, but these will be due to the pre-existence of band-like imperfections (see Pan *et al.*, 1983).

Unlike isothermal shear bands, such as slip-lines in rigid perfectly plastic materials and other bands in rate-dependent and rate-independent materials, the triggering mechanism for adiabatic shear bands is distinctly different. The trigger for adiabatic shear bands involves thermal softening as an essential feature. Therefore, the evolution and morphology of adiabatic shear bands are distinctive, as stated earlier. Despite these differences, there are some features common to adiabatic shear bands and other bands. For example, shear bands frequently follow the trajectories of slip-lines.

The patterns of adiabatic shear bands are often periodic, consisting of a network of two families of narrow bands. There are some similarities between an adiabatic shear band network and a theoretical slip-line field. For example, Fig. 2.1 shows the similarity between a slip-line field network and a shear band network formed in a ballistic impact (Backman and Finnegan, 1973). The active family of shear bands of the two families permits material flow, which is directed inwardly in this case. If the projectile impacts obliquely, then a set of shear bands appears on the compressively loaded side only.

Orthogonal metal-cutting produces two major shear bands. The primary one is located between the metal piece and the chip, whereas the secondary shear zone is on the interface between the chip and the rake of the tool. The primary shear band should coincide with one of the slip-lines. However, the secondary shear band may not always follow the trajectory of the other family of slip-lines. In fact, frictional shear is significant for the secondary shear band in metal-cutting (see Chapter 10).

Patterns, similar to those in metal-cutting, can be seen in explosive welding. The details are discussed in Section 9.6. Many other examples of adiabatic shear band patterns have been observed in

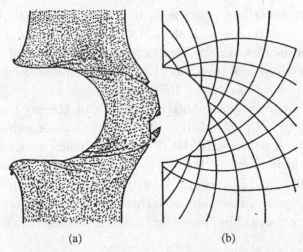

(a) (b)

Fig. 2.1. Comparison of observed shear patterns from a ballistic impact with a theoretical slip-line field model: (a) section of a 2024-T4 aluminium alloy plastic impacted by a rigid sphere; (b) theoretical pattern for impact. After Backman and Finnegan (1973).

various areas of impact dynamics, for example, planar impacts (Section 9.1), fragmentation (Section 9.2), cratering (Section 9.3) and ignition of explosives. The width and spacing of the adiabatic shear band pattern can only be understood by the mechanisms pertinent to adiabatic shear bands, namely their coupled thermomechanical nature (see Sections 7.2, 9.1 and 9.3). The patterns of slip-lines in slip-line fields are determined almost exclusively by boundary conditions, unlike adiabatic shear bands.

2.2. Deformed Bands

We have seen that adiabatic shear bands are a thermomechanical phenomenon. They are a form of flow localization caused by the destabilizing effect of thermal softening. The thermal softening occurs because of the conversion of the plastic work done to a local temperature rise. If the strain rate is high enough, locally, then it is possible for the temperature to increase appreciably over a small zone. This can cause a rapid decrease in flow stress, which in turn can cause a band of shear localization to form.

Adiabatic shear bands are formed in many materials by many different dynamic processes. Materials with a tendency to form shear bands are those of low density, for example, aluminium and titanium. However, they have also been reported in denser metals such as steels and uranium alloys (see Irwin, 1972).

The early experimental work of Tresca and Massey has already been referred to in Chapter 1. In both cases hot crosses were observed to form between punches. In the 1970s shear bands were referred to as either 'deformation' or 'transformation' bands by Backman and Finnegan (1973) and Rogers (1974, 1979). Deformation bands may be regarded as zones of intense plastic shear only, whereas transformation bands are zones of intense shear in which a phase transformation has occurred.

Deformed shear bands are characteristic features of aluminium and aluminium alloys that have been subjected to impact or dynamic loading of some kind, but other metals, particularly non-ferrous ones, also exhibit them. After the onset of strain localization, very steep strain, strain rate and temperature gradients are produced.

Wingrove (1973a,b) impacted an aluminium alloy with a flat-nosed projectile. The impact velocity was 140 m/sec. Wingrove showed that the sharp edges of the projectile acted as sites for shear band initiation.

Leech (1985) investigated adiabatic shear band formation in 7039 aluminium alloy impacted with a flat-nosed projectile as a function of alloy hardness and impact velocity. For all hardnesses the plugging failure mode was encountered. The shear bands formed are shown in Figs 2.2, 2.3 and 2.4. For a starting hardness of 80 VPN a fairly wide zone of intense shear can be seen with a width of about 90 μm. With harder material, the shear bands are narrower, having a width of about 20 μm. Figure 2.5 shows a scanning electron microscope view of a plug fracture surface showing a characteristic knobbly surface. Such fracture surfaces have been reported in steels (see Bedford *et al.*, 1974). This type of fracture surface indicates that localized melting could have occurred. Microhardness traverses of shear bands are shown in Fig. 2.6. Leech points out that the important observation to be made here is the strength differential between the matrix and the shear zone. It can be seen that the hardness differential increases

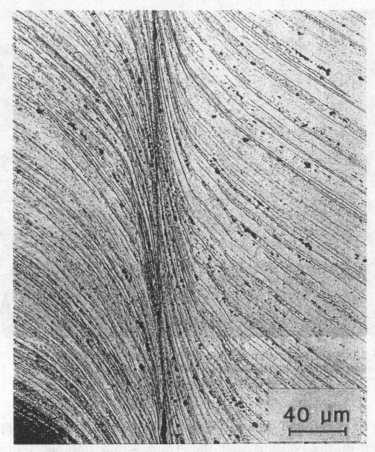

Fig. 2.2. Detail of a shear band formed in a plate of hardness 80 HV, impact velocity 311 m/sec, magnification × 388. After Leech (1985) published by permission of the American Society of Metals and the Minerals, Metals and Materials Society.

with matrix hardness, which indicates that the mechanical properties of the band are dependent on those of the matrix.

2.3. Transformed Bands

There have been numerous reports of transformed bands formed in steels and titanium alloys. The explanation for the formation of a transformation band in steel was given by Zener and Hollomon

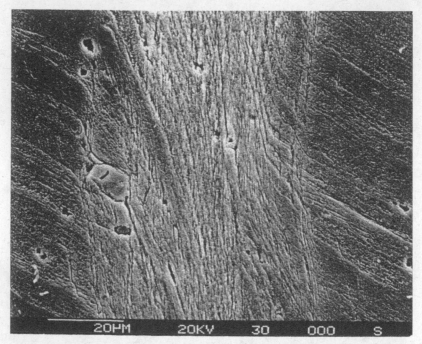

Fig. 2.3.   Scanning electron micrograph of a shear band formed in a plate of hardness 120 HV, impact velocity 315 m/sec. After Leech (1985).

(1944), who carried out extensive punching experiments. They observed during plugging that the strain, strain rate and temperature increased dramatically. On microexamination of sheared-out plugs, the material in the shear zone was seen to have white etching bands when etched in nital. From hardness measurements it was concluded by Zener and Hollomon that the white etching constituent was martensite. These researchers estimated that the thin band of material had undergone a shear strain of about 100. Also they showed that a shear strain of 5 would have produced a local temperature rise of about 1000°C. In impact studies, temperatures of this order are produced in micro- or milliseconds and then the band is quenched by the surrounding material. A temperature of 1000°C means that in a carbon steel, ferrite is transformed to face-centred cubic austenite, which if quenched quickly forms martensite. Clearly, the actual transformation temperature and kinetics of the

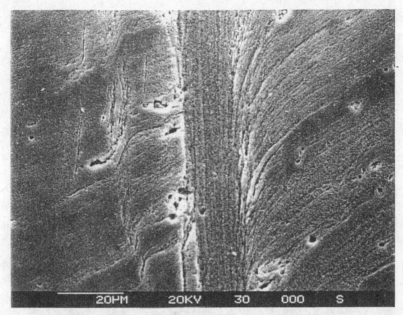

Fig. 2.4. Scanning electron micrograph of a shear band formed in a plate of hardness 150 HV, impact velocity 321 m/sec. After Leech (1985).

phase changes $\alpha \rightleftharpoons \gamma \rightarrow$ martensite will be very sensitive to the process under consideration.

Transformed shear zones or bands have also been observed in wire ropes (as mentioned in Chapter 1) as well as shattered hammer heads (Bhambi, 1979) and failed armament hardware (Samuels and Lamborn, 1978).

Typical transformed shear bands in a steel in an impacted plate are shown in Fig. 1.1; similar bands have been observed in AISI 1040 steel quenched and tempered at 400°C (after Rogers, 1983). Rogers and Shastry (1981) carried out microhardness traverses of shear bands and discovered that the hardness of a band increased linearly with carbon content (see Fig. 2.7). Rogers and Shastry carried out microhardness transverses of adiabatic shear bands for a number of steels (Fig. 2.8). The central area shows the extreme hardness of the bands, but on either side of a band are regions of material, which have undergone various degrees of deformation strengthening and thermal softening. Figure 2.9 shows a microhardness traverse of a 1040

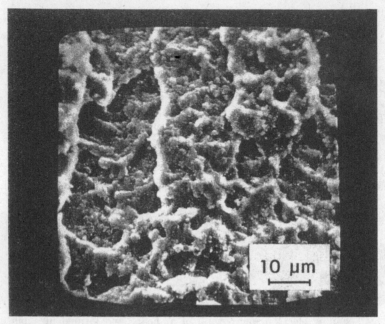

Fig. 2.5.   Scanning electron microscope view of the fracture surface of a plug showing the knobbly surface. After Leech (1985).

quenched and tempered steel. The heat produced in the transformed band was enough to cause a slight softening in the zones immediately adjacent to the band; these are heat-affected zones (HAZ). A little further away from the shear band the hardness begins to increase again over a short width, and these zones were regarded as deformation zones by Rogers and Shastry (1981). It should be remembered, though, that this quenched and tempered steel is metastable and the provision of a little heat will change the microstructure significantly.

Affouard and co-workers (1984) used the split Hopkinson compression bar technique to study the shear banding properties of four martensitic steels. A number of the steel cylinders fractured along the shear bands formed. Affouard *et al.* measured the microhardness across the bands and discovered like Rogers and Shastry that the band hardness was markedly higher than the matrix material.

In a detailed experimental study of shear bands in a chromium armour steel, Derep (1987) observed that shear bands between 0.1

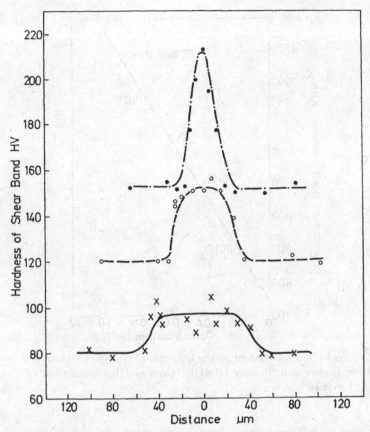

Fig. 2.6.   Microhardness profile of the shear bands shown in Figs 2.2, 2.3 and 2.4. After Leech (1985).

and 0.2 mm in width were formed. On either side of the shear band itself the microstructure consisted of martensite, but the adiabatic shear band consisted of equiaxed grains of δ-ferrite (this is the high temperature form of ferrite) together with laths of martensite. This indicates that this particular alloy was heated to a very high temperature, almost melting it locally. Giovanola (1988a,b) studied the structure of adiabatic shear bands formed in a steel using scanning electron microscope techniques and other fractographic tools. Amongst his many observations he saw ductile voids in the shear bands, indicating that final failure was along the shear bands, but by a ductile mode. The work of Giovanola indicates that the microstructures of

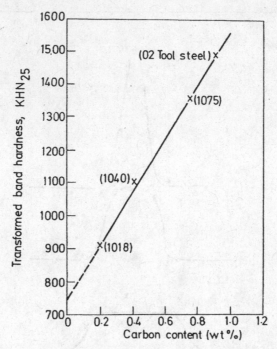

Fig. 2.7.   Effect of carbon content on the hardness of transformed bands in several steels. After Rogers and Shastry (1981), reproduced by permission of Plenum Publishing Corporation.

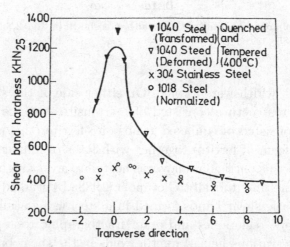

Fig. 2.8.   Profile of microhardness across adiabatic shear bands in several steels. After Rogers and Shastry (1981), reproduced by permission of Plenum Publishing Corporation.

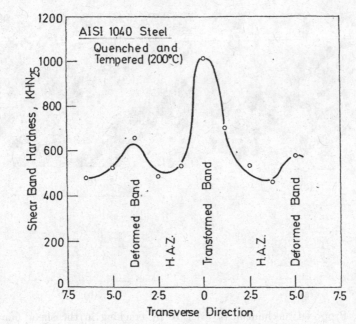

Fig. 2.9. Profile of microhardness across a transformed shear band in 1018 steel quenched and tempered at 200°C. After Rogers and Shastry (1981), reproduced by permission of Plenum Publishing Corporation. HAZ: Heat-affected zone.

shear bands can be very complex, and this will be returned to in Chapter 3.

Numerous researchers have observed adiabatic shear bands in titanium alloys. For example, Wulf (1979) carried out high strain-rate compression on titanium and some titanium alloys using a modified Hopkinson bar technique. Wulf tested three materials: commercial purity (CP) titanium, Ti–6%Al–4%V and Ti–8%Al–7%Mo–1%V. He obtained adiabatic shear bands in all three materials tested at the higher strain rates.

Projectile impact experiments were carried out by Grebe and co-workers (1985) using titanium and Ti–6%Al–4%V alloy targets. These ballistic tests showed that the measured shear band widths varied between 1 and 10 μm for both materials. In the alloy a fracture mode was deduced from a large number of microscopic observations, as shown in Fig. 2.10.

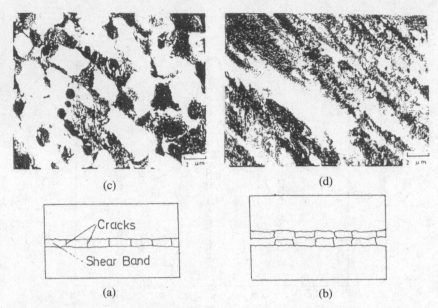

Fig. 2.10. Proposed mechanism of fracture by cracking in the shear band of TI–6%Al–4%V. After Grebe *et al.* (1985).

At 882°C, titanium undergoes a phase transformation from the low temperature α (close-packed hexagonal) phase to the high temperature β (body-centred cubic) phase. It is also possible with titanium alloys to form martensite, depending on composition and cooling rate. As observed by Timothy (1987), a transformed shear band is often partitioned into two regions: a central zone accompanied by significant microstructural modification and an outer zone with a decrease in shear deformation as the undeformed material is approached. Both zones are of comparable dimensions. Martensite has been observed in shear bands in Ti–6Al–4V alloy (see Ansart, 1986). Figure 2.11 shows transformed bands in a titanium alloy and hot-rolled commercial purity titanium (after Timothy and Hutchings, 1984a).

Not all shear bands formed in titanium are transformation bands though. Meyers and Pak (1986) carried out a detailed microscopic analysis of shear bands formed in commercial purity titanium by projectile impact. The shear bands formed had a small equiaxed grain

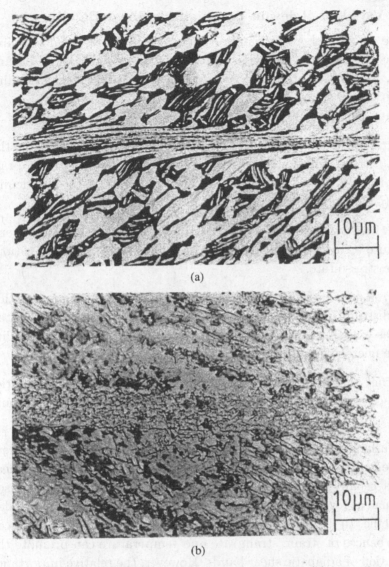

(a)

(b)

Fig. 2.11. Transformed bands in (a) a titanium alloy and (b) hot-rolled commercial purity titanium. After Timothy and Hutchings (1984a).

size indicative of recrystallization, whilst the material adjacent to the shear band had a high dislocation density. It is clear that in these experiments the temperature would not have reached 882°C locally because there was no evidence of a body-centred cubic structure.

2.4. Variables Relevant to Adiabatic Shear Banding

There are a number of variables that affect adiabatic shear banding. These variables can be divided into four categories: material parameters, the stress state, material microstructure and external loading conditions.

An incomplete list of these variables is listed for reference.

Material parameters: density ρ, specific heat c, thermal conductivity λ, thermal diffusion κ, rate of strain hardening $Q = (\partial \tau / \partial \gamma)$, rate of thermal softening $P = -(\partial \tau / \partial \theta)$ and rate of strain-rate hardening $R = (\partial \tau / \partial \dot{\gamma})$.

Stress state: shear stress τ, shear strain γ, shear strain rate $\dot{\gamma}$, temperature θ, hydrostatic pressure p and multiaxial characterizations of the stress state.

Material microstructure: size, shape, spacing, orientation, population and distribution of second-phase particles, inclusions, precipitates, etc., as well as texture, porosity, imperfections and thermal stability of the microstructure.

External loading conditions: disturbances and discontinuities imposed externally, their amplitude, sharpness distributions and energy.

Variables related to external conditions are strongly dependent on the individual case. General experience shows that adiabatic shear bands usually nucleate primarily at stress discontinuities imposed externally, for example, at the sharp periphery of a flat-ended projectile. In addition, there is some feeling that externally induced disturbances of strain, strain rate and temperature can promote the formation of adiabatic shear bands. However, the relative importance of their amplitudes, distribution and so forth is still unclear and under intensive study (see Chapter 8).

The role of the microstructure in adiabatic shear banding is also not well understood. Some effects of the microstructure have been reflected in some macroscopic mechanical variables; for example, the rate of strain hardening, Q, indicates, in a way, precipitation hardening. In addition, pockets of large inhomogeneities or imperfections constitute macroscopic imperfections such as geometrical defects, which should be considered as external disturbances.

There are two main aspects concerned with the effects of microstructure on adiabatic shear banding: nucleation and the eventual morphology of the shear bands.

In dynamic torsion tests (Giovanola, 1987; Marchand and Duffy, 1988) and plugging in ballistic impact (Woodward, 1984, 1990) it was found that adiabatic shear bands nucleate at several isolated locations, although the stress state should be comparatively homogeneous. This may imply that the nucleation of shear bands is affected by the microstructure to some extent. But in these tests there was no direct evidence for effects due to microstructure.

Bedford *et al.* (1974) outlined evidence for several examples of the effects of microstructure on the early occurrence of adiabatic shear bands. An illustrative example has been given by Wingrove and Wulf (1973). It was found that a hardened and tempered 1%C–1%Cr steel containing a fine distribution of small carbides in a martensite matrix was less susceptible to transformed shear bands than a 1%C steel hardened and tempered to the same hardness but with a fully martensite structure. Other examples (described in Sections 2.2 and 2.3) show the tendency of different microstructures to favour either deformed or transformed shear bands.

Shockey and Erlich (1981) and Shockey (1986) reported that there was no positive proof that microstructural heterogeneities were responsible for adiabatic shear bands. The samples used in the contained exploding cylinder technique had no identifiable externally induced imperfections; hence, they were suitable for material study. Examination of micrographs of the initiation region and the paths of adiabatic shear bands did not reveal any identifiable microstructural effects even though inclusions were positioned close to the bands. The shear bands extended without any tendency to favour either the hard tungsten particles or the soft matrix in the 7% tungsten/3% nickel iron alloy.

A strong influence of the texture on adiabatic shear bands was reported by Shockey and Erlich (1981); shear failure appeared to occur along the weaker planes in the rolling texture. So, perhaps, the question is what type of microstructure will influence adiabatic shear banding and how severe will this influence be? The microstructures may fit those wavelengths of disturbances sensitive to adiabatic shear bands.

The early stage of inhomogeneous shearing beyond instability is different from the process that produces fully developed shear bands. Perhaps the microstructure has some influence on early inhomogeneous shearing, but the eventual morphology of adiabatic shear bands is dominated by thermomechanical mechanisms. Some authors have postulated that intense shearing and the elevated temperature within shear bands may destroy the active microstructures responsible for the nucleation of inhomogeneous shearing. Perhaps the more fundamental feature is that adiabatic shear bands are almost always observed in shear or compression. Therefore, the microstructure is not as influential in failure under these stress states as it is in tension in which voids and cracks form. Therefore, it is assumed that the microstructure may have secondary effects on adiabatic shear banding in relation to other major parameters, such as stress state and material parameters. If microvoids or cracks do occur within a shear band, they provide a further softening mechanism, which accelerates the localization. This feedback effect of microdamage on shear banding cannot be ignored.

It has been clarified that adiabatic shear bands usually occur in shear or a compressive stress state, whereas a tensile stress state usually suppresses them. However, the effects of hydrostatic pressure on adiabatic shear banding are still unclear. Figures 2.12 and 2.13 illustrate the effects of hydrostatic pressure on the stress–strain relation. It appears that normal compressive stress and hydrostatic pressure can both retard shear fracture. But how these stresses affect the susceptibility of materials to adiabatic shear banding is unclear. Do these stresses just provide a suitable stress state for adiabatic shear bands to occur, do they accelerate their occurrence or instead do they retard their occurrence? These questions remain unanswered (also see Section 6.1).

The effect of multiaxial stress states on adiabatic shear bands is not obvious, and little experience has been gained of these effects. The study of this question is left largely to numerical simulations (see Section 8.4).

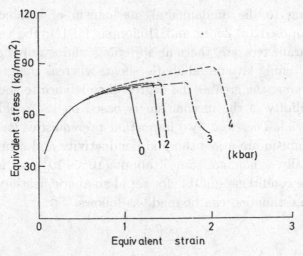

Fig. 2.12. Effect of superimposed hydrostatic pressure on torsional behaviour of a steel. After Osakada and co-workers (1977).

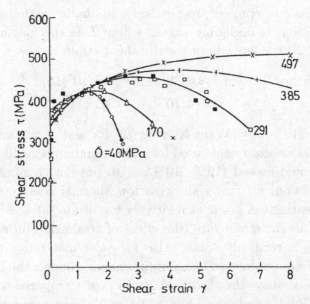

Fig. 2.13. Shear stress–shear strain results for resulphurized low-carbon steel where σ is the normal stress on the shear plane. After Walker and Shaw (1969).

According to the fundamental mechanism of adiabatic shear banding proposed by Zener and Hollomon (1944), the two major material parameters are thermal softening and strain hardening. Thermal softening favours adiabatic shear, whereas the greater the strain hardening the greater the degree of resistance to shear banding. The validity of the mechanism is based on the rapid heating of the material. Therefore, two important prerequisites for the proposed mechanism are a low thermal conductivity and a high strain rate. Generally, a nominal rate of about $(10^2 - 10^3)/$sec is enough to guarantee conditions suitable for rapid adiabatic heating for most metals. The estimation can be made as follows:

$$\frac{\rho c \dot{\theta}}{\tau \dot{\gamma}} \approx 1 \tag{2.1}$$

$$\frac{\lambda \frac{\Delta \theta}{(L/2)^2}}{\tau \dot{\gamma}} \ll 1. \tag{2.2}$$

These expressions represent, respectively, adiabatic heating and that heat conduction is negligibly small, where L is the gauge length. From Eq. (2.2) the lower limit for the shear strain rate is:

$$\dot{\gamma} \gg 4\lambda \, \Delta\theta / \tau L^2 \approx 4(10^1 - 10^2)\Delta\theta/10^8(10^{-3})^2$$
$$\approx 4(10^{-1} - 10^0)\Delta\theta$$

where $\lambda \approx (10^1 - 10^2)\,$W/m K, $\tau \approx 10^8\,$Pa and $L \approx$ mm, which are typical values for metals used for the estimation. Provided $\Delta\theta \approx 100\,$K, $\dot{\gamma}$ should exceed $(10^2 - 10^3)/$sec to prevent significant heat loss. In addition, for materials with low thermal conductivities λ adiabatic heating can occur at relatively low strain rates. Above the lower limit for the strain rate, the effect of strain rate on adiabatic shear bands is relatively small. Also for most materials, including some good heat conductors like aluminium and copper, the heat loss can be ignored above the threshold strain rate. Compared to copper ($\lambda \approx 400\,$W/m K) and iron ($\lambda \approx 82\,$W/m K), titanium has a very low conductivity of $\lambda \approx 19\,$W/m K. Thus it is easy to understand why titanium is so sensitive to adiabatic shear banding.

After heat conduction λ, another thermodynamic parameter is the volumetric specific heat ρc, which has slight variations in metals because of the almost constant molar heat capacity $c \approx 6\,\mathrm{cal/K}$ at elevated temperatures. Certainly, from Eq. (2.1) it can be seen that a lower value of ρc can lead to a higher temperature and a consequent increased tendency to adiabatic shear banding. This may happen at temperatures lower than the Debye temperature in metals, i.e. at cryogenic temperatures, and for some light- and heat-sensitive materials.

In engineering, the general experience is that metals with the same composition but a higher strength or hardness caused by a difference in heat treatment can be prone to adiabatic shear bands. For example, high-strength quenched and tempered steels frequently exhibit adiabatic shear bands compared to their lower strength counterparts. This empirical rule is extremely important in industrial applications and will be explained later in Section 6.4.

It is impossible to list the effects of each individual parameter. Therefore, before embarking on a sophisticated discussion of the mechanisms of shear banding in Chapters 6, 7 and 8, we will use dimensional analysis to provide a general view of these effects. The dimensional analysis is mainly limited to the stress state and material parameters.

In one-dimensional simple shear, the independent and determining parameters of the stress state and the characterization of the material are:

	Parameter	Dimension	Implication
γ	Shear strain		Strain state
$\dot{\gamma}$	Shear strain rate	T^{-1}	Rate state
τ	Shear stress	$ML^{-1}T^{-2}$	Stress state
ρ	Density	ML^{-3}	Inertia
c	Specific heat	$L^2T^{-1}\theta^{-1}$	Heat temperature
κ	Thermal diffusion	L^2T^{-1}	Heat transfer
$(\partial\tau/\partial\gamma)$	Strain hardening	$ML^{-1}T^{-2}$	Strain hardening
$(\partial\tau/\partial\dot{\gamma})$	Strain-rate hardening	$ML^{-1}T^{-1}$	Strain-rate hardening
$-(\partial\tau/d\theta)$	Thermal softening	$ML^{-1}T^{-2}\theta^{-1}$	Thermal softening

T, M, L and θ are the four independent dimensions representing time, mass, length and temperature. According to Buckingham's

π theorem in dimensional analysis, five independent dimensionless parameters can be formed. One choice is as follows:

strain $\quad\quad\quad\quad\quad\quad\quad\quad\quad\quad \gamma$

effective Prandtl number $\quad Pr = \tau/\rho\dot{\gamma}\kappa \approx 10^6$

strain hardening $\quad\quad\quad N = \dfrac{(\partial\tau/\partial\gamma)}{\tau} \approx 10^{-1}$

thermal softening $\quad\quad\quad S = \beta(-\partial\tau/\partial\theta)/\rho c \approx 10^{-1}$

rate sensitivity $\quad\quad\quad\quad m = \dfrac{\dot{\gamma}}{\tau}(\partial\tau/\partial\dot{\gamma}) \approx 10^{-2}$

where β is the fraction of plastic work converted into heat. The approximate numbers are typical for most metals. From these dimensionless parameters it is possible to find some specific combinations of thermal diffusion and rate effect, thermal softening, volumetric specific heat, etc.

Amongst these are $1/Pr \approx 10^{-6} \approx 0$, which implies that heat diffusion has a negligible effect, i.e. approximately adiabatic conditions. Also the more usually defined Prandtl number $Pd = (\partial\tau/\partial\dot{\gamma})/\rho\kappa = Prm \approx 10^4$ leads to a similar adiabatic approximation.

Here, primarily, two dimensionless parameters N and S are comparable. Their balance provides a useful dimensionless number B:

$$B = S/N = \beta\frac{(-\partial\tau/\partial\theta)\tau}{\rho c(\partial\tau/\partial\gamma)}.$$

B describes the balance of the coupled thermomechanical process due to a strain increment $d\gamma$ because B can be rewritten as

$$B = \frac{\beta\tau\, d\gamma(-\partial\tau/\partial\theta)\, d\theta}{\rho c\, d\theta(\partial\tau/\partial\gamma)\, d\gamma}$$

$$= \frac{\begin{array}{c}\text{plastic work resulting from } d\gamma\\ \times\text{ thermal softening resulting from } d\theta\end{array}}{\begin{array}{c}\text{temperature rise due to plastic work}\\ \times\text{ strain hardening resulting from } d\gamma\end{array}}.$$

Later, in Chapter 6, it will be shown that $B = 1$ corresponds to the maximum of the adiabatic stress–strain curve.

Also, instead of N the strain hardening index $n = \gamma N$ is more useful in practice. Therefore, as well as γ, the four compound dimensionless parameters that appear frequently in adiabatic shear banding discussions are:

(a) *Effective Prandtl number* $Pr = \tau / \rho \dot{\gamma} \kappa$, which represents the ratio of the rate effect to the heat dissipation.
(b) *The coupled thermomechanical number*

$$B = \beta(-\partial \tau / \partial \theta)\tau / \rho c(\partial \tau / \partial \gamma),$$

which represents the balance of the temperature effect with the mechanical behaviour.
(c) *The strain hardening index*

$$n = \gamma \left(\frac{\partial \tau}{\partial \gamma} \right) \Big/ \tau.$$

(d) *The strain-rate hardening index*

$$m = \dot{\gamma} \left(\frac{\partial \tau}{\partial \dot{\gamma}} \right) \Big/ \tau.$$

More implications of these dimensionless numbers, especially Pr, will be given later in Chapters 6 and 7.

One feature of these material parameters should be kept in mind: they cannot be kept constant under normal conditions, when the material within the adiabatic shear band experiences an unusually large deformation and temperature. Some variations in these parameters are given in the appendices as a reference, but the conditions under which these data were obtained will not be the same as that occurring in adiabatic shear bands. Therefore, the estimations and values listed above are all illustrative and the data in the appendices are just for reference.

2.5. Adiabatic Shear Bands in Non-Metals

Adiabatic shear bands do not occur exclusively in metals. There have been a number of reports of adiabatic shear bands occurring in polymers and rocks under high pressure. But, so far, there have been no observations of adiabatic shear bands in glasses or ceramics. One may wonder whether adiabatic shear bands can occur in these materials under high pressure. However, under normal conditions most ceramics and glasses are too brittle to accumulate enough plastic shear strain for adiabatic shear to occur.

Generally speaking, polymers may be more prone to the occurrence of adiabatic shear bands because of their lower heat conductivity than most metals. Approximately, polymers have typical values of thermal diffusivity of about $\kappa \approx 10^{-7} \, \mathrm{m^2/sec}$ and specific heats $c \approx 2 \times 10^3 \, \mathrm{J/kg \, K}$. Hence, the thermal conductivity $\lambda = \rho c \kappa \approx 2 \times 10^3 \times 2 \times 10^3 \times 10^{-7} \approx 10^{-1} \, \mathrm{W/m \, K}$. Whereas even for titanium, perhaps the most sensitive metal to adiabatic shearing, the values of heat dissipation are $\kappa \approx 8 \times 10^{-6} \, \mathrm{m^2/sec}$ and $\lambda \approx 19 \, \mathrm{W/m \, K}$, which are two orders of magnitude higher than those for polymers. Clearly, adiabatic shearing should be an important mode of deformation and failure in polymers.

However, polymers are fundamentally very different from metals. The variations in commercial grades having nominally the same composition with the same name are considerable. The variations are not only from batch to batch of the same polymer, but also can be due to chemical and physical aging (Walley *et al.*, 1989a,b). This is a common experience when testing polymers. Therefore, any confirmation, comparison and application of the mechanical properties of polymers should be done very carefully. In addition, the values for heat diffusivity and conductivity are only illustrative. Certainly, there is a necessity to increase fundamental knowledge of the mechanical and thermal properties of polymers. For this reason Table 2.1 lists some fundamental data for various polymers. It would not be surprising at all if different values for the parameters listed in Table 2.1 were found in different reports. The flow stresses of polymers may range between ten and hundreds of megapascals, depending on composition, processing and even moisture. The strain-rate sensitivities of these polymers are generally within the range 5–15 MPa per decade of strain rate over the testing strain rate range of 10^{-2} to 10^3/sec. In the region of strain rate of about 10^3 to 10^4/sec, most polymers tested show a decrease in stress and exhibit thermal softening. However, it must be underlined that this

Table 2.1.

Primary polymers studied

Full name	Polymer	Commercial type	Specific heat Temperature (K)	Specific heat Value ($J\,g^{-1}\,K^{-1}$)	Melting temperature (K)	Heat of fusion ($J\,g^{-1}$) Sample	Heat of fusion ($J\,g^{-1}$) 100% cryst.	Formula mass
Nylon 6	N6	Akulon M244H (Natural) from AKZO (Holland)	298 313 (T_g) 313 483	1.49 1.58 2.41 2.63	483	92 ± 1 (dry)	230	113
Nylon 66	N66	Maranyl A100 (Natural) from ICI	298 323 (T_g) 323 533	1.45 1.48 2.22 2.65	533	95 (dry)	300 190	226
Polycarbonate	PC	Makrolon 2800 from Bayer AG	298 418 (T_g) 418 560 (max. tabulated)	1.20 1.70 1.89 2.21	403 (608 in equilibrium)	*ca.* 1	132	254
	Noryl	731 Grey from GE (US)	298 373 (T_g) 413	1.45 1.88 2.02	413 (decomposes)	*ca.* 0	—	—
Polybutylene teraphthalate	PBT	Valox 325 from GE (US)	298 383 (T_g) 463	1.57 2.22 2.27	496	52	Not known	204
Polyvinylidene difluoride	PVDF	Kynar 730 from Pennwalt ($T_g = 233$ K)	298	1.32	483	Not determined because of risk of damage from HF	106	64

(Continued)

Table 2.1.   (*Continued*)

Secondary polymers studied

Full name	Polymer	Commercial type	Specific heat		Melting temperature (K)	Heat of fusion ($J\,g^{-1}$)		Formula mass
			Temperature (K)	Value ($J\,g^{-1}\,K^{-1}$)		Sample	100% cryst.	
Polytetra-fluorethylene	PTFE	298	0.90	1.02	605	39 ± 1	82	50
		450	1.15	1.27	604			
		603	1.35	1.39	($T_g = 220\,K$)			
Polymethyl-methacrylate	PMMA (Perspex)	298		1.37	Decomposes	—	—	100
		378 (T_g)		1.69				
		378		2.03				
		550 (max. tabulated)		2.44				
Polystyrene	PS	298		1.07	513	*ca.* 0.5	84	118
		373 (T_g)		1.38				
		373		1.64				
		510		1.95				
Polyethylene	PE	298	1.55	2.20	Fully crystalline: 415 (HDPE)	114 ± 3 (HDPE)	293	14
		390	2.31	2.49	HDPE: 387	180 ± 10 (MDPE)		
		413	2.58	2.57	MDPE: 395			
					($T_g = 237\,K$)			
Polyethersulphone	PES	298		1.37	Amorphous	—	—	232
		503 (T_g)		2.11	($T_g = 483\,K$)			
Polyvinylchloride	PVC (Darvic)	298		0.94	565	180	176	62.5
		354 (T_g)		1.11				
		354		1.42				
		380		1.57				

Data collated by Walley *et al.* (1989a).

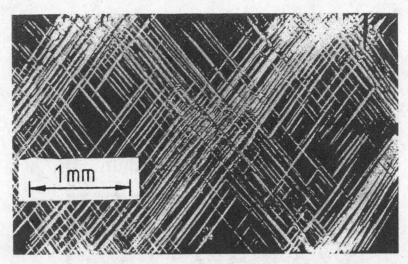

Fig. 2.14. Section cut from a sample of polystyrene that has been deformed just past yield at 22°C in a plane strain compression test viewed between crossed polars. Strong shear bands have been nucleated at the corners of the dies. Drawing of a photograph after Bowden (1970).

is quite a general picture of the thermal and mechanical properties of polymers.

Perhaps Bowden (1970) was the first researcher to report an observation of localized shear bands in homogeneous compression (hence excluding geometrical softening) in polymers. In this case the polymers were polystyrene and PMMA. As the strain rate increased, the shear bands changed from diffuse to sharp shear bands with a width of $1\,\mu$m (Fig. 2.14). The bands were observed at strains of about 4%. The final strain in the shear band of polystyrene was approximately 2.

Later, Winter (1975) reported an observation of adiabatic shear bands in PMMA and compared them with shear bands observed in titanium. Figure 2.15 shows the obvious damage to PMMA subjected to impact by a flat-ended punch. This is similar to the subsurface pattern of adiabatic shear bands in metals (see Section 9.3). The similarity between the two types of shear localization led Winter to think that they were governed by the same mechanism, namely adiabatic heating. Moreover, the quasi-static indentation of a flat-ended punch into PMMA had a serrated curve of load versus displacement (see

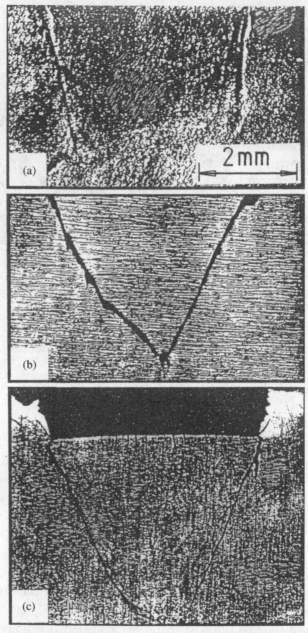

Fig. 2.15. Damage sites in PMMA produced by loading with a flat-ended rod 4 mm in diameter, (a) and (b) by impact, (c) quasi-statically. The impact veloc-ities for the damage sites in (a) and (b) were 69 and 100 m/sec respectively and the length of the rod was 7 mm. After sectioning the specimens were crazed in acetone for 30 sec. Drawings based on photographs after Winter (1975).

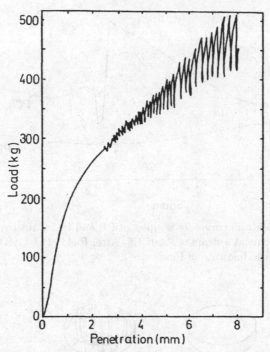

Fig. 2.16.   Load versus displacement curve obtained during loading of PMMA by a flat-ended indenter of diameter 2.5 mm. The loading rate was 0.1 mm/min. After Winter (1975).

Fig. 2.16), which is very similar to that obtained by Basinski (1957) in cryogenic tension tests of metals. The serrated curve of stress versus strain in cryogenic tension was confirmed as being caused by adiabatic heating (see Section 1.3).

Generally speaking, there are two groups of polymers. One group has lower thermal conductivities and latent heats of fusion, for example $E = 7.2 \, \text{kJ/kg}$ for PC. The other group has higher values of these parameters, for example $E = 95 \, \text{kJ/kg}$ for polypropylene (PP) (see Field *et al.*, 1984). The polymers that Bowden (1970) and Winter (1975) tested, i.e. PMMA and PS, all belong to the first group. In addition, PC and PTFE are in the same group. These polymers mainly fail by shear banding, whereas PP, high density polyethylene (HDPE), nylon and PVC belong to the second group, which exhibit mainly bulk deformation without shear bands.

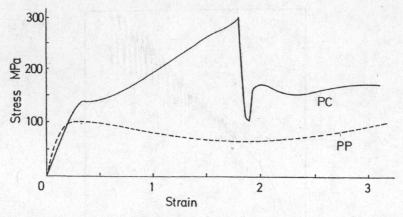

Fig. 2.17.   Stress–strain curves for samples of PP and PC. Catastrophic failure of the PC sample occurs at a strain of about 1.8. After Field *et al.* (1984), reproduced by permission of the Institute of Physics.

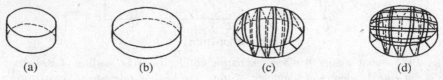

| (a) | (b) | (c) | (d) |

Fig. 2.18.   Shear and cracking sequence observed during impact on PC, PS or PMMA. The sequence was similar for each of the materials, although the duration and time between stages differed: (a) initial undeformed disc, (b) disc at completion of plastic flow stage, (c) failure by parallel shear cracks, (d) continued failure by perpendicular shear-cracks. Drawing based on those of Swallowe *et al.* (1986).

Close examination reveals the difference between the two groups more clearly. Figure 2.17 shows the stress–strain curves for samples of PP and PC. PP exhibits quite a stable stress–strain curve, indicating uniform bulk plastic deformation, whilst the curve for PC shows catastrophic failure occurs at a strain of about 1.8 corresponding to failure by shear bands.

Figure 2.18 illustrates the sequence of failure resulting from adiabatic shear bands in samples of PC, PS or PMMA subjected to impact loading. Firstly, there is only bulk plastic deformation. Then, at some critical compressive strain, parallel shear bands occur, which

Table 2.2.

	A^* natural strain	Time of cracking	Max. temperature of plastic deformation (K)	Max. observed crack temperature (K)	Latent heat (kJ/kg)	
·HDPE	0.9	—	≪200	—	120	
PP	>2.5	—	230	—	95	Group 2
Nylon 6	1.0	—	400	—	130	
PVC	0.7	—	450	—	60	
PS	0.01	250	<200	550	—	
PMMA	0.04	400	<200	530	—	Group 1
PTFE	0.25	150	<200	600	37	
PC	1.1	20	<200	700	7.2	

A^* represents the strain at which cracking occurs or the observed maximum plastic strain.

become connected by perpendicular bands. Eventually, the sample was covered by a network of shear bands and this then failed completely. The timescale for this sequence is different for different materials. For example, shear failure occurs explosively in PC. Table 2.2 compares the strain, cracking time, temperature and latent heat for various polymers. The difference between the two groups of polymers is very clear and the timing for shear failure is different in polymers of the first group. Figure 2.19 illustrates shear banding after drop-weight tests on PC.

To give a clear picture of what a network of adiabatic shear bands in polymers look like in reality, Figs 2.20 and 2.21 show a specimen of PC and PS, respectively, after impact tests. The high-speed photographs reveal the progress of the formation of adiabatic shear bands. The pictures of Figs 2.20 and 2.21 show a transient measurement of the temperature on the surface of the tested sample made with a thin temperature-sensitive sheet (see Section 4.5). The similarity between the patterns of the bands shown in Figs 2.20 and 2.21 is further proof that the mechanics of adiabatic heating is responsible for band-like deformation.

Fig. 2.19.   Specimens after deformation in the drop-weight apparatus. Failure of
1 mm thick PC was by shear banding. After Walley *et al.* (1989a).

Adiabatic shear bands are known to form in ceramics and rocks
under certain conditions. At high hydrostatic pressures, the ductility
of ceramics and rocks is greatly increased as observed by von Karman
(1911). Under these conditions adiabatic shear bands are possible and
are assumed in some cases.

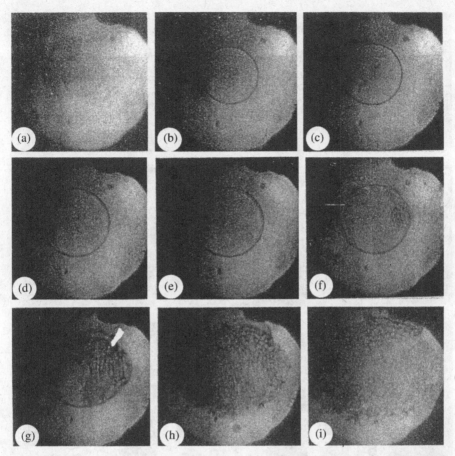

Fig. 2.20.   Impact on a 5 mm diameter, 1 mm thick disc of PC. Times after first frame: (a) 0, (b) 161 μsec, (c) 188 μsec, (d) 241 μsec, (e) 268 μsec, (f) 275 μsec, (g) 281 μsec, (h) 288 μsec, (i) 295 μsec. After Swallowe *et al.* (1986).

Orowan (1960) proposed that adiabatic shearing under extreme pressures leading to shear melting should be the mechanism for seismic faulting in the earth's mantle.

Grady (1974, 1977, 1978) and Murson and Lawrence (1979) calculated temperature histories for four minerals for which both compression and release-wave measurements were available: SiO_2, MgO, $CaCO_3$ and Al_2O_3. Grady (1980) described some shock wave properties of these materials with a model based on inhomogeneous deformation together with a transient thermal state (see Fig. 2.22).

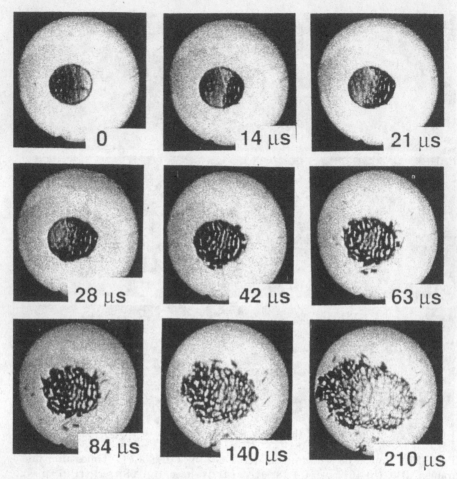

Fig. 2.21. Impact on a disc of PS; times from impact are shown. After Walley *et al.* (1989b).

Kondo (1987) proposed a so-called 'skin-model' to estimate the heterogeneous temperature distribution in the compaction of ceramic powders by shock waves. The hot skin is the localized shearing zone (Fig. 2.23). For further details on the compaction of powders, see Blazynski (1987).

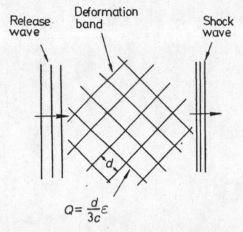

Fig. 2.22. Illustration of a uniform distribution of deformation bands with nominal separation 1, after passage of a shock wave. Based on a diagram after Grady (1980).

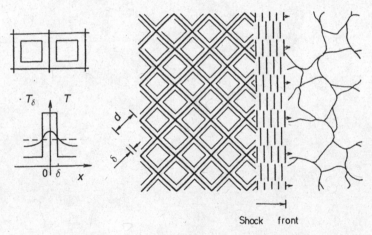

Fig. 2.23. Schematic illustration for the skin model in the compaction of a ceramic powder by a shock wave. Based on a diagram after Kondo (1987).

Chapter 3

Fracture and Damage Related to Adiabatic
Shear Bands

Adiabatic shear bands occur commonly in dynamic compression, which tends to suppress void nucleation and growth as well as cracking. A tensile stress state favours fracture by void growth and cracking and usually prevents a sufficient accumulation of plastic strain for adiabatic shear banding to occur. Thus, although adiabatic shear bands occur in compressive stress fields and are not associated with void nucleation and growth, it should be emphasized that adiabatic shear bands are commonly the precursors to fracture. Also materials that contain shear bands but have not yet failed that are sensitive to successive loadings, particularly if the bands are transformed bands. This is because transformed shear bands are more brittle than the original matrix and provide a potential passage for brittle cracks. In other words, adiabatic shear bands are the weaker areas in the material. These facts are significant for materials that may have been subjected to repeated impact loadings, such as armour or certain fabricated components (Rogers, 1983).

3.1. Adiabatic Shear Band Induced Fracture

Adiabatic shear bands usually act as precursors to brittle or ductile fracture. Figure 3.1(a) shows a polished cross-section through a penetration hole in an electroslag remelted 4340 steel plate. The

65

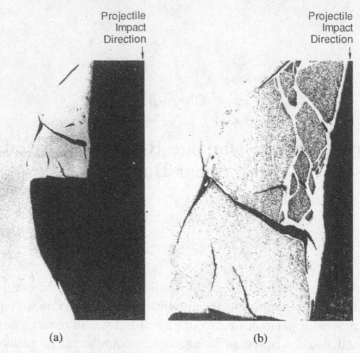

(a)   (b)

Fig. 3.1.   Planar microcracks and adiabatic shear zones near the plugged region in steel armour. (a) Polished section through projectile hole; (b) etched close-up view of (a). After Shockey (1985).

direction of impact is shown in the figure. There are a few visible cracks near the periphery of the hole. Additionally, some fragments seem to have been formed. When the same area was etched, a network of white etching transformed shear bands became visible (Fig. 3.1(b)). Importantly, the majority of the cracks shown in Fig. 3.1(a) lie within these shear bands. Also a number of areas, which are almost fragments, are clear. These fragments have been completely encircled by white etching bands. These pictures show that adiabatic shear bands play a significant role in fracture under impact loading.

The mechanisms for the occurrence of these brittle and ductile fractures within or near the adiabatic shear bands that usually appear in compressive or shear loadings will now be studied.

From the initiation of adiabatic shear bands to complete fracture is not a single simple process. Giovanola (1987) reported that according to his observations made from high-speed photographs of thin-walled tubular specimens, individual independent shear bands nucleate at several locations and then coalesce with each other, finally forming some well-defined steps in the fracture path. Similar steps have also been observed in plugging by Woodward and co-workers (1984). Adiabatic shear bands initiate at several points around the periphery of the projectile, then propagate towards the rear of the target, as well as joining circumferentially, eventually forming that special mode of fracture called plugging.

Now let us examine the fractography of shear band induced fracture (after Huang, 1987). The fracture surface of mild steel subjected to dynamic torsion testing exhibits typical parabolic dimples (see Fig. 3.2). Within the dimples can be seen particles, which indicate the initiation sites of these dimples. This is a clear indication of

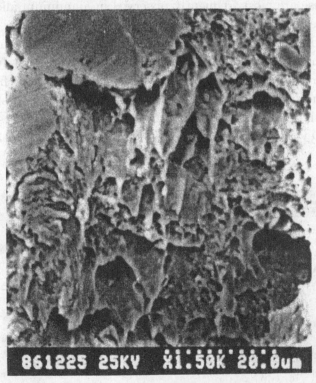

Fig. 3.2.   Parabolic dimple patterns due to adiabatic shear banding in low-carbon steel. After Huang (1985).

ductile shear fracture. On the same fracture surface there are some
smooth areas, which are larger than those with the parabolic dimples.
These areas are caused by mutual rubbing of the surfaces. Even for
high-strength steel tested in dynamic torsion, ductile fracture surfaces with voids appear (see Hargreaves and Hoegfeldt, 1976), as a
result of adiabatic shear bands. These researchers also reported that
the size and spacing of the voids on the fracture surface increase with
increasing strain rate. Brittle fracture is usually observed in transformed structures associated with adiabatic shear banding. Hence,
brittle fracture should occur after the shear band has been formed
and after it has cooled down. In this case the cracks may occur parallel to the shear band or transverse to its length, depending on the
local stress state.

Perhaps the most novel feature of adiabatic shear band induced
fracture is the clusters of small knobbly regions (Fig. 2.5, after Leech,
1985). These are most frequently observed in aluminium alloys, but
are also seen in steels. The mechanism for this feature may be twofold. The high local temperature within the shear band transforms
the band metal to liquid and this liquid is almost immediately cooled
by the surroundings. Alternatively, this could result from frictional
heat produced by the two fractured surfaces subjected to shearing.
Either of these mechanisms can produce high temperatures. There
is further evidence, though, of localized melting. Figure 3.3 is a

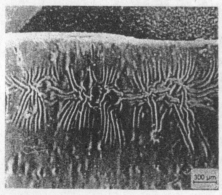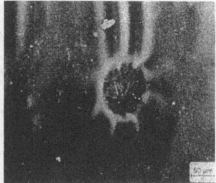

Fig. 3.3.   Solidified liquid film with shrink holes. After Hartmann *et al.* (1981),
reproduced by permission of Plenum Publishing Corporation.

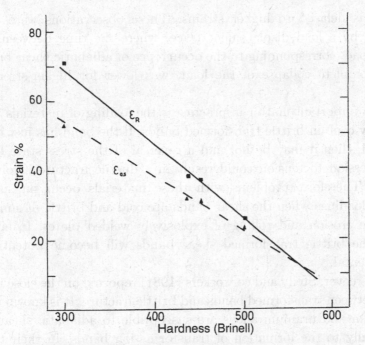

Fig. 3.4. Strain at fracture ϵ_R and strain at the start of adiabatic shearing ϵ_{as} versus the initial hardness in 28 CND 8 steel. After Affouard *et al.* (1984), reproduced by permission of the Institute of Physics.

scanning electron micrograph of a shear surface, which clearly has a solidified liquid film.

Affouard and co-workers (1984) reported a very interesting practical relation between adiabatic shear bands and fracture. Four martensitic steels were tested under dynamic compressive loading with a split Hopkinson pressure bar. Microscopic observations at various stages of shear deformation and fracture were performed to determine the critical strains for the initiation of adiabatic shear bands as well as band-induced fracture. The results are plotted in Fig. 3.4, which shows the critical strains for shear banding and fracture versus the initial hardness of the steels. It was found that the difference between the two corresponding strains decreased rapidly with increasing initial hardness of the steels, though both strains decrease with increasing initial hardness. These results imply that once adiabatic shear bands form in high-strength steels, fracture may ensue rapidly, although the onset of shear banding in high-strength steels is delayed to higher strains. These observations were confirmed by a load–displacement trace, where the time between the peak load, corresponding to the occurrence of adiabatic shear bands and complete collapse of the load, was lower for higher-strength steels.

An important matter in practice is the loading of materials that already contain brittle transformed bands. If the band has just been formed, then it may be hot and a reversal of the stress state from compression to tension could result in a ductile fracture involving voids. Therefore, problems with these materials occur on subsequent loadings when the shear bands are cold and brittle. Examples are the erosion and rolling of explosively welded plates. In either case the brittle transformed shear bands will become potentially detrimental.

Moreover, Stelly and co-workers (1981) reported on the close relation between transformed bands and brittle fracture. It is known that titanium and uranium alloys are susceptible to adiabatic shearing, especially to the formation of transformation bands. In their tests using exploding cylinders and shells made with these materials, it was found that fracture was clearly related to the shear bands and the metals exhibited small elongations. In the exploding cylinder tests there is a tensile stress in the test piece, though high pressure also acts on the inner surface of the wall.

3.2. Microscopic Damage in Adiabatic Shear Bands

It has been generally assumed that fracture induced by adiabatic shear bands is the consequence of the accumulation of microscopic damage that occurs in adiabatic shear banding because of the large plastic deformations with a band. In particular, metals rich in second-phase particles, or with a large number of inclusions, should produce an apparent deformation-induced anisotropy, a consequence of which is significant microdamage. Now, the occurrence of adiabatic shear bands may be influenced by microdamage.

Firstly, let us examine the relation between adiabatic shear bands and cracks as well as voids. The most straightforward case is a sharp crack within a transformation band (Fig. 3.5). This is a

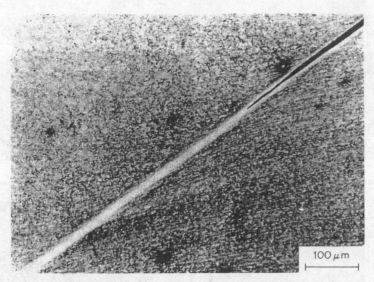

Fig. 3.5.   Adiabatic shear band and an associated crack in a martensitic steel deformed in a dynamic compression test. From Dormeval (1987), courtesy Elsevier Applied Science Publishers Ltd.

typical brittle fracture showing a sharp crack tip extending along the transformation band. This situation often, but not always, happens in brittle transformed bands. Backman and Finnegan (1973) photographed several types of brittle cracks associated with transformed bands (Fig. 3.6). Apart from cracks that are completely confined to the shear band, there are various intersections of microcracks with the transformed shear bands. Though the microcracks extend into the matrix to some extent, they are all closely related to the transformed bands. This indicates that brittle transformed bands are responsible for microcracks. The complexity of individual cases may be due to differences in the stress states that initiate fracture or to the properties of the transformed structure and matrix.

However, ductile microvoids form in transformed bands as well. Figure 3.7 shows this very clearly. Here a number of nearly equiaxed voids have formed within the shear band of a uranium alloy. The void shape indicates that they formed after shearing has ceased but while the band remained hot. In addition, the stress state responsible

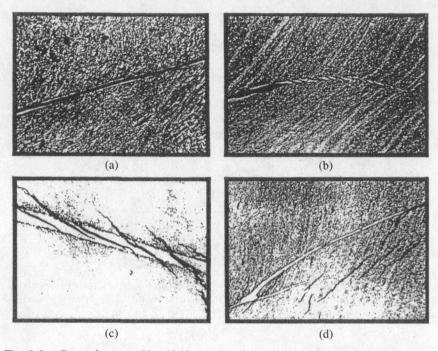

(a)

(b)

(c)

(d)

Fig. 3.6.   Several types of brittle fractures of transformed bands. Drawings based on photographs after Backman and Finnegan (1973).

for void formation should be tensile and normal to the band. At the upper left-hand side of the picture, the voids are almost linked together to form a ductile rupture. Figure 3.8 shows a similar but slightly different picture. In Fig. 3.8(a), instead of equiaxed voids, elliptical voids have formed within the transformed band within the titanium alloy plate, indicating the difference in the stress state under which the voids initiated. Me-Bar and Shechtman (1983) also reported these two types of voids. For comparison, Fig. 3.8(b) also shows a brittle crack formed in the transformation band of the titanium alloy. From the photographs shown in Figs 3.6 to 3.8, it is clear that the morphology of microdamage is strongly affected by the stress state and the current state of the band, i.e. whether it is hot or cold. However, in either case the shear bands are the origin of the cracks and voids.

Several reports have indicated that the microdamage associated with adiabatic shear bands consists of a number of microvoids and

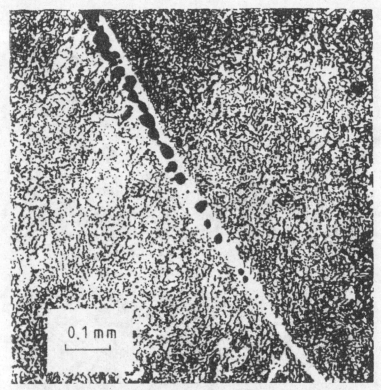

Fig. 3.7.   Section of a fractured shear band in a U-2Mo alloy. After Irwin (1972).

microcracks (see, for example, Johnson *et al.*, 1983a,b; Giovanola, 1987; Huang, 1987; Xu *et al.*, 1989). In these cases, significant cavitation occurs both in and outside the shear band. In their metallographic examination of the microdamage in a less ductile metal, S-7 tool steel, at shear strain rates of 10^2/sec, Johnson *et al.* (1983a,b) found that eventual fracture occurs in relation to a shear band and they observed a large number of voids in a distinct shear band of about 100 μm in width. While the void density in the shear band was high, voids of a similar size were also observed in the remainder of the gauge length of the thin-walled tubular specimen. Johnson and co-workers noticed that in a few cases there were voids in undeformed metal. Although it is not certain that the voids observed in the shear band were all induced by the shear deformation, the higher density

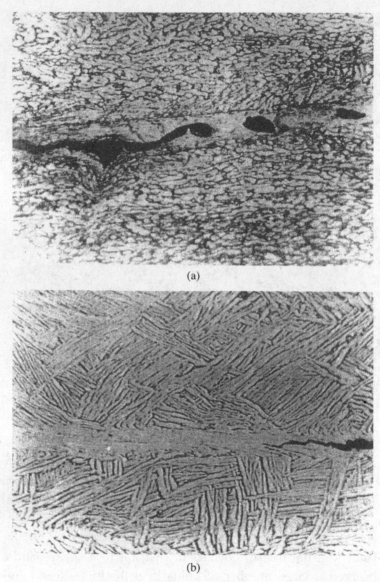

(a)

(b)

Fig. 3.8.   Adiabatic shear bands and associated voids and cracks in a titanium plate under impact loading. Courtesy Xian Mengmei.

of voids there does imply that the increase of microdamage results from adiabatic shear banding.

Huang (1987) and Xu *et al.* (1989) clarified these points. Tests were also performed in dynamic torsion but at high strain rates of

10^3/sec using a split Hopkinson torsional bar. The material tested was more ductile, a hot-rolled low-carbon steel ($20^\#$). They sectioned the specimen and then examined the distortion and rotation of the grains. In addition, the microcracks on the sectioned surface were counted. Xu *et al.* (1989) assembled the photographs as functions of distance along the gauge length (Fig. 3.9). This diagram shows clearly the accumulation of microdamage that accompanies grain distortion. In particular, the microdamage becomes extremely severe near the core of the shear band, i.e. in those regions subjected to greater shear

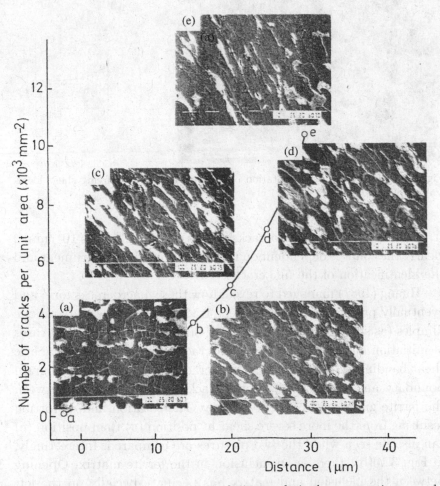

Fig. 3.9.   Relation between the number of microcracks in the cross-section normal to the shear band and the distance. After Xu *et al.* (1989).

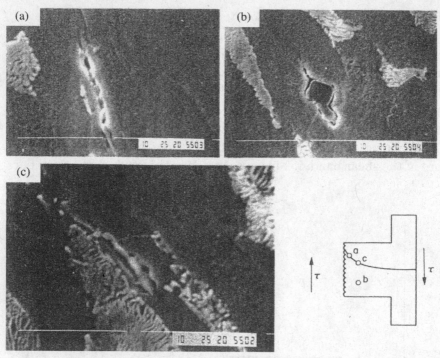

Fig. 3.10.   Initiation and orientation of microcracks accompanying shear localization. After Huang (1987).

strain. The number of microcracks in this case is as high as $10^4/\text{mm}^2$;
clearly the count is dependent on the resolution of the instrument and
the identification of the fine cracks.

Huang (1987) managed to reveal how these macrocracks form and
eventually produce the ductile shear fracture surface with parabolic
dimples (as shown in Fig. 3.2). Figure 3.10 shows three typical cases
of initiation and orientation of microcracks accompanying adiabatic
shear banding. Figure 3.10(a) shows cracking along the ferrite grain
boundary and Fig. 3.10(c) shows cracks at the interface between
the ferrite grain and a pearlite colony. The rotation and distortion
resulting from the more severe shear at position (a) than position (c)
can also be seen when the two pictures are compared. Interestingly,
in Fig. 3.10(b), there is an inclusion in the ferrite matrix. Opening
between the inclusion and matrix has begun, especially on the left
upper and right lower areas of the inclusion, due to the driving shear

orientation. Additionally, a second crack has begun to extend along the direction of alignment of the grains at the same position. From these micrographs it is a little clearer how the occurrence of micro-cracks is strongly influenced by the microstructure of the material and how the accumulation of a large amount of microdamage may result in eventual fracture. It would obviously be very interesting to compare the effects of shear bands and microdamage in pure and single-phase metals with the above. Furthermore, it still remains unclear whether this kind of microdamage is responsible for the occurrence and formation of adiabatic shear bands as a mechanism of softening.

Further observations have been made using a transmission elec-tron microscope on the microstructural changes occurring in adia-batic shear bands (see, for example, Me-Bar and Shechtman, 1983; Meyers and Pak, 1986; Xu *et al.*, 1989). As these observations were made on very small areas, fundamental features were revealed, such as dislocation substructures. Although these features are not directly related to microdamage, such as microvoids and microcracks, they should illustrate the underlying microscopic mechanisms contribut-ing to adiabatic shear banding as well as microdamage.

Me-Bar and Shechtman (1983) and Meyers and Pak (1986) examined adiabatic shear bands formed in ballistic tests in targets of Ti–6%Al–4%V and commercial purity titanium. In both cases transformed shear bands were formed with a width of 3–40 μm in Ti–6%Al–4%V and 1–10 μm in titanium. The grains were nearly equiaxed and were very fine, ranging from 0.05 to 0.3 μm in grain size. Both spherical and elliptical voids were formed in the bands (Me-Bar and Shechtman, 1983). More importantly, the diffraction patterns of the shear bands in the as-received alloy plate are rings (see Fig. 3.11). These rings correspond to the close-packed hexagonal phase and indi-cate that there is a very fine structure with almost no preferred ori-entation. These observations also revealed that the transformation products in the shear bands contain a high density of dislocations. Also a high dislocation density was observed outside the shear bands.

Xu *et al.* (1989) performed a very detailed examination of the microscopic morphology of shear bands and microdamage. The

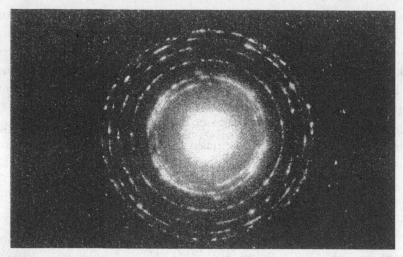

Fig. 3.11.   A diffraction pattern from a shear band. After Me-Bar and Shechtman (1983).

samples were taken from fractured thin-walled tubular specimens tested with the split Hopkinson torsion bar. The specimens sustained a comparatively well-defined torsion loading and lower strain rates of 10^3/sec, unlike those of previous researchers as mentioned above, which were about 10^5/sec. The material was hot-rolled low-carbon steel and only deformed bands were formed. The thin transmission electron microscopy (TEM) foil was made approximately perpendicular to the eventual fracture surface. Figure 3.12 shows that the grains with the shear bands were distinctly elongated. Grains with original sizes of tens of millimetres become long strips with widths of fractions of a micron. However, the elongated grains retain their clear crystallographic characterization (see Fig. 3.14). From these observations, it may be assumed that microscopic shear results from crystallographic slip on a dominant slip system within individual grains. Figure 3.13 reveals that the density of dislocations is very high. In particular, within grains the tangled arrangement of dislocations tends to align in accord with the shear. It might be anticipated that the large strains and stress concentrations resulting from dislocation pile-ups (as shown in Fig. 3.13) at grain boundaries and phase boundaries within the shear band are responsible for the initiation of microcracks, such as those shown in Fig. 3.11.

Fig. 3.12. TEM foil micrograph showing the microscopic and crystallographic nature of the elongated grains within a shear band. After Xu *et al.* (1989).

Fig. 3.13. The high density of dislocations and their tangled arrangement in the elongated grains within a shear band. After Xu *et al.* (1989).

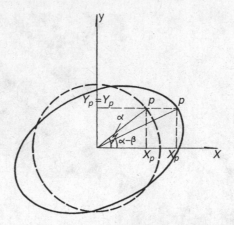

Fig. 3.14.   Schematic of rotation and stretching in large simple shear.

Unfortunately, for such an important problem as shear band induced damage and fracture in engineering practice, there is not at present a detailed model or theory. Such a model is required badly, even if it only gives estimations. Certainly, the problem is much more difficult than that of a shear band itself. The following illustration shows some of the difficulties involved in the kinematics of extremely large shears.

Figure 3.14 shows the change in the configuration in simple shear, which consists of rotation and stretching. When we examine the orientation of a microcrack it is important to take into account shear and rigid body rotation. Let P denote a point on a circle of radius R in the original configuration. Then when subjected to simple shear

$$x = X + \gamma Y,$$
$$y = Y.$$

P moves to (x_P, y_P) on an ellipse in the current configuration. The major and minor axes should represent the principal stretching, i.e. the following must be satisfied:

$$r_P^2 = x_P^2 + y_P^2 = X_P^2 + 2\gamma X_P Y_P + (1 + \gamma^2) Y_P^2$$
$$d\gamma_P = 2(X_P + \gamma Y_P) dX_P + 2\{\gamma X_P + (1 + \gamma^2) Y_P\} dY_P = 0.$$

The tangent and normal of the principal stretching in the original configuration then should be:

$$\frac{dY_P}{dX_P} = -\frac{X_P + \gamma Y_P}{\gamma X_P + (1 + \gamma^2)Y_P}$$

$$\tan\alpha = \frac{Y_P}{X_P} = -\frac{dX_P}{dY_P} = \frac{\gamma X_P + (1 + \gamma^2)Y_P}{X_P + \gamma Y_P}$$

therefore,

$$\tan\alpha = \frac{1}{2}\{\gamma \pm \sqrt{\gamma^2 + 4}\}.$$

The orientation of the principal stretching in the current configuration is

$$\tan(\alpha - \beta) = y_P/x_P = \frac{\tan\alpha}{1 + \gamma\tan\alpha}$$

where β is the rotation of the principal stretch and

$$\tan\beta = \gamma/2.$$

Accordingly, the principal stretch is

$$1 + \epsilon = \frac{a}{R} = \sqrt{x_P^2 + y_P^2}/R = \sqrt{1 + \gamma\tan\alpha}.$$

The following illustrate the corresponding orientations.

y	$\alpha°$	$(\alpha - \beta)°$	$\beta°$	ϵ
0	45	45	0	0
1	58.3	31.3	26.6	0.62
4	76.7	13.3	63.4	3.24
10	84.4	5.64	78.7	9.1

Clearly at large shear strains, such as $\gamma \sim 10$, the principal stretch is comparable to the shear strain and nearly parallel to the shear (5.64°).

This simple exercise illustrates the difficulty of correlating the orientation of cracks to the stretching or the stress state. For instance, the maximum tensile stress remains at an inclination of 45° to the shear direction, but the maximum stretching attained may not. If

$\gamma = 10$, it would only be $5.64°$ in the shear direction. In fact, the orientation of the maximum stretching rotates in an element with increasing shear strain, from $45°$ to become normal and parallel to the shearing in the original and current configuration, respectively. If a microcrack appears at some time and is treated as a marker, its orientation in the final configuration can be examined by taking the distortion and the rotation of the element into account.

3.3. Metallurgical Implications

As we have seen, the initiation and propagation of fracture is often the result of the formation of adiabatic shear bands at high strain rate. Backman (1969) has listed three ways in which the termination of deformation can occur: (a) tensile stress wave failure leading to spalling, (b) large strain failure after general yielding and (c) inhomogeneous shear fracture. In ballistic impact it is clear that more than one of these modes may be present in the termination of deformation.

Much research has been carried out using stepped projectiles (see Wingrove, 1973a,b, for example) to study the strain gradient associated with the projectile. Wingrove used 1-cm-thick steel plates. Careful examination of the plates showed that shear bands were only observed after the projectile had penetrated a distance of about 1 mm. Cracks developed at the back surface of the plate later in the process and further penetration caused a plug to be detached from the plate.

As observed by Rogers (1974) in steels subjected to dynamic loading conditions, temperatures can reach in excess of $1000°C$ in the centre of the shear band in microseconds and be quenched subsequently forming a brittle band. These brittle bands can fail in ordnance because of the number of stress waves released. Abbott (1960) observed a reflected tensile wavefront interact with the conical adiabatic shear zone beneath a flat-faced projectile, detaching the cone.

Andrew *et al.* (1950) and Beetle and co-workers (1971) have provided detailed fractographic evidence of adiabatic fractures in steels. Andrew and co-workers carried out impact compression tests on steels of varying compositions. In a higher carbon steel (greater than 0.8% C) fracture occurred usually on a plane at $45°$ to the compression axis, but occasionally on planes parallel to the axis. The former fracture surfaces were observed to be smooth due to rubbing, whilst the other surfaces were rough.

Using specimens from exploded silicon–manganese steel cylinders, Beetle and co-workers observed that the radial fracture mode consisted of cleavage fracture. Further from the centre of the cylinder the authors observed a large number of small equiaxed dimples ($<1\,\mu$m). Beetle *et al.* concluded that there were two fracture modes.

If fracture is ductile it will have occurred during the formation of the adiabatic shear band when the band was hot. Brittle fractures occur after deformation in a cool band. Fracture occurs under these conditions normal to the tensile axis. When bands fracture during shearing, the voids are normally very elongated. Sometimes brittle transformed bands may form in a steel without fracture. This is a very dangerous situation because on subsequent loading fracture will occur along the bands.

Giovanola (1988a,b) carried out an extensive investigation into the shear banding in a 4340 steel in pure shear. He used high-speed photography and a number of microscopic facilities. Giovanola found that the shear band phenomenon occurred in two stages over widths of $60\,\mu$m and $20\,\mu$m, respectively. Strain rates approaching 1.4×10^6/sec were calculated and strains of 17 were measured in the band. The total energy dissipated per unit area of the shear band was about $0.18\,\text{MJ/m}^2$.

From the detailed experimental work it was possible to develop a knowledge of fracture in shear bands. Figure 3.15 shows how the strain develops in the shear band as a function of time. Figure 3.16 shows the strain-rate history during the shear test. The strain rate gradually increases during homogeneous deformation to reach about 10^4/sec. During the first localization of deformation, the strain rate jumps to a value of about 10^5/sec and then reaches a value of 1.4×10^6/sec during the second localization. The shear stress, shear strain and temperature of the material within the shear band can be constructed from all the collected data, and this adiabatic diagram is shown in Fig. 3.17.

It was observed that each specimen failed on a number of parallel planes connected by well-defined steps. Individual shear bands

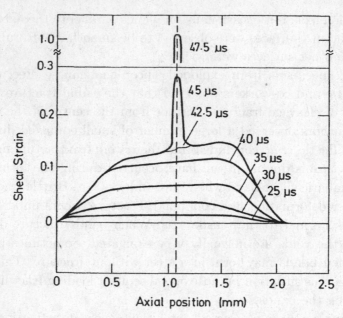

Fig. 3.15. Development of shear strain with time in a shear band. After Giovanola (1988a).

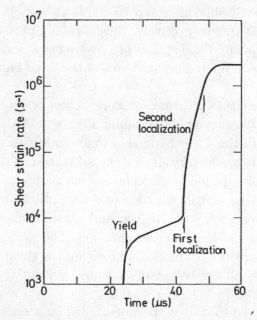

Fig. 3.16. Two-step shear localization in a 4340 steel. After Giovanola (1988a).

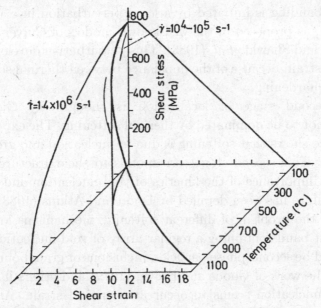

Fig. 3.17. The shear stress, shear strain and temperature history of 4340 steel. After Giovanola (1988a).

nucleate at several locations and on different planes around the circumference. From scanning electron microscope studies, it was concluded that the fracture surface consists of three main features: (1) large areas of smeary holey material, (2) patches of microvoids and (3) tongues of highly smeared material covered partially by knobbles. All three features are the result of the mechanism of fracture under shear of the material. Void patches are formed when unbroken ligaments between the two sliding planes break under tensile loading parallel to the shear plane. This mechanism has also been suggested by Dormeval and Stelly (1981).

It has been confirmed that large strains accumulate in the shear band material, which in turn points to strong evidence of high local temperatures. High-speed photographs suggest that the two stages of localization may well correspond to a deformed band and a transformed band in situations of combined shear and compression. The evidence points to white etching bands in this material resulting from the high temperatures caused by opposite surfaces rubbing against each other.

Shear banding is initiated by a local perturbation in a uniform strain field, as proposed in the well-known models of Culver (1973), Bai (1981) and Shawki *et al.* (1983). This perturbation develops with increasing strain because of the imbalance between thermal softening and strain hardening.

The second stage of localization is thought by Giovanola (1988a,b) not to be dominated by thermal softening. The experimental evidence shows that softening is due to nucleation and growth of microvoids. This second stage corresponds to shear fracture of the band. The importance of the kinetics of void nucleation and growth in shear failure has been detailed by Dodd and Atkins (1983). They considered the problem of different softening mechanisms for a different shear band containing a regular array of void nucleation sites, which could be second-phase particles, inclusions or grain boundaries.

From the work of Goods and Brown (1979) and others it is clear that void nucleation seems to occur at a critical strain. At larger strains γ_c, voids will have nucleated along the plastically deforming zone. From the work of Rice and Tracey (1969) the growth rate of a single void has the form

$$\dot{H}_T = \dot{\gamma}\left(a + b\frac{\sigma_H}{\sigma_Y}\right) \quad \text{for } \gamma > \gamma_c \qquad (3.1)$$

where a is related to the amplification of the growth rate of the void relative to the strain rate of the matrix (Le Roy *et al.*, 1981), b is a constant dependent on geometry, σ_H is the hydrostatic stress and σ_Y is the yield stress. Integration of Eq. (3.1) with respect to time gives

$$H_T = (\gamma - \gamma_c)\left(a + b\frac{\sigma_H}{\sigma_Y}\right) \quad \text{for } \gamma \geq \gamma_c. \qquad (3.2)$$

Once we have an assumed constitutive equation, it is possible to derive an equation for γ_i, the geometrical instability strain in shear derived in the absence of any consideration of thermal softening effects.

It is clear that any induced hydrostatic pressure will tend to suppress void nucleation to higher strains and therefore under these circumstances there is a greater chance of thermoplastic shear. This is somewhat similar to the case in compression testing. According to Giovanola (1988a,b), there has been undue emphasis on white etching bands and their width and structure and this has overshadowed important studies on the macro- and micromechanisms of adiabatic shear banding.

Timothy and Hutchings (1984b) studied the initiation and growth of microfractures along adiabatic shear bands in Ti–6%Al–4%V. This alloy was subjected to ballistic impact. There were two distinct types of fracture. In the rims of the indentations the voids grow during the late stages of loading or at the start of unloading when the lip is raised above the level of the rest of the free surface. In these lip zones, fracture occurred by growth and coalescence of voids along the shear bands. At the base of the impact craters, even though the material is subjected to a compressive stress, field and void growth is unlikely although any microvoids present may elongate.

Proposed mechanisms for fracture in this material are shown in Fig. 3.18. In Fig. 3.18(a), when the void nucleation sites within the shear band are widely separated, voids grow under the combined action of tension and shear. At a critical void separation coalescence occurs, the surface exhibiting ductile dimples. When the void nuclei are closely spaced, voids coalesce almost immediately. The model of Timothy and Hutchings (1985) is simple and is encompassed by the detailed model of Dodd and Atkins (1983).

3.4. Effects of Stress State

When a material is subjected to a tensile load the onset of necking normally precludes shear banding because of the lack of accumulation of plastic work. Despite this drawback, adiabatic shear band failures can occur in cryogenic tensile tests where the thermal properties of the material are very different (see Basinski, 1957).

In ballistic impact experiments, Zener (1948) showed qualitatively the effects of superimposed tension and compression on the planes of maximum shear. For example, when a large blunt-nosed projectile impacts a thin plate, radial tensile stresses are set up by the bulging of the target, which produces an inverted conical plug. In

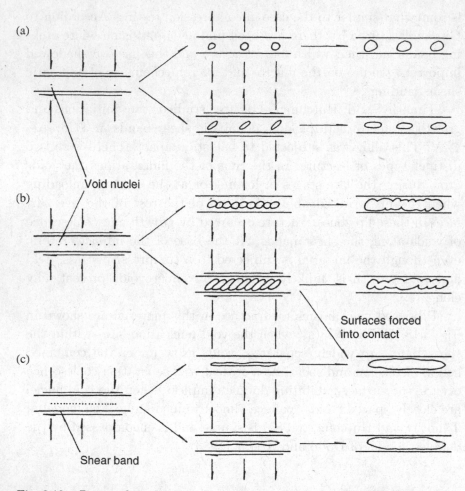

Fig. 3.18.   Proposed mechanisms for fracture in a titanium alloy. (a) Void nucleation sites widely spaced, (b) void nucleation sites more closely spaced, (c) void growth from more closely spaced nucleation sites. After Timothy and Hutchings (1985).

impact tests on thick plates with blunt noses as well as hemispherically ended projectiles, a dead metal cap is created underneath the projectile in the target and frictional forces are important in determining the orientation of the maximum shear planes.

In the explosive fragmentation of cylinders, shear bands are observed on the inner wall. If we consider that there is an initial compressive hoop stress in the interior of the cylinder wall and a tensile hoop stress in the outer part of the wall, then the tensile stress wave reflected by the outer surface causes failure along the shear bands.

Fracture has been observed to occur under the action of tensile stresses whether the fracture is cleavage or void nucleation and growth. Fracture under compressive loading occurs along one or more normally well-defined shear planes.

In ballistic impact and other loading situations, where the stresses are predominantly compressive or shear, more plastic work can be accumulated without necking, which can lead to adiabatic shear banding. Dodd and Bai (1985) observed that the adiabatic shear band width is not very sensitive to stress state. However, the microscopic nature of fracture along the bands will be sensitive to stress state. If for example a shear band is formed in a compressive stress field, the voids or microcracks within the band will tend to weld up as deformation proceeds. However, as shown by Giovanola, final fracture may still be by void growth along hot or warm shear bands.

If a band contains a number of void nucleation sites, then the critical strain at which voids will nucleate will depend on the hydrostatic pressure as illustrated by Dodd and Atkins (1983). For high hydrostatic pressures, void nucleation will be delayed. However, for tension, void nucleation, growth and coalescence can be rapid, leading to a rapid catastrophic failure, as shown in Fig. 3.18(c).

Although stress state does not affect the shear band width significantly, it can affect the mode of failure profoundly, as for example we can see from the work of Timothy and Hutchings (1985) on Ti–6%Al–4%V.

Backman and co-workers (1986) studied the scaling of adiabatic shear banding in aluminium alloy targets impacted with steel spheres. No significant differences were found in the shear band width no matter what the size of the indenter. A post-mortem counting of fragments suggested a linear scaling factor for the number of fractures and therefore the number of shear bands, since fracture occurred along the bands. However, there was no evidence of any changes in the mode of fracture with scale and this confirms that the fractures were intimately connected with the adiabatic shear bands.

Further Reading

Xi, Y. and Meyers, M. A. (2012) 'Nanostructural and microstructural aspects of shear localization for materials'. In *Adiabatic Shear Localization: Frontiers and Advances* (eds. Dodd, B. and Bai, Y.) pp. 111–171, Elsevier, London.

Chapter 4

Testing Methods

The first three chapters of this book have provided a general background to adiabatic shear bands and some features that are specific to them. It is well known that the phenomenon of adiabatic shear banding appears in a variety of fields in mechanical engineering and materials technology. In addition, the unusual characteristics of shear bands make the phenomenon impossible to describe or identify in terms of materials data commonly available in handbooks and from conventional testing techniques. Furthermore, the practical and engineering experiments in which adiabatic shear bands do occur, such as high-speed impacts, generally cannot be used for the systematic study of the occurrence and formation of shear bands, identification of the susceptibility of materials to adiabatic shear banding and so forth. This is because, in such cases, usually a number of phenomena with different physical origins occurs simultaneously and they are impossible to differentiate. For example, in penetration, adiabatic shear banding, cracking, spalling, etc., can all occur in the projectile and target. Moreover, the stress states involved in these cases are too complex to distinguish the effects of stress states and material behaviour. Therefore, there is a necessity to design and establish specific testing methods and techniques for the study of adiabatic shear banding.

4.1. General Requirements and Remarks

Before studying the details of the technical aspects of these testing methods and techniques, it is very important to clarify some requirements that are quite general but specific to the study of adiabatic shear bands.

Firstly, the testing techniques should be able to create a high enough strain rate in the tested materials to form adiabatic shear bands. Figure 4.1 shows some common tests in stress–strain rate space. More importantly, the stress state in the gauge length of the tested material should be well defined and comparatively simple; for example, simple shear and uniaxial compression. This is very important, particularly when the data are used in practice or in comparison with other tests. For the time being, various split Hopkinson bar techniques, flyer plate impact tests and several specially designed high-speed testing machines can meet the requirements.

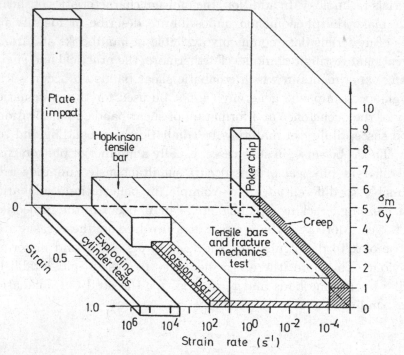

Fig. 4.1.   Regions of stress, strain and strain rate attained in various mechanical tests. After Curran *et al.* (1987).

However, the above are only the essential and necessary requirements. The crucial requirement may be to achieve a relatively large and homogeneous plastic shear deformation within a defined gauge length. As described in Chapters 1 and 2, the occurrence of adiabatic shear bands is mainly due to thermal softening outweighing the plastic hardening of the material. However, the increase in the temperature depends on the plastic deformation work done on the material. Therefore, a knowledge of the amount and accumulation of plastic shear deformation is necessary for the study of adiabatic shear bands.

It is necessary to avoid geometrically inhomogeneous deformation defects such as necking or barrelling to achieve a relatively large and uniform plastic shear deformation at high strain rates together with a well-defined and simple stress state. This constitutes the basic prerequisite for the experimental study of shear bands.

Conventional materials testing equipment provides records of load and displacement or strain. In dynamic tests monitoring the strain rate is also necessary. Fortunately, the simultaneous recording of the average stress, strain and strain rate is possible using the split Hopkinson bar techniques (see Sections 4.2 and 4.3). Although these mechanical variables averaged over the whole gauge length are necessary for the study of adiabatic shear bands, a shear band is a localized area of deformation within the gauge length. Once the shear band has formed, the average deformation variables, the nominal strain and nominal strain rate do not provide enough information to describe the plastic deformation fully. Hence, transient recordings, in particular the transient recordings of the whole deformation field, both inside and outside the intensely strained narrow zone, are required. This requires not only the temporal response of the instruments to resolve the advance of shear localization, but also two spacial scale resolutions, a coarse one for the general deformation field and a fine one for the localized zone.

Usually, three types of testing techniques are commonly used in laboratories. These are the dynamic torsion, dynamic compression and contained exploding cylinders. Each technique has its own advantages and disadvantages and each test covers different strain-rate ranges and stress conditions. Some general remarks will be made here concerning the testing techniques, but the more detailed specifications and key procedures will be described under separate headings.

Torsion tests have a considerable number of advantages over other tests in the study of large dynamic plastic flow and relevant instability and localization phenomena. It is for this reason that it has been the most commonly adopted test for the study of adiabatic shear banding (see Lindholm *et al.*, 1980).

Torsional loading produces a simple shear deformation with no macroscopic volume change and, therefore, it provides the true effect of shear deformation in materials. However, one point should be taken into account. Due to the nature of simple shear, a rigid rotation of the material element occurs. Therefore, when considering the large plastic simple shear deformations, the rigid body rotation should also be taken into account. Furthermore, microscopic dilation due to microvoid nucleation resulting from local non-shear stresses at inhomogeneities is not precluded by the requirement of no macroscopic volume change. Nevertheless, simple shear deformation, provided by torsion tests, is a convenient tool in experimental studies.

In a thin-walled tubular specimen tested in torsion, a nearly homogeneous state of shear deformation can be obtained fairly easily along the gauge length and across the wall thickness. This mode of homogeneous shear deformation can greatly reduce the amount of data processing required for tests. However, the flanges attached to the two ends of the thin-walled tube should be designed carefully and the flange corners and the testing section should be machined carefully so that a state of homogeneous shear can be guaranteed in the gauge length.

Under these conditions, because there is no change in the cross-section of the thin-walled tube, homogeneous shear deformation can be maintained to very large plastic strains. Simple shear overcomes problems with macroscopic geometrical instability such as diffuse and localized necking in tension and barrelling in compression, and therefore it is a particularly important test in the study of shear localization.

In high strain-rate tests, such as in the split Hopkinson bar technique, the torsional version has distinct advantages over the compression machine. When elastic torsional waves propagate along long cylindrical rods, they do so without suffering any geometrical dispersion. However, elastic compression waves can distort severely as they propagate along the rod. A long bar subjected to compressive wave loading cannot be loaded in an exactly one-dimensional state. The surface of the rod is stress free, whilst near the centre an element of the bar will be in a multiaxial stress state. This geometrically induced geometrical dispersion can only be minimized by reducing the ratio of the bar diameter to the elastic wavelength. For the torsional split Hopkinson bar system this restriction is not necessary. Moreover, there will be no friction on the interface of the thin-wall tube specimen and the bars, as occurs in dynamic compression tests. Finally, the average strain rate and average strain can be controlled accurately in the torsional Hopkinson bar case, unlike the other tests such as cylinder impact or explosive ring expansion. All of these features make the data processing in dynamic torsion much easier than in the other tests.

Nevertheless, torsion tests have disadvantages. The torsional deformation mode is not straightforward for engineers in industry because it is not obvious how results can be used. Also, test-piece machining is more complicated than that of the solid cylinders used in compression testing. Another aspect of dynamic testing is the rise time of the loading pulse. The rise time for torsional loading usually ranges between 20 and 40 μsec, because of the limitation of the mechanism of torque release. Therefore, the strain rate in loading is limited, although shortening the gauge length can increase the strain rate.

Compression tests have advantages as well as disadvantages. The most attractive aspect of dynamic compression tests is that they are straightforward. Dynamic loading in various applications, industrial and military, is almost always due to dynamic compression, at least initially. Such features as tensile cracking, spallation and shear banding can be traced back to impulsive compression, impact compression or compressive shock waves. Therefore, the results obtained from dynamic compression tests can be linked directly to specific applications; hence this type of test tends to be more readily accepted by engineers in industry. Another feature of dynamic compression that is often overlooked is compression under plane strain conditions. The importance of the plane strain stress state is that it can provide pure shear deformation, i.e. shear without any rotation in the deformed configuration. This can make the examination of morphological changes of a material element during the course of deformation easier to follow. Of course, the two end surfaces of the plane strain compression test piece are nearly subjected to plane strain conditions. Finally, because of the direct impact nature of these tests, the strain rates tend to be higher than those achieved in torsion tests. However, the drawbacks of dynamic compression have already been alluded to.

The contained exploding cylinder is a special testing procedure. It is not a conventional material test because it has its origin in practical studies in explosion dynamics. Furthermore, generally speaking, the requirements for conventional tests such as a well-defined stress state and uniform deformation within a gauge length are not realized in this technique. However, if Fig. 4.1 is examined carefully it is possible to see that the technique used in contained exploding cylinders has a specific place in the stress, strain and strain-rate space — this is the left-front corner (the region of ultra-high strain rate and large plastic deformation). Interestingly, these two features are amongst the prerequisites for adiabatic shear banding. Therefore, contained exploding cylinder tests are very important in the study of adiabatic shear bands.

4.2. Dynamic Torsion Tests

Making use of shear tests to study adiabatic shear bands certainly seems to be the most straightforward approach. However, to create a truly uniform shear deformation, without bending, induced axial stresses, etc., is by no means an easy task. Therefore, experimentalists have been researching for many years how to improve specimen design so that ideal shear deformation can be achieved.

Punching a piece of metal with a hard punch is one of these tests (Fig. 4.2(a)). However, it has been discovered that the results of tests are strongly influenced by the clearance between the punch and die

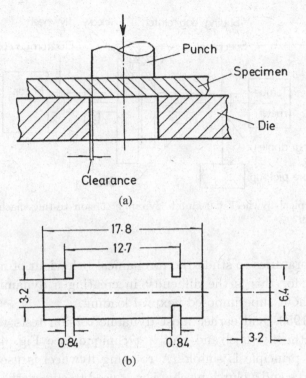

Fig. 4.2. Sketch of punching (a) and double shear (b) tests.

and the sharpness of the edges of both punch and die. Thus, this test configuration is rather complicated in the derivation of material behaviour.

The double shear test (Fig. 4.2(b)), fitted to an hydraulically operated testing machine or a split Hopkinson bar apparatus, is another attempt to minimize the non-uniform stress and strain in the testing section. However, it is only when the plastic deformation is considerable that the non-uniformity of deformation decreases. To avoid substantial bending of the specimen, the gauge length must be very small; however, the degree of bending in these specimens has not been assessed. See Campbell and Ferguson (1970), Harding and Huddart (1979) and Ruiz (1991) for experimental results using this technique.

The torsion test has obvious advantages over this notched shear test. However, before the 1960s dynamic torsion tests were seldom

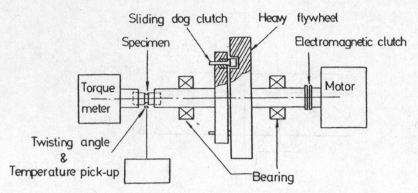

Fig. 4.3.   Typical flywheel and clutch dynamic torsion testing machine. After Vinh *et al.* (1979).

used in laboratories to study the mechanical behaviour of materials. The reason for this is the difficulty in creating a dynamic torque loading without superimposed flexural loading.

In the 1950s and earlier most dynamic torsion tests were conducted by the flywheel and clutch technique (see Fig. 4.3). The operational principle is simple. A rotating flywheel is used as the loading source and a clutch mechanism is used to engage the flywheel with the specimen. In the 1970s, Culver (1972) used a lathe to perform dynamic torsion tests and studied the susceptibility of materials to adiabatic shear banding. Kobayashi (1987) also modified a lathe (Fig. 4.4) and investigated carefully the formation and growth of adiabatic shear bands in titanium alloys and aluminium–lithium alloys. The advantage of this lathe-type torsion testing machine is the ability to control rotation speed easily. To avoid undesirable axial and flexural loadings, the design of the clutch is very important. For example, in Kobayashi's torsion lathe, a ramp clutch was adopted to give a smooth engagement whilst at the same time minimizing the induced axial and flexural loadings. The lathe type of torsion testing machine is easy to develop, and according to Culver and Kobayashi this type of machine can produce maximum shear strains of 3 and shear strain rates of between 1 and 300/sec. However, it should be borne in mind that before testing, the axial and flexural loading must be monitored.

Impact torsion testing machines have been under development for many years (for example, see Sakui *et al.*, 1966). Lindholm

Dynamic Torsion Testing Machine

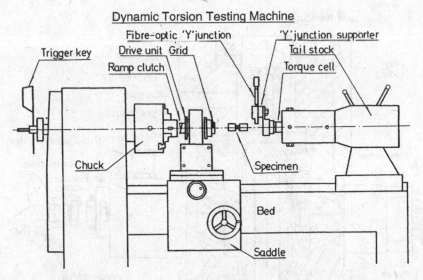

Fig. 4.4. Schematic of the torsion testing lathe. After Kobayashi (1987).

and co-workers (1980) developed an advanced hydraulic actuator, which they used in a high-speed torsion testing machine. Although a hydraulic testing machine can only provide strain rates up to $10^2/\text{sec}$, this type of machine can be used in tests with large plastic deformations. A hydraulic torsional machine can typically achieve rotational displacements of about 3.3 radians. Therefore, provided that the radius and gauge length of the specimens are 7 mm and 3 mm respectively, the corresponding shear strain is about 7.

Figure 4.5 shows the cross-section of a torsional actuator. The concept behind the operation of this machine is quite simple. It consists of a low-mass hydraulically driven torsional actuator and shaft, which coupled to the torsional test piece by a very stiff housing. To provide the machine with the maximum torsional stiffness, the entire loading system is contained in a simple cylindrical housing. The sections of the housing are rigidly bolted together.

Application of a hydraulic pressure between the rotary vanes and those fixed to the housing creates a torque, which acts on the specimen. The special design of this machine is to use two stacked, single-vane sections. This design utilizing these two stacked vane sections may induce a lateral moment on the shaft. In this particular machine this was compensated for by an area of bushing.

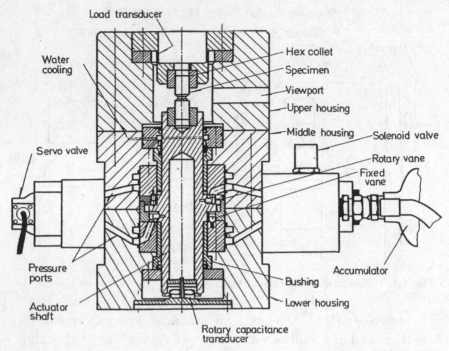

Fig. 4.5.   Typical hydraulic actuator dynamic torsion testing machine. After Lindholm *et al.* (1980).

For higher strain rates an open loop was used in the machine. A fast-opening solenoid valve is incorporated. A hydraulic pressure differential of 22 MPa provides a high rotational velocity. The maximum design torque is 339 Nm. This machine is representative of the second type of dynamic torsional testing apparatus.

The third type of dynamic torsion testing technique is the split Hopkinson torsion bar apparatus. This is a relatively inexpensive experimental set-up and can reach high strain rates of the order of 10^3/sec.

The principle of the torsional Hopkinson bar apparatus is completely different from the previous two machines. The wheel-clutch technique makes use of rotary kinetic energy to cause torsional deformation in the specimen. The hydraulic actuator takes advantage of the potential energy of the pressurized medium. Compared to these two mechanisms, the torsional Hopkinson bar is simple: a sudden release of a stored torque acts on the specimen. However, the design of the mechanisms for storing and releasing the torque were proposed in the late 1960s.

Baker and Yew (1966) and Duffy and co-workers (1971) are amongst the first researchers to develop split Hopkinson torsional bar apparatuses. After numerous improvements over the last 20 years, the current version of the torsional split Hopkinson bar apparatus in common use is different from the original.

Figure 4.6 shows a general view of this kind of apparatus. A schematic layout of the apparatus is shown in Fig. 4.7. In this machine, a short thin-walled tubular specimen is sandwiched between two long elastic bars, called the input (or loading or incident) bar and the output (or transmitted) bar, respectively. Before dynamic testing, torque is stored in the part of the input bar between the rotating head and a clamp. The torque is provided by a hydraulic actuator. The stored torque is monitored by strain gauges attached to the bar. A dynamic torsion test is started by simply releasing the clamp. When this is done, half of the stored torque travels down the bar and loads the specimen. The torsional stress pulse is then partially reflected along the input bar and partially transmitted into the specimen. At the interface of the specimen and the output bar the wave is subjected to a further transmission and reflection and a transmitted wave enters the output bar. Figure 4.8 shows a schematic Lagrangian diagram of the waves and typical torque traces recorded by the input and output bar strain gauges. The reflections of the waves within the short sandwiched specimen lead to a quasi-static state being established within the specimen and therefore a uniform stress state is reached within the gauge length. Quasi-static equilibrium is established in two reflections and the time for equilibrium is about $t \approx 4 \, l/c \sim 4 \times 2\,\mathrm{mm}/3\,\mathrm{mm}/\mu\sec \approx 3\,\mu\sec$. Thus for a test lasting 1 msec, the non-equilibrium time is extremely short.

The shear stress, shear strain rate and shear strain can all be calculated according to the recordings of the incident torque wave M_i and the reflected torque M_r recorded at the input strain gauges and the transmitted torque M_t, recorded at the output gauge:

$$\tau = \frac{M_i + M_r + M_t}{2sr_m} \approx \frac{M_t}{sr_m} \qquad (4.1)$$

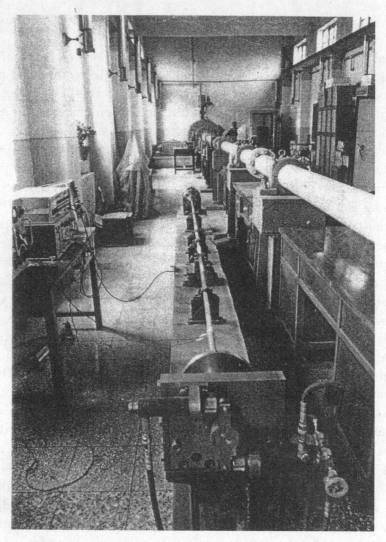

Fig. 4.6.   Split Hopkinson torsional bar apparatus. Courtesy L. T. Shen

where s is the cross-sectional area of the specimen and r_m its mean radius. For thin-walled tubular specimens, $s \approx 2\pi r_m t_s$ where t_s is the wall thickness.

The equality $M_t = (M_i + M_r + M_t)/2$ used in formula (4.1) illustrates the assumed quasi-static equilibrium conditions. The average strain rate over the gauge length L can be calculated as the difference of the linear velocities at the two ends of the tube

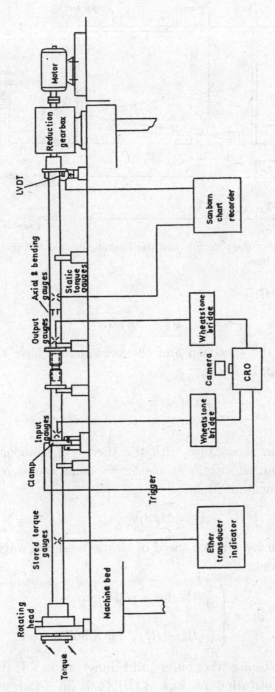

Fig. 4.7. Schematic layout of a split Hopkinson torsional bar apparatus.

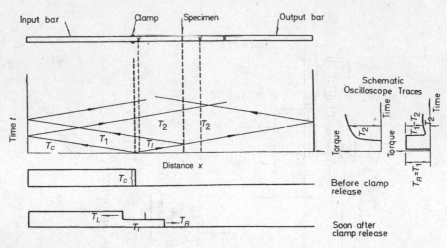

Fig. 4.8.   Schematic of the principle and the Lagrangian $x - t$ diagram of split Hopkinson torsional bars.

specimen, A and B:

$$\bar{\dot{\gamma}} = r_m(\dot{\varphi}_A - \dot{\varphi}_B)/L. \tag{4.2}$$

The interface of the specimen and the bars should have the same angular velocity $\dot{\varphi}$, hence

$$\dot{\varphi}_A = \dot{\varphi}_i - \dot{\varphi}_r, \tag{4.3}$$

$$\dot{\varphi}_B = \dot{\varphi}_t. \tag{4.4}$$

In addition, in accordance with the theory of torsional elastic waves, the angular velocity in an elastic torsional bar is related to the torsional momentum M by

$$\dot{\varphi} = M/J\rho c \tag{4.5}$$

where ρ is the density, c is the speed of elastic torsional waves and J is the polar moment.

$$J = \int 2\pi r^3 \, dr = \begin{cases} \dfrac{\pi}{2}R^4 & \text{for a solid bar} \\[2mm] \dfrac{\pi}{2}(R_o^4 - R_i^4) & \text{for a hollow rod} \end{cases} \tag{4.6}$$

where R_o and R_i are the outer and inner radii of the tube, respectively. Substitution of Eqs. (4.3)–(4.6) in (4,2) gives the following expression for the average strain rate:

$$\bar{\dot{\gamma}} = \frac{r_m}{L} \frac{(M_i - M_r - M_t)}{(J\rho c)_b} = \frac{2(M_i - M_t)}{L(J\rho c)_b} \qquad (4.7)$$

where $(J\rho c)_b$ is called the torsional impedance of the elastic bar. Although the reflected wave propagates in the opposite direction to the incident wave, the particle velocity of the reflected wave has the same direction as that of the incident wave. The average shear strain can then be obtained by simply integrating (4.2):

$$\bar{\gamma} = \int_0^t \bar{\dot{\gamma}} dt.$$

As we have seen, the torsional split Hopkinson bar is a very elegant testing technique and it is worth enquiring why this simple testing technique only appeared relatively recently. Also it is valuable to find the improvements made in the technique in recent years.

The most important thing is to store the torque and then to release it as soon as possible when the test initiates. The introduction of the friction clamp into the apparatus was a major breakthrough, which made dynamic torsion testing a widespread technique.

Baker and Yew (1966) were the first to apply the friction clamp mechanism. The force provided to the clamp to hold the stored torque was provided by a hydraulic jack. To conduct a test, part of the clamp was knocked out by an air gun, which unavoidably introduces flexural and axial loading into the system. Campbell and Lewis (1969) adopted the fracturing of the release mechanism in the friction clamp (see Fig. 4.9(a)). The clamp consists of two symmetrical separating shoes with the same curvature as the input bar. A high-strength steel bolt is tightened through the two shoes, to keep the torque stored in the loading part of the input bar. The bolt should be pre-notched at its centre. When a test is performed, the stored torque is released by tightening the notched bolt until it fractures. Clearly, this small device greatly simplified the operation of the stored torque release method and facilitated the spread of the test technique. Figure 4.9 shows various types of friction clamps and notched bolts reflecting the advances made by investigators in this area. Efforts have been concentrated predominantly in the minimization of the induced bending

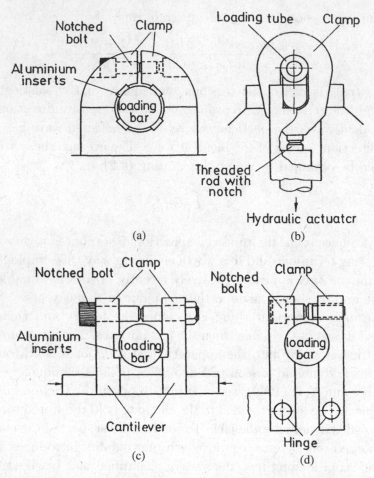

Fig. 4.9.   Various types of friction clamp. After Kobayashi (1987).

and axial loads and the shortening of the rise time of the loading torsional wave.

Adjustment of the friction clamp and proper fracture of the notched bolt are two significant advances which help to obtain a satisfactory torsional wave loading. However, the expertise of the experimentalist is still a very important factor. Some further important factors are: symmetrical contact between the loading bar surface and the two clamp shoes; the finish of the contact surfaces of the loading bar and the two clamp shoes; the method used to tighten and fracture the bolt; the choice of bolt material to give a quick fracture; the mass of the hinges and clamp. A well-designed and machined friction clamp can provide a rise time of 10–20 μ sec (see Fig. 4.9(c), (d)). More usually though a rise time of between 20 and 40 μ sec is obtained

Another important aspect of Hopkinson bar tests is the machining of the specimen and its design. The specimen cannot be solid because of serious inhomogeneities in plastic torsion. Thin-walled tubes are used to provide a homogeneous state of stress. There are a variety of designs for thin-walled tubular specimens, as shown in Tables 4.1 and 4.2. The specimens used in Hopkinson bar tests usually have vertical flanges to minimize their influence on the stress wave profiles. However, the re-entrant corners of the specimen where the gauge length of the thin-walled tube joins the flanges may become the site for stress concentration. Then, if the strain is concentrated there this can completely spoil the test. Therefore, particular attention should be paid to the design and machining of this part of the specimen. The surface finish of the gauge length of the specimen is also important; this is particularly so in the study of adiabatic shear bands. If there are slight fluctuations in the wall thickness then this can influence the position of a shear band. On the other hand, this can be an advantage to investigators who wish to study the continual development of shear bands, because they can predict the initiation sites of bands.

Impedance matching of the bars and the specimen is necessary to avoid undesirable signals

$$(J\rho c)_{\text{bar}} = (J\rho c)_{\text{flange}}.$$

The dimensions of the flange are decided by this requirement.

The method by which the specimen is attached to the split Hopkinson bars is also of importance. In most experimental set-ups, specimens are usually attached to the input and output bars by epoxy resin adhesive, cement or brazing to enhance the smooth propagation of waves. However, this technique is time-consuming; for example, while waiting for the adhesive to cure. There are alternative techniques for attachment for example, by using an optional attachable head. Duffy and co-workers used hexagonal sockets with set screws (see Marchand and Duffy, 1988). To ensure the sockets

Table 4.1. Hollow specimens.

Investigators	r_o (mm)	r_i (mm)	r_i/r_o	L (mm)	Shape of gauge section
Calvert (1955)	3.94	3.18	0.81	3.81	
Hodierne (1963)	4.76	3.18	0.67	3.18	
Tsubouchi and Kudo (1968)	7.00	6.00	0.86	4.0 2.0	
Nicholas and Camplell (1972)	7.24 7.11 6.99	6.35 6.35 6.35	0.88 0.89 0.91	9.53 9.53 9.53	
Bitans and Whitton (1971)	3.97 6.73	2.38 5.79	0.60 0.89	3.18 3.18	
Culver (1972)	7.16 9.53	6.35 8.89	0.89 0.93	3.18 4.78	
Bailey *et al.* (1972)	4.45	3.18	0.72	3.18	
Vinh *el al.* (1979)	25.5	25.0	0.98	1.00	
Lindholm *et al.* (1980)	7.29	6.45	0.89	3.18	

and flanges fitted, a set of 12 small set screws in each socket were used to hold the flanges of the specimen against the driving faces of the hexagonal sockets.

It is clear that the split Hopkinson torsional bar apparatus is very suitable for the study of adiabatic shear banding. However, one of its main drawbacks is the relatively small shear strain attained by it (a shear strain of about 1). Table 4.3 lists the three types of dynamic torsional testing machines and the corresponding characteristics.

Table 4.2. Thin-walled tubular specimens.

Investigators	r_o (mm)	r_i (mm)	r_i/r_o	L (mm)	Shape of gauge section
Baker and Yew (1966)	5.35	4.76	0.90	25.4	No flange
Campbell and Lewis (1969)	7.76	7.63	0.98	1.02	Vertical flange
	7.89	7.56	0.96	2.04	
Duffy et al. (1971)	8.28	7.77	0.94	2.54	
Campbell and Dowling (1970)	7.11	6.35	0.89	1.27	Glue
Nicholas and Lawson (1972)	6.99	6.35	0.91	1.27–12.7	R = 0.8 mm
Nicholas and Campbell (1972)	6.99	6.35	0.91	6.35	
Tsao and Campbell (1973)	5.33	4.83	0.91	3.18 6.35	
Stevenson and Campbell (1974)	8.34	7.94	0.95	1.27	Vertical flange
Clyens and Campbell (1974)	5.28	4.77	0.90	1.27– 6.35	
Eleiche and Duffy (1975)	8.04	7.53	0.94	1.27 2.54	
Eleiche and Campbell (1976a,b)	8.32	7.94	0.95	2.54	$\theta \cong 40°$
Chatani and Hosei (1978)	7.00	6.30	0.90	10.0	Vertical flange
Senseny et al. (1978)	8.15	7.77	0.95	2.54	
	8.33	7.77	0.93	2.54	
Costin et al. (1979)	5.15	4.75	0.92	2.50	

4.3. Dynamic Compression Tests

Partly because of the ease of accessibility of dynamic compression tests, this technique is the most widely used testing technique in laboratories. The test is used to characterize various dynamic properties of materials subjected to impact loadings such as the dynamic yield and flow stresses and the dependence of flow stress on strain rate and

Table 4.3.   Special features of torsion testing machines.

Type of machine	System	Strain rate (/sec)	Max. strain	Remarks
Flywheel and clutch	Simple	Static to 1000	6	Unavoidable vibration in engaging clutch
Hydraulic machine	Complicated	Static to 500	7	Discontinuity in measurable strain rate
Torsional Hopkinson bar	Simple	500–10,000	1.3	Difficulty of releasing clamp

temperature. Dynamic compression tests have been used extensively to study the onset of adiabatic shear bands. Industrial processes such as upsetting and forging, which are akin to dynamic compression, can produce adiabatic shear bands.

Generally, there are two types of dynamic compression testing machines, the conventional machines and the split Hopkinson compression bar technique. The former techniques normally cover strain rates less than 10^2/sec, whereas the Hopkinson bar technique can reach strain rates of 10^3/sec.

In the context used here, by conventional testing machines we mean: the drop hammer, pendulum-type machine, flywheel machine and pneumatic and oil pressure machines. In fact, although we call these conventional machines, they are all specially designed for material testing under dynamic conditions. One of the major requirements for these testing machines is the achievement of constant or nearly constant strain rate. However, not all of the apparatus can meet this requirement. For example, in drop hammer testing it is normally impossible to control the strain rate.

Amongst the special machines, the cam plastometer was designed originally by Orowan (1950) and is capable of maintaining a constant strain rate. Later, Hockett (1967a,b) and Suzuki *et al.* (1968) developed and improved the apparatus. In Hockett's machine, the loading part of the cam is logarithmic. Hence, if the cam speed is constant, the true strain rate will be constant. These machines have large capacities; for example, the cam plastometer designed by Suzuki *et al.* has a maximum natural strain of 0.8.

The types of compression tests that can be carried out on these machines resemble fairly closely industrial upsetting and forging operations. Doraivelu *et al.* (1981) and Semiatin and Lahoti (1983) are among the investigators who examined the occurrence of adiabatic shear bands in upset metal specimens. Doraivelu *et al.* applied a double-action hydraulic press, a friction screw press and a high velocity gravity drop hammer to encompass the range of mean strain rates of 0.6, 11 and 470/sec. In addition, a tubular furnace was mounted on the press bed for hot upsetting tests. During the process, the force, the displacement and the temperature were monitored with strain gauges, linear velocity transducers (LVDTs) and capacitance displacement gauges (for strain rates between 11 and 470/sec). Thermocouples were used to monitor temperature during the process. More importantly, to reduce barrelling to a minimum, a 0.1 mm thick Teflon sheet was used as a lubricant up to temperatures of 400°C and different compositions of glass were used as lubricants at higher temperatures. By using these techniques, these researchers studied the patterns and morphology of shear cracks in upset forged samples.

One important mode of forging is side-pressing or diametral compression tests. The test configuration creates a so-called plane strain stress state in the specimen. This can be particularly useful for the study of adiabatic shear bands because of the pure shear state in the specimen with a superimposed hydrostatic pressure. Therefore, it is suitable for the characterization of materials in relation to shear banding. Figure 4.10 shows a schematic representation of the mechanism of shear band formation in the side-pressing of circular bars.

The split Hopkinson compression bar technique, also termed the Kolsky bar technique, was introduced by Kolsky in 1949. The essential principle of this technique was outlined in the last section in the discussion of the torsional split Hopkinson bars. However, in compression the formulation becomes more straightforward. The expressions for the nominal compressive stress, strain rate and strain in the specimen, similar to formulae (4.1), (4.2) and (4.8) for torsion are:

$$\sigma = (\sigma_i + \sigma_r + \sigma_t)\frac{s_b}{s_s} \approx \sigma_t\frac{s_b}{s_s} \tag{4.8}$$

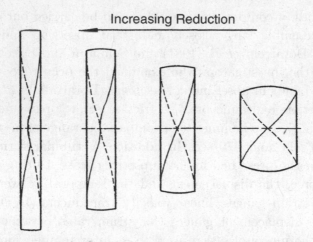

Fig. 4.10.   Schematic representation of the mechanism of shear band formation in non-isothermal side-pressing. After Semiatin and Lahoti (1983).

where s_b and s_s are the cross-section of the bar and specimen, respectively.

$$\dot{\epsilon} = \frac{2(\sigma_i - \sigma_t)}{\rho c_c L} = \frac{-2c_c \epsilon_r}{L} \qquad (4.9)$$

where ρ and c_c are the density and the speed of one-dimensional elastic compressive stress waves, respectively.

$$\epsilon = \int_0^t \dot{\epsilon} dt. \qquad (4.10)$$

To study adiabatic shear banding with split Hopkinson compressive bars, some modifications have been made to facilitate observations.

Wang *et al.* (1988) combined a thermostatic box with the split Hopkinson bars, then conducted dynamic compression tests at low environmental temperatures of $-90°$ and $-190°$C. Affouard *et al.* (1984) wanted to determine the critical strain for the onset of shear banding and to allow further microscopic observations on the tested specimens both before and after fracture. Therefore, these researchers used hardened steel chocks with different lengths positioned around the specimen. With this equipment they could stop plastic deformation at predetermined levels during the dynamic tests. Some of their results are shown and discussed in Section 6.4.

Wulf (1979) used a further type of modified split Hopkinson bar to achieve strain rates as high as 10^4/sec to facilitate the study of adiabatic shear failure. In these tests, the incident bar was removed and a hardened steel projectile with the same diameter as that of the transmitted bar impacted the specimen directly. The stress was recorded by strain gauges as usual mounted on the transmitted bar. However, the variation in the specimen length was measured by the position of the rear of the projectile in a coaxial capacitor. Graphite grease was applied to hold the specimen on the transmitted bar as well as to act as a lubricant.

The split Hopkinson compression bar technique has also been modified to conduct punching experiments. The technique was adopted by Harding and Huddart (1979) and was then developed by Hartmann et al. (1981) to study adiabatic shear banding and plugging. This punching technique was also used by Meyer and Manwaring (1986) and Tian and Bai (1985) to study banding and plugging. In these tests the modifications in the specimen configurations are shown in Fig. 4.11. In effect this technique is the study of shear deformation using the advantages of the compression Hopkinson bar technique. These advantages are higher strain rates and ease of measurement of the load and the deformation velocity. However, although these tests are relatively easy to set up and carry out, it is difficult to obtain accurate and reliable data (see Section 4.2). Moreover, in the design and performance of these tests, particular attention should be paid to guarantee the strength and rigidity of the punch and the annular die, the clearance between the two parts as well as the sharpness of the edges.

Dynamic compression tests are a very useful and straightforward testing method. Various modifications based on compression tests provide access to other tests, such as punching and split Hopkinson tensile bars. For the study of adiabatic shear banding, dynamic compression tests are very helpful but care should be taken in correlating the observed shear bands with the state of stress or strain because of barrelling and friction effects at the ends.

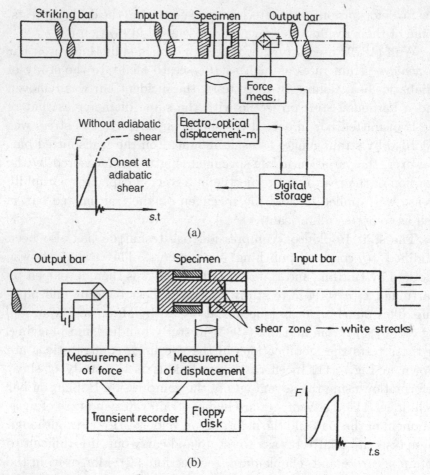

Fig. 4.11.   Configuration of punching tests fitted into a split Hopkinson pressure bar set-up. (a) After Hartmann *et al.* (1981) and (b) after Meyer and Manwaring (1985).

4.4. Contained Cylinder Tests

Contained cylinder tests, also called contained exploding cylinders (CEC technique), are able to provide even higher strain rates of about 10^4–10^6/sec, as is the case with planar impact tests. However, in these tests large plastic strains can be achieved, $O(1)$, due to the apparent shear deformation involved. These two characteristics make the technique, though not ideal for studying adiabatic shear banding, a unique and powerful tool for measuring the kinetics of

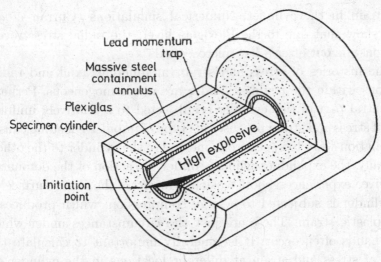

Fig. 4.12. Exploding cylinder experiments for studying shear band kinetics. After Curran *et al.* (1987).

microdamage, including adiabatic shear banding, occurring under similar loading conditions. In addition, the experimental set-up is closely related to practical applications such as fragmentation of shell rounds. Also the results and data obtained by using these kinds of tests can be used in engineering practice quite straightforwardly.

Figure 4.12 shows a cutaway of the experimental configuration. It consists of an explosive-filled specimen cylinder and a massive containment annulus surrounding it. The internal detonation of the filled explosive induces a compressive wave which propagates out along the testing cylinder. Under this impulsive loading, the cylinder expands extremely quickly. The loading compressive wave passes through the Plexiglas layer and then enters the steel containment. From the interface of the Plexiglas layer and the steel containment, a reflective compressive wave propagates back and retards the expansion of the cylinder due to the refractive wave reflected from the outermost free surface of the device in the later stages of testing. The Plexiglas layer between the expanding cylinder and the massive steel containment annulus acts as a buffer to soften the impact and reduces the effect of the reverberating stress waves.

The high explosive filling the cylinder and the Plexiglas layer are the major factors that govern the magnitudes and histories of stress and strain in the cylinder. Numerical simulations (Curran *et al.*, 1987) show that due to the Plexiglas layer, the radial stress reverberations die out after a few microseconds.

The histories of stress and strain at different axial and radial positions within the testing cylinder are not homogeneous. Perhaps the central part of this cylinder is subjected to a relatively uniform state of stress and strain because this type of impulsive loading causes a detonation wave to sweep from one end of the cylinder to the other. Generally, the expanding cylinder, under the action of the detonated explosive, experiences hoop tension. However, the inner surface of the cylinder is subjected to a large compression, which produces a large plastic strain. These are just the circumstances under which shear bands often occur. It is generally important to calculate the states of stress and strain at different locations in the cylinder by numerical simulation.

This experimental technique has been used for two types of study. The first is the study of the relation between the adiabatic shear instability strain and material properties (Staker, 1980). The second type of study is the examination of the kinetics of this kind of microdamage (Curran *et al.*, 1987).

In the first study, the determination of the shear instability strain is the major aim. Because of the configuration of the right cylinder, the average radial and hoop strains can be obtained by measurements of the inner diameter (ID) and the outer diameter (OD) made before and after the detonation experiments as

$$\epsilon_R = \ln(t_f/t_o) \quad \text{and} \quad \epsilon_H = \ln\left\{\frac{(\text{ID} + \text{OD})_f}{(\text{ID} + \text{OD})_o}\right\}$$

where the subscripts R, H, f and o denote the radial, hoop, final and original values, respectively, and t is the thickness of the specimen. The axial strain is a rough average over the whole length of the cylinder, i.e. $\epsilon_A = \ln(L_f/L_o)$, where L is the length of the cylinder. It was found that in this configuration adiabatic shear bands usually appear on the inner surface on a plane parallel to the axial direction of the cylinder but at 45° to the radial and hoop directions. Staker (1981) applied the post-test measurement of the values of ϵ_R and ϵ_H

to determine the critical adiabatic instability strain. Section 6.4 will discuss these results in detail.

Curran *et al.* (1987) summarized their experimental study of adiabatic shear bands, for this contained exploding cylinder. They reported that the adiabatic shear bands formed in this testing cylinder usually nucleate at points on the inner surface and grow outwards to form 'halfpenny'-shaped shear bands in the plane of the maximum resolved shear strain. Also some shear bands originate within the cylinder wall. Curran and co-workers thought that the shear bands have a geometry similar to macroscopic dislocations with edge and screw components. The number of shear bands and their distribution provide information on the nucleation and growth of shear bands and this data is used to predict fragmentation (see Section 9.2).

4.5. Transient Measurements

The devices for conventional transient measurements in dynamic tests are, for example, the load cell, displacement or velocity transducers and strain gauges. The histories of the nominal stress, strain and strain rate of specimens utilizing the split Hopkinson technique, whether torsional or compressive, are all deduced from recordings made at two stations of strain gauges attached to the elastic input and output bars. The details of this method have been given in Section 4.2.

All the above transient measurements are made for homogeneous deformation where the nominal values of the mechanical variables represent the true state of the material element undergoing deformation. As soon as adiabatic shear bands form, the strain rate and strain obtained from the split Hopkinson bar records are only an average value taken over the highly localized shear band and the region outside the shear band. Therefore, the results do not reveal the true shear localization process. Hence, for the study of the formation of adiabatic shear banding, local and transient measurements of the local strain, strain rate and temperature are necessary.

It has been realized for many years that both post-mortem observations and recordings of histories of the nominal stress, strain and strain rate do not provide enough information to study adiabatic shear banding. Moreover, the difficulties involved in making transient and local measurements are unique. For example, if we imagine that an adiabatic shear band with a width of tens of microns suddenly appears within tens of microseconds, somehow and somewhere within several millimetres of gauge length, there will clearly be difficulties in observation and measurement.

The above-mentioned difficulties raise the specific requirements for the detection and recording system for the study of adiabatic shear banding. Briefly, these requirements are:

Response time. The response of the instruments should be quick enough not only for the whole high strain-rate deformation, but also the transition period from homogeneous to inhomogeneous deformation, i.e. the formation of a shear band.

Field of view. We require a large field of view of the whole gauge length so that it is possible to identify where and when inhomogeneities occur. Moreover, we need a small field of view to observe the inner structure of the shear band, bearing in mind that we do not necessarily know where the small field of view will be located.

Spacial and temporal resolution. Perhaps, for both time and space we need two resolutions. A coarse resolution for the whole deformation process and field and a fine one for the localization process and area. The two variables for which these types of measurements have been achieved are the shear strain and the temperature. These were recorded in inhomogeneous shear by Giovanola (1987, 1988a,b) and Marchand and Duffy (1988).

Using a high-speed camera, Giovanola (1987) recorded the shear deformation of a grid on the surface of a torsional specimen. He studied thin-walled tubes of 4340 steel sandwiched between two torsional bars. A fine square grid, with a $100\,\mu$m $\times$ $100\,\mu$m line spacing, was etched on the outer wall of the 2-mm-long gauge length. High-speed framing photographs were taken with an interframe time of $2.5\,\mu$ sec. The measurements show that the spacial resolution ($100\,\mu$m) is enough to study homogeneous deformation, but it is too coarse to distinguish the structure of the shear band since the shear localizations observed by Giovanola were less than $100\,\mu$m in width.

Marchand and Duffy (1988) adopted a different approach for the examination of shear deformation fields. They used three static cameras, taking short-exposure simultaneous photographs of the shear band, to reveal features of adiabatic shear banding. They used 35 mm single lens reflex cameras with standard 50 mm focal length lenses and 110 mm extension tubes and obtained some very interesting results (see Section 7.1, Fig. 7.3). A high-speed flash source was used and the light was guided to the observed area by means of optical fibre tubes. The flash duration of the light was less than 2 μsec. The light was triggered by the strain gauge mounted on the input bar and the trigger was recorded as part of the stress–time recording. To guarantee the quality of the photograph, the surface of the thin-walled tube specimen was given a fine polish, which was followed by a very light etch, which dulled somewhat the mirror finish. A grid of 98 lines per centimetre was deposited on the outside surface of the specimen. Thus the transient measurements of Giovanola (1987) and Marchand and Duffy (1988) have roughly the same spacial ($\approx 100\,\mu$m) and temporal ($\approx 2\,\mu$sec) resolutions. Generally, the spacial resolution in the localized area needs further improvement, although the grid lines parallel to the axis of the tube specimen can provide some information on shear banding. Perhaps two photographic systems with different fields of view are required with different spacial resolutions. For example, it should be possible to encompass the gauge length (2 mm) and the band width (0.1 mm).

No direct measurement of the history of localized shear strain has been reported in the literature so far. As we will see in Chapters 7 and 8, the formation of adiabatic shear bands can be marked sensitively by an abrupt increase in local shear strain rate and therefore this type of direct measurement may be worth pursuing.

The transient measurement of the temperature increase of material during plastic deformation can be traced back to the 1970s. Moss and Pond (1975) and Dao and Shockey (1979) measured the inhomogeneous temperature increase of materials subjected to plastic elongation. Moss and Pond (1975) used a copper-doped germanium semiconductor crystal to measure the increase in temperature at the surface of copper samples due to plastic work done during elongation. The working principle is based upon the measurement of infrared radiation emitted by the surface of the test-piece, which is heated by the fraction of plastic work carried out on it (see Section 6.1). This technique possesses several advantages, which arc cspecially helpful in the study of shear banding.

This is a non-contact technique of temperature measurement, therefore the detector does not influence the temperature measured. This is of crucial importance for the study of adiabatic shear bands because a shear band is very narrow and the heat is limited although the temperature can be high. Also, in this case the thermal inertia involved in the contact detector need not be considered. The second advantage of the technique is its rapid response time, normally less than 100 nsec. According to the requirements of transient measurement, this can meet the need for shear localization. The third advantage is that the area measured can be small, about 20 to $100\,\mu m$. Although the spot is still not as small as the shear band width, 10 to $100\,\mu m$, this spacial resolution is still good compared to conventional thermocouples.

Costin and co-workers (1979) started using this technique to examine the local temperature of adiabatic shear bands in 1018 cold-rolled steel (CRS). They used an indium antimonide crystal to detect the increase in temperature on the surface of thin-walled tubular specimens tested on torsional split Hopkinson bars. Since the observation spot is $1\,mm^2$ in surface area, the results underestimate the actual rise in temperature within the shear band, which is about $200\,\mu m$ wide.

After a number of improvements (see Duffy, 1984; Hartley *et al.*, 1987; Marchand and Duffy, 1988), this technique has become quite sophisticated in the detection of the transient and local temperature in the study of adiabatic shear banding. The following description will be based on a version of the technique given by Marchand and Duffy (1988).

The infrared detectors are a 12-element system constructed by Judson Infrared. A linear array of 12 individual indium antimonide cells are mounted in a Dewar flask and cooled with liquid nitrogen to 77 K. The detectors are used in a photovoltaic mode with a response time of less than $1\,\mu s$. Therefore, the temperature measurements can be performed simultaneously at 12 spots on the surface within the gauge length of the thin-walled tubular specimen.

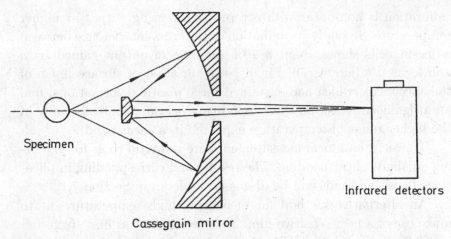

Fig. 4.13. Schematic diagram showing Cassegrain mirror used in the detector-mirror system. After Marchand (1988).

The infrared radiation emitted from the heated surface of the specimen should be focused onto the detectors by a mirror system. Instead of the gold-plated mirror used by Hartley *et al.* (1987) on an off-axis position, Marchand and Duffy (1988) employed a reflecting objective of the Cassegrain type (Fig. 4.13). The advantage of mirrors over lenses is that they avoid chromatic aberrations due to lens refraction. Since infrared detectors respond in a broad range of wavelengths, mirrors are necessary. Furthermore, the Cassegrain system, consisting of two precisely aligned confocal paraboloid mirrors, does not suffer from spherical aberration. Moreover, because of the accessibility of its focal point and the system's relatively low obscuration ratio and diffraction effects, it can provide a good image quality over a very small field of view. This is particularly useful in the study of shear banding. The Cassegrain objective used was from Ealing Electro-Optics with a magnification of 15 and a focal length of 13 mm. Therefore, the corresponding object spot is 35 μm wide with 11 μm spacing. This can cover about 0.5 mm of the field of view with 12 cells.

Other key steps in employing this technique are calibration and identification of cell interactions. In their preparatory experiments, Marchand and Duffy (1988) clarified the following facts. Firstly, the calibrations of all cells were found to be almost identical. The calibration is non-linear with a rapidly increasing output at higher temperature. Secondly, examination of the cross-interference between adjacent cells shows about a 10% increase in output gained by a non-focused adjacent cell. Finally, the variation of surface finish of the specimens, which had sustained large plastic deformations, had an insignificant effect on the temperature measurement especially in the higher range of temperature expected in a shear band.

Transient and local measurements are powerful tools for exploring adiabatic shear banding. The results and corresponding implications for shear bands will be discussed in detail in Section 7.1.

An alternative method for measuring high temperatures is to make use of a heat-sensitive film. This technique was first suggested by Coffey and Jacobs (1981). It has subsequently been applied in the study of hot spots in polymers and explosives; for example, see Swallowe *et al.* (1984, 1986). Compared to the above-mentioned infrared method, this technique is much easier and cheaper with a large field of view but less spacial resolution due to the size of the grains ($\approx 0.2\,\text{mm}$) on the film. Certainly, the accuracy of the results obtained by this technique is not as good as the infrared method. However, because of its convenience and large field of view it is suitable for the observation of the distribution of adiabatic shear bands.

The heat-sensitive film is a kind of transparent sheet like that used on overhead projectors. The acetate sheet is coated with a heat-sensitive layer, which darkens, on exposure to heat. However, the sheet does not darken under natural or artificial light or infrared radiation. Also it is not darkened by tearing, stretching or impact up to a peak pressure of $1\,\text{GPa}$. The darkening produced by heating is stable for at least a year at room temperature. To carry out a measurement a piece of film is just placed on the sample with its sensitive side in contact with it. By comparing the darkened photograph with a calibration curve, it is possible to deduce estimates of the temperature of the deformed sample. The darkening is, in fact, a function of both temperature and time. So use of the correct calibration is of crucial importance. Figure 4.14 shows calibration curves used by Swallowe *et al.* (1986).

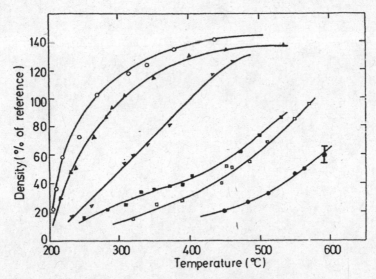

Fig. 4.14.   Calibration curves for a heat-sensitive film. Density is measured relative to a standard neutral density filter. Times represent the contact time of the film with the heated plate. Error bar shown is typical of all points. Contact times (μsec): (●) 30; (□) 50; (■) 140; (▽) 330; (▲) 440; (○) 600. After Swallowe *et al.* (1986).

Further Reading

Meyer, L. W. and Pursche, F. (2012) 'Experimental methods'. In *Adiabatic Shear Localization: Frontiers and Advances* (eds. Dodd, B. and Bai, Y.) pp. 21–109, Elsevier, London.

Chapter 5

Constitutive Equations

An equation that describes the macroscopic response of a material to applied loads is defined as a constitutive equation. As observed by Malvern (1969), it is not realistic to have one equation for a material for all possible combinations of loads, strains, strain rates and temperatures. The conventional approach to these problems is to formulate equations that describe an ideal material response, which approximates to the behaviour of a real material. Equations describing Bingham solids and Newtonian viscous behaviour are examples of this type.

Under circumstances where temperature changes can be ignored and the deformation is quasi-static so that inertial effects are negligible, the most general form of constitutive equation for plastic deformation is that the stress tensor is some function of the strain increment tensor. The precise formulation of the equations depends upon the properties of the material as well as the loading conditions.

The classical work on constitutive equations is discussed at length by Malvern (1969) and there is no need to study the background theory here. Rather it is necessary to understand the physical origins of the equations that are used in research so as to be able to implement appropriate equations where necessary.

One of the most frequently used relations between stress and strain is a simple power law due to Ludwik (1927):

$$\sigma = K\epsilon^n \qquad (5.1)$$

where σ and ϵ are the true stress and strain, respectively, and n is the strain hardening or work hardening exponent; the n value usually lies in the range $0 \leqslant n < 1$. This type of equation is very useful in the study of problems with large plastic deformations. For example, in many metal-forming problems a simple power law hardening equation is quite adequate for obtaining a measure of strain hardening in what are often non-uniform strain fields (see, for example, Johnson and Mellor, 1973).

5.1. Effect of Strain Rate on Stress–Strain Behaviour

The general effect of increasing the strain rate on the stress–strain curve is to move the entire curve to higher stresses. Pure metals appear to be affected more markedly than alloys. Along with a general elevation of the stresses, the fracture strain normally decreases with increasing strain rate.

Valuable compilations of data on the effect of strain rate on the yield and flow stresses of many metals and alloys have been presented by Suzuki *et al.* (1968), Lindholm and Bessey (1969), Eleiche (1972) and Campbell (1973).

Figures 5.1(a) and (b) show the effects of strain rate on the stress–strain curves in torsion for commercial purity aluminium (Tsao and Campbell, 1973) and titanium (Eleiche and Campbell, 1974). In only two tests did fracture occur; these tests are shown with an F at the end of the curves. For both materials the flow stress increases with increasing strain rate.

Strain-rate sensitivity data are normally presented as yield stress or flow stress at a given strain versus the logarithm of the strain rate. Figure 5.2 shows typical behaviour. Three zones of behaviour have been recognized, but not all need to be present for each metal. Zone I corresponds to the region of low strain rate where the flow stress is insensitive to strain rate. This region is often referred to as the thermal region because plastic flow is not thermally activated (see Harding, 1987). For this first zone of low strain rate, the strain-rate sensitivity parameter $\beta = (\partial\sigma/\partial \log \dot{\epsilon}^p)_{\epsilon^p, \theta}$ approaches zero.

Over a fairly wide range of strain rates, the strain-rate sensitivity parameter β is often constant; this is Zone II in Fig. 5.2.

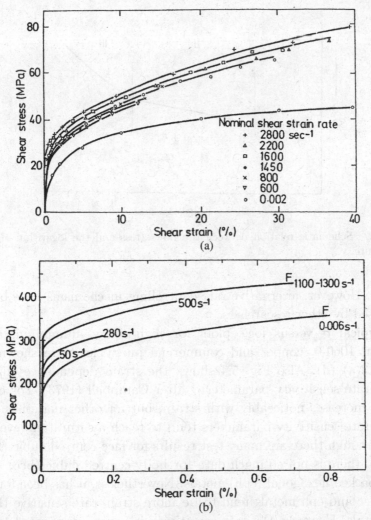

Fig. 5.1. Effects of strain rate on the shear stress–shear strain response for (a) aluminium (after Tsao and Campbell, 1973) and (b) titanium (after Eleiche and Campbell, 1974).

This region is one of thermal activation in which dislocations require thermal energy to overcome obstacles, such as precipitates, by cross-slip (see Seeger, 1955). Region III may be one in which phonon drag on dislocations is the rate-controlling mechanism and increases rapidly with increasing strain rate. This phonon drag mechanism was suggested by Ferguson and co-workers

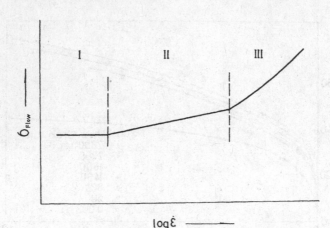

Fig. 5.2.   Schematic relation between the flow stress and the logarithm of the strain rate.

(1967). However, alternative rate-controlling mechanisms have been proposed by other researchers.

Typical σ versus $\log \dot{\epsilon}$ plots for high-purity aluminium, aluminium 1060-0, copper and commercial purity lead are shown in Figs 5.3(a)–(d). Also Fig. 5.4 shows the strain dependence of the strain-rate sensitivity parameter β after Campbell (1973). For copper, β increases noticeably with strain, but for other materials the strain-rate sensitivity parameters tend to reach a saturation level.

Although there are many test results for face-centred cubic (fcc) metals, there is not so much data for body-centred cubic (bcc) and close-packed hexagonal (cph) metals. Nevertheless, it has been found that bcc and cph metals tend to be more strain-rate sensitive than fcc metals. However, the rate sensitivity is more complicated than for fcc metals, as has been discussed at length by Campbell (1973). The differences in strain-rate sensitivity must be ascribed to different dislocation mechanisms.

The strain and strain-rate sensitivity of the flow stress for mild steel, beryllium and titanium–6% aluminium–4% vanadium are shown in Figs 5.5(a)–(c). Negative rate sensitivity is observed in mild steel tested at elevated temperatures.

Constitutive equations for materials that take account of strain rate were first proposed by Ludwik (1909), who introduced the

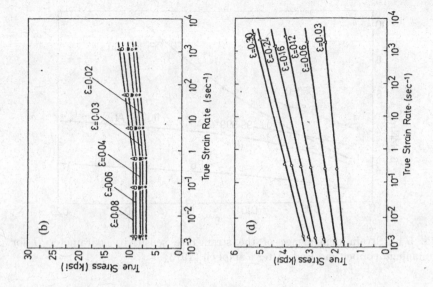

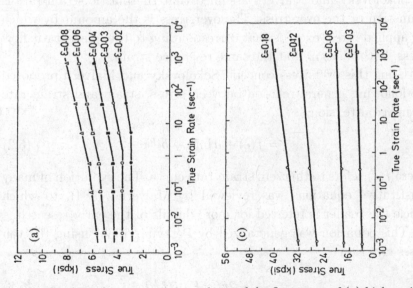

Fig. 5.3. Strain and strain-rate dependence of the flow stress of (a) high-purity aluminium, (b) aluminium 1060-0, (c) oxygen-free high thermal conductivity (OFHC) copper and (d) commercially pure lead. Data collected by Campbell (1973).

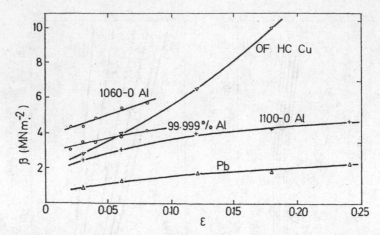

Fig. 5.4.   Strain dependence of the strain-rate sensitivity parameter β for aluminium, copper and lead. After Campbell (1973).

concept of overstress. This concept was used by Sokolovsky (1948) and Malvern (1951).

Sokolovsky and Malvern assumed that the plastic strain rate is a function of the overstress. The overstress is the amount by which the applied stress exceeds that corresponding to the quasi-static flow stress, both measured at the same reference strain.

Using this overstress concept, Sokolovsky and Malvern proposed the following general relation between stress, strain and strain rate for uniaxial tension:

$$\sigma = f(\epsilon) + a\ln(1 + b\dot{\epsilon}^p) \qquad (5.2)$$

where $f(\epsilon)$ refers to this zero strain-rate curve. The evolution of many constitutive equations was reviewed by Malvern (1984), to which article the reader is referred for more details of the early research.

This equation was generalized by Perzyna (1966) using the von Mises criterion

$$\dot{\epsilon}_{ij} = \frac{\dot{s}_{ij}}{2G} + 2r\langle\varphi(F)\rangle\partial f/\partial\sigma_{ij} \qquad (5.3)$$

where s_{ij} are the components of the stress deviator tensor, G is the shear modulus, v is Poisson's ratio, σ_m is the hydrostatic pressure, δ_{ij} is the Kronecker delta, r is a material constant, φ is a function of

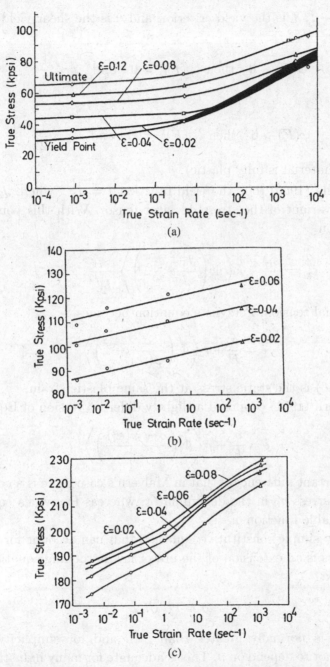

Fig. 5.5.   Strain and strain-rate dependence of the flow stress of (a) mild steel, beryllium 1-400 and (c) titanium–6% aluminium–4% vanadium. After Campbell (1973).

$F = f/k - 1$, f is the yield criterion and k is the shear yield stress. Now if

$$\varphi(F) \leqslant 0 \quad \text{then} \quad \langle \varphi(F) \rangle = 0 \quad \text{and} \quad v \leqslant \frac{1}{2}$$

and if

$$\varphi(F) > 0 \quad \text{then} \quad \langle \varphi(F) \rangle = F \quad \text{and} \quad v = \frac{1}{2}$$

(i.e. the material is fully plastic).

Perzyna (1966) took the yield function $f = J_2^{1/2}$ where J_2 is the second invariant of the stress deviator tensor. With this von Mises assumption

$$\dot{\epsilon}_{ij} = \frac{\dot{s}_{ij}}{2G} + r \left\langle \Phi \left(\frac{J_2^{1/2}}{k} - 1 \right) \right\rangle \frac{s_{ij}}{J_2^{1/2}}, \quad i \neq j. \tag{5.4}$$

For uniaxial tension, Perzyna's equation becomes

$$\dot{\epsilon} = \frac{\dot{\sigma}}{E} + r \left\langle \Phi \left(\frac{\sigma}{\varphi(\epsilon^p)} - 1 \right) \right\rangle \tag{5.5}$$

where $\varphi(\epsilon^p)$ is the static stress at the same plastic strain.

Malvern (1951) proposed a slightly different version of Eq. (5.5):

$$\dot{\epsilon} = \frac{\dot{\sigma}}{E} + r \left\langle \Phi \left(\frac{\sigma - \varphi(\epsilon^p)}{\sigma_0} \right) \right\rangle. \tag{5.6}$$

The important difference is that in Malvern's case there is a constant reference stress σ_0 in the denominator, whereas Perzyna's equation has a variable function φ.

A very simple constitutive equation that can account for strain-rate effects is an extension of the power law hardening equation due to Ludwik:

$$\sigma = k\epsilon^n \dot{\epsilon}^m \tag{5.7}$$

where m is normally greater than zero and for simplicity m is assumed not to depend on n. This is adequate for many industrial circumstances but there are limitations in this approach. For example, for steels there is evidence that the strain-rate exponent m decreases as the strain increases (see Hosford and Caddell, 1983; Saxena and Chatfield, 1976).

5.2. Strain-Rate History Effects

Klepaczko (1968) was one of the first researchers to investigate the effects of strain-rate history on the stress–strain behaviour of metals. Aluminium torsion specimens were tested either at (i) the lowest rate, and unloaded and reloaded at the highest rate or (ii) the highest rate, and unloaded and reloaded at the lowest rate. In both cases, on reloading, the flow stress differed from constant rate tests.

Figure 5.6 shows that strain-rate changes can have significant effects on the stress–strain curve. Klepaczko found that continued straining at the new rate caused the flow stress to tend towards the corresponding constant rate test and in consequence he deduced that the material had a 'fading memory' of the earlier deformation.

Similar types of tests were carried out by Campbell and Dowling (1970) on copper, niobium and molybdenum (see Campbell, 1973). Similar effects were found for copper as had been found by Klepaczko, but for bcc metals the strain-rate history or jump tests can have very different effects. For these metals there is an overshoot when the strain rate is changed from the lower to the higher values, as shown in

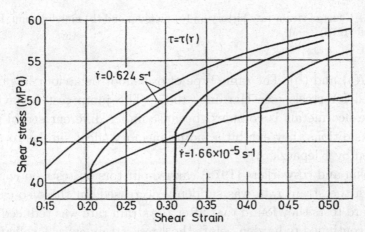

Fig. 5.6.   The results of strain-rate change for three initial values of strain: initial strain rate is $\dot{\gamma}_i = 1.66 \times 10^{-5}$/sec and the strain rate on reloading is $\gamma_r = 0.624$/sec. After Campbell (1970).

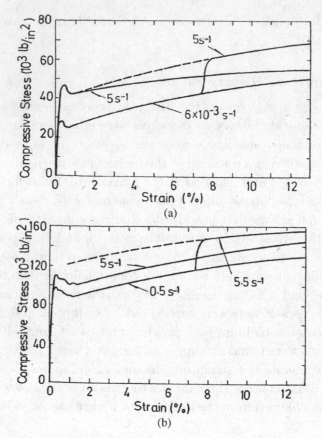

Fig. 5.7.   Effect of strain-rate history on (a) niobium and (b) molybdcnum. From Campbell (1970).

Figs 5.7(a) and (b). The same type of overshoot behaviour has been observed for steels (see Harding, 1987). The jump test behaviour of these bcc metals is complex, because under different strain-rate jump conditions they exhibit a fading memory effect similar to that observed by Klepaczko.

Lipkin and co-workers (1978) carried out torsion tests on copper in which the strain rate was suddenly decreased. In these so-called downward tests they found that after the strain rate was reduced the copper continued to harden before the stress–strain curve levelled off.

Subsequently, strain-rate history effects have been investigated for a number of different materials by Eleiche and Campbell

(1976a,b), Campbell and co-workers (1977) and Lipkin and co-workers (1978). Much of this work has been reviewed by Duffy (1979) and Hartley and Duffy (1984).

Shirakashi and Usui (1970) proposed a constitutive equation that includes an hereditary integral allowing for strain-rate changes in upward and downward tests in compression.

Campbell and co-workers (1977) considered jump tests in torsion from one strain rate $\dot{\gamma}_1$ to a second strain rate $\dot{\gamma}_2$ after a prestrain to $\gamma = \alpha$ in copper (note that $\dot{\gamma}_2 > \dot{\gamma}_1$). They found that the stress increase $\Delta\tau$ above the low rate curve depended upon the strain increment $\gamma - \alpha$ but not on the prestrain α.

Campbell and co-workers proposed the following constitutive equation, taking into account the different strain rates. For a strain jump from $\dot{\gamma}_1$ to $\dot{\gamma}_2$ they proposed:

$$\tau = f_1(\gamma) + f_2(\gamma, \dot{\gamma}_1) + f_2(\gamma - \alpha, \dot{\gamma}_2) - f_2(\gamma - \alpha, \dot{\gamma}_1) \qquad (5.8)$$

where $f_1(\gamma)$ is the quasi-static flow stress, $f_2(\gamma, \dot{\gamma})$ represents the overstress and the terms in $(\gamma - \alpha)$ disappear when $\gamma \leqslant \alpha$. For copper with a constitutive relation of the form

$$\tau = A\gamma^n\{1 + m\ln(1 + \dot{\gamma}/B)\} \qquad (5.9)$$

then the constitutive relation according to Eq. (5.9) is

$$\tau = A\{1 + m\ln(1 + \dot{\gamma}_1/B)\}(\gamma)^n \\ + \{mA\ln[(B + \dot{\gamma}_2)/(B + \dot{\gamma}_1)]\}(\gamma - \alpha)^n.$$

Of course, it is possible to develop more and more complicated constitutive equations, but it is clearly going to be difficult to take into account upward as well as downward tests with the same equation.

Klepaczko and Chiem (1986) derived a theory for jump tests in fcc metals, which includes two rate sensitivities of the stress. This approach will be returned to later in the chapter because of the importance of temperature changes.

Bodner and Partom (1975) used an incremental constitutive equation to allow for history effects. The plastic strain rate is given, at a given temperature, by

$$\dot{\epsilon}^p = \dot{\epsilon}^p(\sigma, H) \tag{5.10}$$

where H is a history-dependent state variable that characterizes the material's resistance to plastic flow. If the von Mises yield criterion is assumed, then the evolution equation is

$$\dot{H} = F(J_2, H) \tag{5.11}$$

where J_2 is the second invariant of the stress deviator tensor. Bodner and Partom assume that H is a function of the plastic work

$$H = H_1 - (H_1 - H_0)\exp(-mW_p). \tag{5.12}$$

The plastic strain rate for uniaxial stress is given by

$$\dot{\epsilon}^p = \frac{2}{\sqrt{3}}\frac{\sigma}{|\sigma|}D_0\exp\left[-\frac{1}{2}\left(\frac{H}{\sigma}\right)^{2n}\left(\frac{n+1}{n}\right)\right] \tag{5.13}$$

where H_0, H_1, m, n and D_0 are material constants. For computational purposes the evolution equation takes the form

$$\dot{H} = \frac{\partial H}{\partial W_p}\frac{dW_p}{dt}. \tag{5.14}$$

The deformation rate is an instantaneous function of the stress and plastic work and there is no quasi-static stress–strain curve for a material.

Bodner and Merzer (1978) used Bodner and Partom's approach successfully to describe the behaviour of copper tested over six decades of strain rate. The equation was found to give an accurate representation of the properties of the copper.

5.3. Effect of Temperature on Stress–Strain Behaviour

Temperature variation can have a profound effect on the mechanical response of metals and other materials. In metals many of the dislocation mechanisms of plastic deformation are thermally activated and are governed by Arrhenius-type activation equations. Therefore, any variation in testing temperature will change the rate at which deformation will occur.

Increasing the testing temperature normally has the same effect on the flow stress as decreasing the applied strain rate, i.e. the level of the stress–strain curve is decreased and the fracture strain is increased.

A further important factor, which appears particularly in bcc metals such as steels, is a transition temperature from brittle to ductile behaviour as described for example by Dieter (1986) and Dodd and Bai (1987). Here a standard notched specimen is struck with a hammer. A ductile material will absorb a large amount of energy during fracture, whereas a brittle material will absorb little energy in fracturing.

Using a cam plastometer, Hockett (1967b) examined the effect of strain rates between 0.05 and 200/sec and temperature on the stress–strain behaviour of aluminium in compression. Figures 5.8(a) and (b) show that as the temperature increases the levels of the stress–strain curve decrease. Figure 5.9 shows the combined effect of temperature and strain rate on the flow stress at a strain of 0.5.

MacDonald and co-workers (1956) investigated the effects of temperature and strain rate on the stress–strain behaviour of a

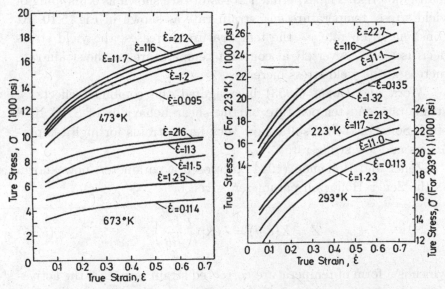

Fig. 5.8. True stress–true strain curves for 1100-0 aluminium: (a) at two temperatures; (b) at lower temperatures than (a). After Hockett (1967b).

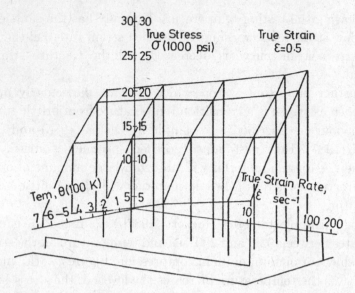

Fig. 5.9. True stress versus logarithm of true strain rate versus temperature at a true strain of 0.5. After MacDonald and co-workers (1956).

fully-killed temper rolled steel (Fig. 5.9). The equipment used was a hydraulic press. The experimental results are shown as a mapping of yield stress, temperature and strain rate as shown in Fig. 5.10. At constant strain rate, as the temperature increases, the yield stress decreases and conversely at constant temperature, as the strain rate increases, the yield stress increases.

Work and Dolan (1953) investigated the combined effects of strain rate and temperature on the shear behaviour of SAE 1018 steel. Some of their results show very large strains for high temperatures and low strain rates.

Zener and Hollomon (1944) proposed a parameter, now known as the Zener–Hollomon parameter, given by

$$Z = Z(\dot{\epsilon}, \theta) = \dot{\epsilon}\exp\left(\frac{U}{R\theta}\right), \qquad (5.15)$$

which is a form of temperature-corrected strain rate. U is the activation energy for plastic deformation, R is the universal gas constant and θ is the absolute temperature (see also Pöhlandt, 1989).

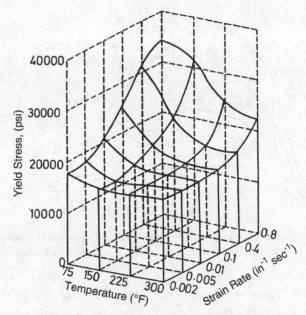

Fig. 5.10. The effect of strain rate and temperature on the upper yield stress of fully Al-killed temper rolled steel. After Work and Dolan (1953).

MacGregor and Fisher (1946) used an alternative approach, which they called the velocity-modified temperature θ_v, where

$$\theta_v = \theta\{1 - A\ln(\dot{\epsilon}^p/\dot{\epsilon}_0^p)\} \tag{5.16}$$

where A and $\dot{\epsilon}_0^p$, are constants, which are related to the activation process for plastic flow. $\dot{\epsilon}^p$ is the actual strain rate. At a given strain, the flow stress is supposed to be a unique function of θ_v. The velocity-modified temperature approach has been used successfully by Oxley and co-workers in their studies of machining and cutting processes (see, for example, Oxley, 1989).

Lindholm (1974) used a Zener–Holloman-like thermally activated parameter given by $\theta^* = \theta\ln(A/\dot{\epsilon})$. Here $\dot{\epsilon}$ is the effective strain rate and A is a material constant. Figure 5.11 shows the dependence of deformation and fracture upon this Zener–Hollomon parameter. Clearly, the boundary between elastic and stable plastic deformation is the yield surface. The boundary between stable and unstable plastic deformation is the maximum in the engineering stress, after which

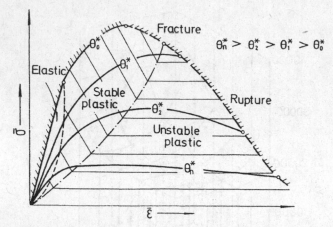

Fig. 5.11. Schematic representation of stress–strain behaviour for metals. After Lindholm (1974) published by permission of the Institute of Physics.

there is either fracture or rupture. The effects of temperature on the stress–strain properties is the inverse of the effects of strain rate.

The most general relation between stress, strain rate and temperature is

$$\sigma = f(\epsilon, \dot{\epsilon}, \theta). \tag{5.17}$$

This type of relation ignores temperature and strain-rate history effects. One example of such a constitutive equation is

$$\sigma = \epsilon^n \dot{\epsilon}^m \exp(-Q/R\theta), \tag{5.18}$$

which clearly emphasizes the Arrhenius thermal energy approach to plastic flow. Vinh and co-workers (1979) found a good agreement between the shear version of Eq. (5.18) and torsion tests on aluminium, copper and mild steel.

A further simple relation, which has been used previously, is

$$\sigma = \epsilon^n \dot{\epsilon}^m \theta^{-v}. \tag{5.19}$$

Often equations of the form given by (5.18) and (5.19) have been used in the literature and provide good fits to experimental data.

Linear temperature dependences of the flow stress have been assumed by many workers. For example, Duffy (1981) used

$$\tau = c(1 - a\theta)(1 + b\dot{\gamma})^m \gamma^n \qquad (5.20)$$

and Kobayashi and Dodd (1988, 1989) used

$$\tau = B\gamma^n \dot{\gamma}^m (1 - a\theta). \qquad (5.21)$$

Linear thermal softening was found to be a good approximation for low-carbon steel, titanium and oxygen-free high conductivity copper for the strain rates considered by Kobayashi and Dodd.

Temperature can in principle be taken into account in Perzyna's formulation in Eq. (5.4) if r and k and the function φ are taken to be functions of temperature. Perzyna (1966) showed that it was possible to correlate a considerable quantity of data by assuming that r and k were temperature sensitive and φ was temperature insensitive.

Klepaczko and Chiem (1986) derived a theory for jump tests in fcc metals. They proposed that there are two separate rate sensitivities of the flow stress. Referring to Figs 5.12(a) and (b), if the strain rate is increased from $\dot{\gamma}_i$ to $\dot{\gamma}_r$ the flow stress exhibits an instantaneous elastic response. This describes path ABCD in Fig. 5.12(a). The same effect would result from a decrease in the temperature as indicated.

It was noted by Klepaczko and Chiem that the value of incremental or jump tests is that they allow the measurement of the instantaneous elastic response $\Delta\tau_s$, as well as the hardening due to the deformation history $\Delta\tau_h$. Therefore, $\Delta\tau_h$ can be associated with the rate sensitivity of the strain hardening. Jump tests allow the determination of the two stress increments as functions of the initial strain γ_i.

Klepaczko (1987) mentions that many constitutive equations do not take into account the fact that the powers n and m are temperature sensitive such as the equations (5.13) and (5.14). Thus, in the more general case:

$$\sigma = B(\theta)\epsilon^{n(\theta)}\dot{\epsilon}^{m(\theta)} \qquad (5.22)$$

and normally n decreases as θ increases and m increases as θ increases.

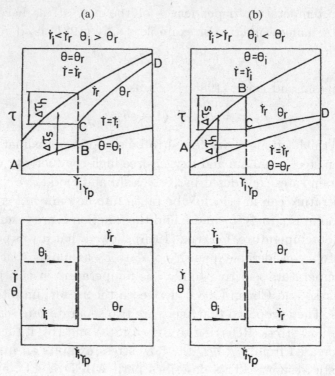

Fig. 5.12.   Typical responses of fcc metals: (a) increase in strain rate from $\dot{\gamma}_i$ to $\dot{\gamma}_r$ or decrease in temperature for T_i to T_r; (b) decrease in strain rate from $\dot{\gamma}_i$ to $\dot{\gamma}_r$ or increase in temperature from T_i to T_r. After Klepaczko and Chiem (1986).

Klepaczko assumes an Arrhenius-type relation to correlate stress, strain rate and temperature:

$$\tau = \hat{\tau} \left\{ \frac{\dot{\gamma}}{\dot{\gamma}_0} \exp\left(\frac{\Delta H}{k\theta_m\theta_h} \right) \right\}^{m(\theta_h)} \tag{5.23}$$

where $\hat{\tau}$ is a threshold stress, $\dot{\gamma}_0$ is a frequency factor, ΔH an apparent activation energy, θ_m is the melting point and θ_h the homologous temperature: $\theta_h = \theta/\theta_m$. The expression within the brackets is the Zener–Hollomon parameter. If we assume that deformation is strain history independent and power-law strain hardening is applicable:

$$\hat{\tau} = c(\theta_h)(\gamma_0 + \gamma)^{n(\theta_h)} \tag{5.24}$$

with

$$c(\theta_h) = c[1 - (\theta_h - p)\exp\{q(1 - \theta_h)\}] \qquad (5.25)$$

where c, γ_0, p and q are material constants.

Substituting, this gives

$$\tau = c(\theta_h)(\gamma_0 + \gamma)^{n(\theta_h)} Z^{m(\theta_h)} \qquad (5.26)$$

where Z is the Zener–Hollomon parameter.

Now for fcc metals:

$$\left.\begin{array}{l} m(\theta_h) = \alpha_m\theta_h, \quad 0 < \theta_h < 0.5 \\[2mm] \text{and} \quad m(\theta_h) = \alpha_m^*(\theta_h - \theta_h^*), \quad 0.5 < \theta_h < 1 \end{array}\right\} \qquad (5.27)$$

where α_m, α_m^* and θ_h^* are non-dimensional constants. For bcc metals the functional relation of n and m on θ is more complicated than for fcc metals. Klepaczko gives as an example the $m(\theta_h)$ expression for a hot-rolled steel (HRS):

$$m(\theta_h) = a\theta_h \exp(-b\theta_h) + \alpha_0\theta_h^r, \quad 0 < m < 1$$

where a, b, α_0 and r are material constants.

For specifying bcc metals 15 constants are needed and 12 are required for fcc metals (this reduces to eight for low-temperature deformation). These equations have been found to fit reasonably accurately the data for many metals.

Figures 5.13(a) and (b) show calculated τ versus γ curves at different temperatures and strain rates. The experimental points are taken from Senseny and co-workers (1978). Of course, there are modifications that could be made to these equations to give a better fit to experimental results. Certainly, for technological purposes, Klepaczko's suggested equations are adequate. However, there are too many undetermined parameters in the equations.

There have been many attempts to base constitutive equations on micromechanisms of plastic flow. See, for example, the reviews by Klepaczko (1988, 1989). Most micro-mechanical links begin with the assumption that dislocation glide is the predominant mode of plastic deformation. The plastic shear strain rate $\dot{\gamma}^p$ is related to the mean

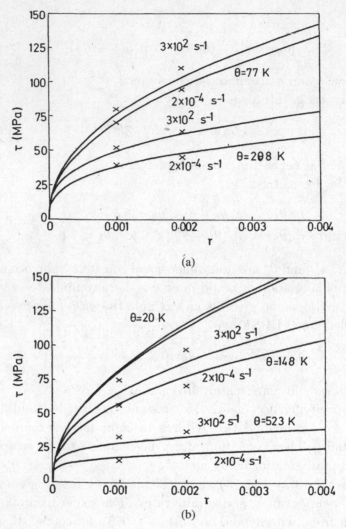

Fig. 5.13.   Calculated τ versus γ curves for aluminium at different temperatures: (a) $\theta = 77$ K and 298 K, for two strain rates $\dot{\gamma} = 2 \times 10^{-4}$ and 3×10^2/sec; (b) $\theta = 20$, 148 and 523 K, for the two strain rates $\dot{\gamma} = 2 \times 10^{-4}$ and 3×10^2/sec. Experimental points are shown by crosses. Plots after Klepaczko (1987).

velocity of dislocations by the well-known Orowan relation

$$\dot{\gamma}^p = b\rho_m V \qquad (5.28)$$

where b is Burger's vector and ρ_m is the mobile dislocation density.

If we consider dislocations to be held up at N points of intersection per unit volume as Seeger (1955) suggested, then plastic flow occurs at a rate governed by the frequency v with which dislocations intersect each other and the area A swept out by a dislocation. The strain rate is then

$$\dot{\gamma}^p = NAbv. \tag{5.29}$$

Now the frequency v is governed by thermal activation and therefore

$$v = v_0 \exp(-U/k\theta) \tag{5.30}$$

where v_0 is the intersection velocity and U is the activation energy. Now the activation energy is reduced by the applied stress during intersection. Then we can write

$$U = U_0 - (\tau - \tau_i)V$$

where τ_i is the internal stress, V the activation volume and U_0 the activation energy for intersection, which is a function of the plastic strain. Substitution of expressions (5.30) and U into (5.29) gives

$$\dot{\gamma}^p = NAbv_0 \exp\left(\frac{-U_0}{k\theta}\right) \exp\left[\frac{(\tau - \tau_i)V}{k\theta}\right]. \tag{5.31}$$

Gilman (1969) discussed this type of approach in detail.

There is a characteristic drag acting on dislocations. Drag is caused by the interaction of moving dislocations with phonons and at lower temperatures electrons (Kocks *et al.*, 1975). Dislocation drag is one of the many forces that can act on a dislocation. Other forces are the lattice Peierls–Nabarro stress, dislocation–dislocation interaction, Orowan looping around precipitates and cross-slip.

The essential problem with these mechanisms is their number. In essence we have a complex dislocation substructure in which there are a number of different mechanisms of deformation. It is almost impossible to correlate the mechanisms with observed macroscopic behaviour.

Because of these obvious difficulties, work has generally been confined to the thermally activated mechanism where correlation with testing temperature can be reasonably close.

Klepaczko and Chiem (1986) introduced the concept of structural evolution in constitutive equations. They suggest that structural evolution depends on strain rate and temperature. So if ρ is the dislocation density

$$\frac{d\rho}{d\gamma^p} = M_{\text{eff}}(\rho, \dot{\gamma}, \theta) \qquad (5.32)$$

where M_{eff} is the effective obstacle multiplication coefficient. Klepaczko and Chiem divide M_{eff} into two portions

$$M_{\text{eff}} = M_g(\rho, \dot{\gamma}) - M_a(\rho, \dot{\gamma}, \theta). \qquad (5.33)$$

M_g is due to defect generation and M_a is due to dynamic recovery. When M_{eff} is known for a material, the plastic strain rate can be derived by integration. For constant strain rate and temperature

$$\gamma^p = \int_{\rho_0}^{\rho} \frac{dx}{M_{\text{eff}}(x, \dot{\gamma}, \theta)} + C_1. \qquad (5.34)$$

Because the equation for the rate of change of dislocation density with strain is non-linear, then this integral is difficult to evaluate. However, the evolution of the microstructure is related to the internal stress via the current dislocation density by a relation of the form

$$\tau_\mu \propto \sqrt{\rho(\dot{\gamma}, \theta)} \qquad (5.35)$$

where τ_μ is the internal stress.

Lücke and Mecking (1973) discussed the contribution of the hardening and softening effects to the constitutive equation because the evolution equation is non-linear and this is still a very active area of research.

One of the most attractive classes of model relates an internal state variable to a single structural parameter. The internal state variables are certain to be microstructural parameters, for example $d\rho/d\gamma$ (Klepaczko, 1988). A simple evolution equation described by Klepaczko is

$$\frac{d\rho}{d\gamma} = A - B(\rho - \rho_0) \qquad (5.36)$$

where A and B are characteristic constants and ρ_0 and ρ are the initial and current dislocation densities. With a relation between the flow stress τ and the current dislocation density it is possible to model material behaviour numerically. However, at this stage it may well be that this approach is too complex for the requirements of a reasonably straightforward thermoviscoplastic constitutive equation. Moreover, there are many undetermined parameters involved.

In a remarkable treatise on the dynamic deformation of metals at various temperatures, Bell (1968) examines, with the aid of many experimental results, a constitutive equation that has a very wide application.

It was observed by Bell that for the many metals he tested the large-strain stress–strain function is parabolic when the stress and strain are expressed in nominal forms. All parabola coefficients were shown to be linearly dependent on the temperature and to be proportional to the zero-point isotropic elastic modulus μ_0 multiplied by a dimensionless universal constant $B_0 = 0.028$.

The stable zero-point elastic shear modulus for all the elements tested could be expressed in the form

$$\mu(0) = \left(\frac{2}{3}\right)^{s/2} \left(\frac{2}{3}\right)^{p/4} A \tag{5.37}$$

where $s = 1, 2, 3, \ldots$ and p is a constant structure factor, which has a value of either unity or zero. A is a universal constant.

The uniaxial stress–strain function proposed is

$$\sigma = \left(\frac{2}{3}\right)^{r/2} \mu_0 B_0 (1 - \theta/\theta_m)(e - e_b)^{1/2} \tag{5.38}$$

where r is what Bell calls a deformation mode index, σ is the nominal uniaxial stress, e is the nominal uniaxial strain and e_b is the intercept on the strain axis: 'r can have integer values 1, 2, 3, 4... and is used to fit experimental data. The only unknown in the equation is r, which was taken arbitrarily to equal unity' (Bell, 1968). Bell showed that it is possible to fit this equation to data for numerous materials and consequently this appears to be an important result, which has been overlooked by most researchers working in this area.

Figure 5.14 shows a typical correlation between experimental data in compression and tension with theoretical predictions for aluminium.

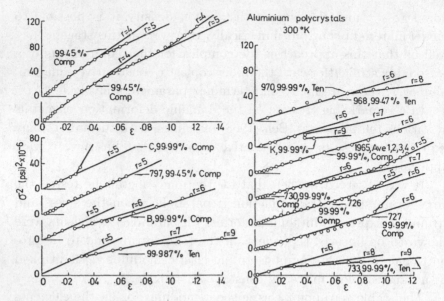

Fig. 5.14.   Tension and compression experiments for polycrystalline aluminium of medium and high purity compared with predictions (solid lines) for the indicated mode indices. After Bell (1968).

Some modern computer codes have used the Johnson–Cook equation and the Zerilli–Armstrong equations. These constitutive models have been found particularly useful in the modelling of the Taylor cylinder impact test.

The Zerilli–Armstrong equations (Zerilli and Armstrong, 1987) are modern constitutive models based on dislocation mechanisms. The equation for face-centred cubic metals is

$$\bar{\sigma} = C_0 + C_2 \bar{\epsilon}^{1/2} \exp(-C_3\theta + C_4\theta \ln \dot{\bar{\epsilon}}) \qquad (5.39)$$

where $\bar{\sigma}$, $\bar{\epsilon}$ and $\dot{\bar{\epsilon}}$ are the equivalent stress, equivalent strain and equivalent strain rate, respectively. C_0, C_2, C_3 and C_4 are constants and θ is the absolute temperature. It is assumed that the initial yield stress C_0 is independent of strain rate and temperature. The corresponding expression for body-centred cubic metals is

$$\bar{\sigma} = C_0 + C_1 \exp(-C_3\theta + C_4\theta \ln \dot{\bar{\epsilon}}) + C_5\bar{\epsilon}^n. \qquad (5.40)$$

The initial yield stress is a function of C_0, C_1, C_3 and C_4.

Although the Zerilli–Armstrong equations are used, it has been found difficult to apply them in practice, therefore the phenomenological Johnson and Cook equation (1983) is sometimes used in preference (see Johnson and Holmquist, 1988). The Johnson–Cook equation is

$$\bar{\sigma} = (A + B\bar{\epsilon}^n)(1 + C\ln\dot{\bar{\epsilon}}^*)(1 - \theta^{*m}) \qquad (5.41)$$

where $\dot{\bar{\epsilon}}^* = \dot{\bar{\epsilon}}/\dot{\bar{\epsilon}}_0$ is a dimensionless strain rate and $\theta^* = (\theta - \theta_{\mathrm{room}})/(\theta_{\mathrm{melt}} - \theta_{\mathrm{room}})$ is the homologous temperature for $0 \leqslant \theta^* \leqslant 1$. The five material constants that must be found are A, B, C, n and m.

5.4. Constitutive Equations for Non-Metals

As pointed out by Lataillade (1989), experimental techniques are advancing very quickly and this makes it possible to examine materials very accurately.

The mechanical properties of polymeric materials and composites based on polymers are very sensitive to testing rate and temperature and therefore jump tests are particularly valuable for ascertaining constitutive relations for these materials.

Also the mechanical properties of non-metallic materials can be sensitive to a superimposed hydrostatic pressure. We know from the pioneering work of Bridgman (1952) that the ductility of metals is dependent on superimposed hydrostatic pressure for high pressures (see also Osakada and co-workers, 1977). Von Karman (1911) showed that rocks and other ceramics become ductile at very large hydrostatic pressures. Bearing these points in mind we can write a general constitutive equation for a material as

$$\sigma = f(\epsilon, \dot{\epsilon}, p, H) \qquad (5.42)$$

where p is the hydrostatic pressure and H represents the history effects of the strain rate and temperature history effects. We know that strain rate and temperature influence the mechanical properties of polymers significantly, whereas hydrostatic pressure influences those of ceramics and history effects will be minimized for the latter.

Ceramics usually have a low fracture strain, and this is normally governed by the defect population and morphology (Davidge, 1979). But of course the yield stress, flow stress and fracture strain can be varied dramatically in synthetic composite ceramics.

Chapter 6

Occurrence of Adiabatic Shear Bands

6.1. Empirical Criteria

Although adiabatic shear bands are very narrow, they are still a macroscopic mode of deformation. The occurrence of adiabatic shear bands signals a mode transition from a homogeneous to a localized mode in continuum mechanics. From this point of view the occurrence of adiabatic shear bands may be regarded as straightforward, if the process of banding itself is not considered.

The essential idea of banding is physically straightforward. When the work increment required for a fixed increment of plastic deformation begins to decrease, there will be a transition in the deformation modes in the deforming system. This idea was proposed by Drucker (1951) in his well-known hypothesis concerning stable and unstable deformation in plasticity theory. The concept is that plastic strain may occur at decreasing stresses beyond a stress maximum, signifying a clear unstable deformation mode (see Fig. 6.1).

Historically, the concept of instability occurring at a peak in the load or stress can be traced back to Considere (1885), who provided a geometrical construction to define the maximum load condition for necking in simple tension. Following the maximum load conditions in tension, Recht (1964), Culver (1973) and other investigators proposed empirical criteria for the onset of adiabatic shearing in accordance with the idea of a maximum shear stress criterion. Because there is no change in cross-sectional area in simple shear, in many ways it is a simpler configuration than tension. However, other factors, such as adiabatic shear localization, can make simple shear of

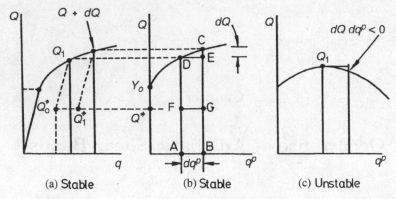

Fig. 6.1. Stable and unstable plastic materials under uniaxial stress. After Malvern (1969).

real materials complex. If we ignore elastic deformation, strain rate and temperature history effects and phase transformations in the tested material, then the shear stress can be written as a function of shear strain γ, shear strain rate $\dot{\gamma}$ and temperature θ:

$$\tau = f(\gamma, \dot{\gamma}, \theta). \qquad (6.1)$$

The increment in the shear stress can then be deduced:

$$d\tau = \left(\frac{\partial \tau}{\partial \gamma}\right)_{\dot{\gamma},\theta} d\gamma + \left(\frac{\partial \tau}{\partial \dot{\gamma}}\right)_{\gamma,\theta} d\dot{\gamma} + \left(\frac{\partial \tau}{\partial \theta}\right)_{\gamma,\dot{\gamma}} d\theta. \qquad (6.2)$$

The three partial differential terms are of great importance for the occurrence of adiabatic shear banding. The physical meanings of each of these terms are as follows:

$\left(\frac{\partial \tau}{\partial \gamma}\right)_{\dot{\gamma},\theta}$ — the isothermal strain hardening rate at constant strain rate

$\left(\frac{\partial \tau}{\partial \dot{\gamma}}\right)_{\gamma,\theta}$ — the isothermal strain-rate hardening rate at constant strain

$-\left(\frac{\partial \tau}{\partial \theta}\right)_{\gamma,\dot{\gamma}}$ — the rate of thermal softening at a constant strain and strain rate

The temperature and rate dependence of these three differential terms can lead to complicated contours of the critical condition for the occurrence of adiabatic shear banding (see, for example, the forging results of Turner, 1988a,b).

The maximum shear stress criterion requires $d\tau = 0$; this is for the point in the process where thermal softening just begins to outweigh strain and strain-rate hardening. Strain-rate hardening is usually appreciable only at high strain rates. Therefore, strain-rate hardening is usually neglected in comparison with strain hardening. Then the criterion $d\tau = 0$ corresponds to

$$\left(\frac{\partial \tau}{\partial \gamma}\right)_{\dot{\gamma},\theta} = -\left(\frac{\partial \tau}{\partial \theta}\right)_{\gamma,\dot{\gamma}} \cdot \frac{d\theta}{d\gamma}. \tag{6.3}$$

For adiabatic deformation, the increase in temperature due to the plastic work is

$$\rho c \, d\theta = \beta \tau \, d\gamma \tag{6.4}$$

where ρ is the material density, c is the specific heat and β is the fraction of plastic work converted into heat.

The fraction β of the plastic work done, which appears as heat is a very important factor in the further investigation of adiabatic shear banding. Therefore, it is worth referring to the original research carried out by Farren and Taylor (1925) and Taylor and Quinney (1934). By using thermocouple measurements on mild steel and copper tensile specimens, Farren and Taylor found that, within the range of their experiments, 86.5–95.5% of the plastic work done on the tensile specimens was converted into heat. The specific percentages were: 86.5% for steel, 90.5–92% for copper, 92–93% for aluminium and 95–95.5% for aluminium single crystals. It was thought that the energy retained in metals during distortion should be a definite fraction of the work done. The fractional energy retained appeared to be a constant for various amounts of distortion, with some variation from metal to metal. Subsequent to this work, it is generally accepted in engineering that between 85 and 95% of the plastic work done is converted into heat.

Later, Taylor and Quinney (1934) argued that it seemed unlikely that it would be possible to increase the amount of latent energy indefinitely by doing cold work. To justify this, they carried out a series of torsion tests. A typical set of experimental results is shown in Fig. 6.2. By assuming that the condition of a metal depends only on the amount of cold work retained in it (latent energy), the results

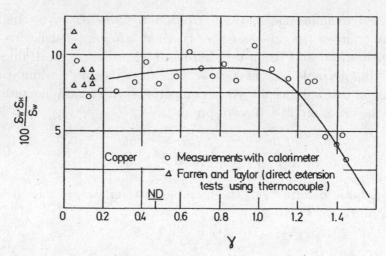

Fig. 6.2.   Percentage of work done that remains latent in specimens subjected to plastic twisting. After Taylor and Quinney (1934).

of Farren and Taylor (1925) were plotted on the same diagram as a comparison. It appears that the ratio of the retained latent energy to the plastic work done on the metal remains unchanged within about 10%. Beyond a certain amount of distortion, the retained latent energy tends to decrease with further distortion and approaches a saturated value.

Therefore, although a constant fraction of the plastic work is assumed to be converted into heat, this is not physically valid. However, it is an acceptable approximation in engineering practice. For simplicity we will take the fraction of plastic work converted into heat, $\beta = 1$.

If we make the assumptions of negligible strain-rate sensitivity and adiabatic deformation, the following criterion for adiabatic shear banding is obtained (see, for example, Culver, 1973), according to expressions (6.3–6.4):

$$\frac{\beta}{\rho c}\left(-\frac{\partial \tau}{\partial \theta}\right)_\gamma \bigg/ \frac{1}{\tau}\left(\frac{\partial \tau}{\partial \gamma}\right)_\theta = 1, \tag{6.5}$$

which is identical to the dimensionless number $B = 1$ (see Section 2.4).

It is clear that materials with a high rate of thermal softening, low strain hardening rate, low density and low specific heat will be prone to adiabatic shear banding. As these are all material parameters, this criterion leads to a type of classification regarding their susceptibility to adiabatic shear banding. Provided the constitutive equation of the material is available, then the criterion can be simplified to a characteristic shear strain γ_i and this can be used as an indication of material behaviour. For example, if power-law hardening is assumed, $\tau = k\gamma^n$, then γ_i is given by

$$\gamma_i = \frac{n\rho c}{-\left(\frac{\partial \tau}{\partial \theta}\right)_{\gamma, \dot{\gamma}}}. \tag{6.6}$$

Table 6.1 lists various predictions of the critical shear strains and the related constitutive equations. The correlation with experimental data appears to be reasonably good and will be discussed later in Section 6.4.

The adiabatic heating principle can also be used to explain phenomena in cryogenic tension. Here instability occurs at maximum axial load $dL = 0$ instead of maximum shear stress $d\tau = 0$. If we make the usual assumption of incompressibility during plastic deformation then the increment of uniaxial tensile strain is

$$d\epsilon = -dA/A$$

where A is the current cross-sectional area. Then it follows that the maximum load criterion in thermoplastic deformation with the adiabatic assumption gives

$$\left\{\frac{\beta}{\rho c}\left(-\frac{\partial \sigma}{\partial \theta}\right)_\epsilon + 1\right\} \bigg/ \frac{1}{\sigma}\left(\frac{\partial \sigma}{\partial \epsilon}\right)_\theta = 1.$$

For isothermal deformation of a power-law hardening solid, $\sigma = K\epsilon^n$, the critical tensile strain is

$$\epsilon_i = \frac{n}{\left\{\frac{\beta}{\rho c}\left(-\frac{\partial \sigma}{\partial \theta}\right)_\epsilon + 1\right\}}. \tag{6.7}$$

Of course, if we can ignore temperature this reverts to the classical $\epsilon_i - n$, which is the condition for diffuse necking (see, for example, Swift, 1952; Backofen, 1972).

Table 6.1.   Criteria for shear instability.

	No.	Constitutive relation	Instability strain	References
Linear strain hardening with yield strength	1	$\tau = A(1 + \alpha\gamma)$	$\gamma_i = \dfrac{\rho c}{\left(-\frac{\partial \tau}{\partial \theta}\right)} - \dfrac{1}{a}$	Bai (1981), Staker (1981)
	2	$\tau = A(1 + \alpha\gamma)\exp(-\beta\gamma)$	$\gamma_i = \dfrac{1}{\beta_1} - \dfrac{1}{a}$ $\beta_1 = \beta + \dfrac{\left(-\frac{\partial \tau}{\partial \theta}\right)}{\rho c}$	Olson et al. (1981)
Power law without yield strength	3	$\tau = A\gamma^n$	$\gamma_i = \dfrac{n\rho c}{\left(-\frac{\partial \tau}{\partial \theta}\right)}$	Culver (1973)
	4	$\tau = A\gamma^n \exp(-\beta\gamma)$	$\gamma_i = \dfrac{n}{\beta_1}$	Olson et al. (1981)
	5	$\tau = A\gamma^n \dot{\gamma}^m \exp\left(\dfrac{W}{\theta}\right)$	$\gamma_i = \left\{\dfrac{n\rho c}{AW}\dfrac{\theta^2}{\dot{\gamma}^m}\exp\left(-\dfrac{W}{\theta}\right)\right\}^{\frac{1}{n+1}}$	Vinh et al. (1979)
	6	$\tau = A\gamma^n \dot{\gamma}^m \theta^{-v}$	$\gamma_i = \left\{\dfrac{n\rho c}{Av}\dfrac{\theta^{v+1}}{\dot{\gamma}^m}\right\}^{\frac{1}{n+1}}$	

(Continued)

Table 6.1.   (*Continued*)

No.	Constitutive relation	Instability strain	References
7	$\tau = A\gamma^n(1+b\dot{\gamma}^m)(1-c_1\theta)$	$\gamma_i = \left\{ \dfrac{n\rho c}{Ac_1}\dfrac{1}{(1+b\dot{\gamma})^m} \right\}^{\frac{1}{n+1}}$	Burns and Trucano (1982)
Power law with yield strength			
8	$\tau = A(1+a\gamma^n)\left(1+b\ln\dfrac{\dot{\gamma}}{\dot{\gamma}_0}\right)$	$\gamma_i = \dfrac{n\rho c(\theta_M - \theta_0)}{A(1+a)} - \dfrac{\gamma_i^{1-n}}{a}$	Lindholm and Johnson (1983)
9	$\times \dfrac{\theta_M - \theta}{\theta_M - \theta_0}$		
Others			
10	$\tau = A + B\log\dot{\gamma}$	$\gamma_i = \dfrac{B\rho C}{\tau|\partial\tau/\partial\theta|}$	Pomey (1966)
11	$\dot{\epsilon} = ct\left(\dfrac{\partial\dot{\epsilon}}{\partial\epsilon} = 0\right)$	$\dot{\epsilon}_c = \dfrac{4\pi(\epsilon - \epsilon_y)}{L^2\tau_y^2} \cdot kc \left(\dfrac{\partial\tau/\partial\epsilon}{\partial\tau/\partial\theta}\right)^2$	Recht (1964)
12	$\tau = H(\bar{\epsilon}^n)F\left(\dfrac{\theta}{\theta_m}\right)$ $= \tau_0\bar{\epsilon}^n F\left(\dfrac{W}{E_m}\right)$		Curran (1979)

(*Continued*)

Table 6.1. (*Continued*)

No.	Constitutive relation	Instability strain	References
	$\rightarrow F\left(\dfrac{W}{E_m}\right) = \left(1 - \dfrac{\alpha W}{E_m}\right)^{1/2}$	$\bar{\epsilon}^p_c = \left[\dfrac{2\rho E m n(n+1)}{\alpha\tau_0(2n+1)}\right]^{1/(n+1)}$	
	$\rightarrow F\left(\dfrac{W}{E_m}\right) = \left(1 - \dfrac{\alpha' W}{E_m}\right)$	$\bar{\epsilon}^p_c = \left[\dfrac{n\rho E m}{\alpha\tau_0}\right]^{1/(n+1)}$	
13	$\dot{\gamma}^p_{xy} = \dot{\gamma}_0 \exp\left(\dfrac{-\Delta H}{K\theta}\right)$	$\left[\dfrac{1}{\tau\tau}\left(\dfrac{\partial\tau\tau}{\partial\gamma p}\right) - \dfrac{1}{C}\left(\dfrac{\partial\tau}{\partial\theta}\right)\right]\dfrac{\dot{\gamma}^p}{m}$	Clifton (1980)
	$\Delta H\left(\dfrac{\sigma_{xy}}{\tau}\right)$	$+\dfrac{k\xi^2}{C} = 0$	
	$m = \dfrac{\partial\ln\sigma_{xy}}{\partial\ln\dot{\gamma}^p_{xy}}$		

Practically, plastic deformation cannot be truly adiabatic: some heat will be conducted away from the deformation zone. As a simplified approach, the loss of heat can be described by a fractional factor b. Then the temperature increase owing to the plastic work can be written as

$$d\theta = b\tau \, d\gamma/\rho c.$$

Correspondingly, the criteria and critical strains in shear and tension may be modified as follows:

$$\left.\begin{array}{c} \frac{\frac{b}{\rho c}\left(-\frac{\partial \tau}{\partial \theta}\right)}{\frac{1}{\tau}\left(\frac{\partial \tau}{\partial \gamma}\right)} = 1, \quad \frac{\frac{b}{\rho c}\left(-\frac{\partial \sigma}{\partial \theta}\right)+1}{\frac{1}{\sigma}\frac{d\sigma}{d\epsilon}} = 1 \\[3mm] \gamma_i = \frac{n\rho c}{b\left(-\frac{\partial \tau}{\partial \theta}\right)}, \quad \epsilon_i = \frac{n}{\frac{b}{\rho c}\left(-\frac{\partial \sigma}{\partial \theta}\right)+1} \end{array}\right\}. \tag{6.8}$$

$b = 1$ and 0 represent the adiabatic and isothermal cases, respectively.

The effect of compressive stress on the occurrence of adiabatic shear is not very clear. However, an examination of the effects of hydrostatic pressure on the shear stress–shear strain curve, even measured at low strain rates, may provide some sort of clue to the problem. Figures 2.12 and 2.13 show shear stress–shear strain curves under different hydrostatic pressures and normal compressive stresses on the shear plane, respectively. Both reveal the same tendency. That is, as the pressure increases the maximum shear stress is delayed to greater strains. This may imply that unstable shear, even adiabatic shear banding, may be retarded by hydrostatic pressure. However, there is a dilemma because of the observation that adiabatic shear bands are favoured in compressive stress states (see Rogers, 1979), and also the fact that there is no volume change in shear deformation.

One possible explanation for this paradox may be the hypothesis that geometrical imperfections (voids) as well as thermal softening can both affect the occurrence of adiabatic shear banding. Consider a thin-walled tubular specimen of gauge length l containing an array of voids of total length w. Furthermore, suppose that the interaction of the voids can be ignored and the void growth rate is pressure dependent:

$$\dot{w} = \dot{\gamma}\left(a - \frac{hp}{\sigma_y}\right) \tag{6.9}$$

where a and b are material parameters. Integration gives the relation between the void length w, the shear strain γ and the hydrostatic pressure:

$$w = (\gamma - \gamma_n)\left(a - \frac{bp}{\sigma_y}\right)$$

where γ_n is the threshold strain for void nucleation. In this case the maximum of the shear load is given by

$$dL_{\mathrm{shear}} = d\{\tau(l - w)\} = (l - w)d\tau - \tau dw = 0.$$

Assuming that there are negligible changes in the strain rate and that deformation is adiabatic the criterion becomes:

$$\frac{dw}{d\gamma}\Big/(l - w) = \frac{1}{\tau}\left(\frac{\partial\tau}{\partial\gamma}\right) + \frac{\beta}{\rho c}\left(\frac{\partial\tau}{\partial\theta}\right). \qquad (6.10)$$

Substitution of the void growth rate, the expression for the void length together with power-law hardening gives a critical shear strain as follows (Dodd and Atkins, 1983):

$$\gamma_i = \frac{n}{\frac{\beta(-\partial\tau/\partial\theta)}{\rho c} + \frac{a - bp/\sigma_y}{l - (\gamma_i - \gamma_n)(a + bp/\sigma_y)}}. \qquad (6.11)$$

This criterion indicates that a compressive hydrostatic pressure retards the occurrence of adiabatic shear bands. If $p \geqslant (a/b)\sigma_y$ the contribution from void growth (that is geometrical softening) is completely suppressed.

The above-mentioned criteria are based on the assumption that the effect of shear strain rate can be ignored. Although most investigations are of this opinion, some such as Recht (1964) think the rate effect cannot be discarded completely.

Recht examined the process of metal-cutting and thought that catastrophic thermoplastic shear would occur on a thin layer where the plastic work transformed into heat and only part of the heat was conducted away to the surroundings. So Recht developed a model based on heat generation and conduction instead of making the adiabatic assumption. It is best to describe his law using dimensional analysis. There are five relevant quantities:

$$[\theta] = \theta, [l] = \text{T}, [\rho c] = \text{M}/\text{LT}^2\theta, [\lambda] = \text{ML}/\text{T}^3\theta \quad \text{and} \quad q = \text{M}/\text{T}^3$$

where $q = \tau\dot{\gamma}l$ and is the heat generation rate per unit area and l is the specimen length. There are only four independent dimensions: θ, T, M and L. Using dimensional analysis, the five quantities must be combined as one dimensionless quantity of order O(1), i.e.

$$\frac{\theta}{q}\left(\frac{\lambda\rho c}{t}\right)^{1/2} \approx 1$$

or

$$\theta \approx q\left(\frac{t}{\lambda\rho c}\right)^{1/2} = \tau\dot{\gamma}l\left(\frac{t}{\lambda\rho c}\right)^{1/2}.$$

Because of the occurrence of adiabatic shear banding, $d\tau = 0$, differentiation gives

$$d\theta = \frac{\tau\dot{\gamma}l}{2}\left(\frac{1}{\lambda\rho c t}\right)^{1/2}dt$$

or

$$\frac{d\theta}{d\gamma} = \frac{\tau l}{2}\left\{\frac{\dot{\gamma}}{\lambda\rho c(\gamma - \gamma_y)}\right\}^{1/2} \tag{6.12}$$

where, as before, the process is assumed to occur at a constant rate of shear strain, $\gamma = \dot{\gamma}t + \gamma_y$, where γ_y is the initial yield shear strain. Equation (6.12) is the counterpart of the adiabatic expression (6.4). Substitution of Eq. (6.12) into criterion (6.3) yields the following criterion, which includes strain rate and heat conduction:

$$\frac{\partial\tau}{\partial\gamma} = -\left(\frac{\partial\tau}{\partial\theta}\right)\frac{\tau l}{2}\left\{\frac{\dot{\gamma}}{\lambda\rho c(\gamma - \gamma_y)}\right\}^{1/2}$$

or

$$\dot{\gamma} = 4\lambda\rho c(\gamma - \gamma_y)\left[(\partial\tau/\partial\gamma)\bigg/\left(\frac{\partial\tau}{\partial\theta}\right)\right]^2\bigg/\tau^2 l^2. \tag{6.13}$$

Recht (1964) suggested that the relative susceptibility of materials to shear localization could be compared by taking the ratio of the characteristic strain rates at the same plastic shear strain:

$$\frac{\dot{\gamma}_1}{\dot{\gamma}_2} = \frac{(\lambda\rho c)_1}{(\lambda\rho c)_2} \left\{ \frac{\left(\frac{\partial\tau}{\partial\gamma}\right)_1 \left(\frac{\partial\tau}{\partial\theta}\right)_2}{\left(\frac{\partial\tau}{\partial\theta}\right)_1 \left(\frac{\partial\tau}{\partial\gamma}\right)_2} \right\}^2 \left(\frac{\tau_2}{\tau_1}\right)^2. \tag{6.14}$$

Recht's pioneering work led to an early understanding of adiabatic shear banding. It is true that unstable shear deformation may occur at a maximum in the shear stress and along a thin zone, but this approach is too simplistic (see Chapter 7).

An alternative approach emphasizes the coupling effects of strain hardening, strain-rate hardening and thermal softening. Combining the maximum shear–stress criterion with the adiabatic assumption and a constitutive equation provides a set of first-order ordinary differential equations with the independent variable γ:

$$\frac{d\dot{\gamma}}{d\gamma} = -\left(\frac{\partial\tau}{\partial\gamma}\right) \bigg/ \left(\frac{\partial\tau}{\partial\theta}\right) - \frac{d\theta}{d\gamma}\left(\frac{\partial\tau}{\partial\theta}\right) \bigg/ \left(\frac{\partial\tau}{\partial\dot{\gamma}}\right) \tag{6.15}$$

$$\frac{d\theta}{d\gamma} = \frac{\beta t}{\rho c} \tag{6.16}$$

$$\tau = f(\gamma, \dot{\gamma}, \theta). \tag{6.17}$$

This system of equations has the following solutions:

$$\dot{\gamma} = \dot{\gamma}(\gamma; a, b)$$
$$\theta = \theta(\gamma; a, b)$$

where a and b are two constants determined by the initial conditions. The solutions give the threshold values of the variables $\dot{\gamma}$, θ and γ, which define the occurrence of adiabatic shear banding. Because the constitutive equation is non-linear, numerical integration of the differential equations is necessary. Until now, it appears that these types of solutions have been neglected. The reason for this may be the uncertainty of the constitutive equation at high strain rates and temperatures. When such constitutive equations become more widely available, the adiabatic shearing criterion obtained will be more accurate and will therefore be more helpful for industrial applications.

As an example, let us consider the following constitutive equation:

$$\tau = A\gamma^n \dot{\gamma}^m \theta^{-v}.$$

Substitution into Eqs (6.15)–(6.17) provides:

$$\gamma_i = \frac{\rho c \left(n + m \frac{d\ln\dot{\gamma}}{d\ln\gamma} \right) \theta}{\beta v \tau}. \tag{6.18}$$

Wang and his co-workers (1988) presented a further example. They carried out compression experiments on titanium. The constitutive equation they used was:

$$\sigma = (\sigma_0 + E\epsilon) \left(1 + g\ln\left(\frac{\dot{\epsilon}}{\dot{\epsilon}_0}\right) \right) \left(1 - \alpha\frac{\theta}{\theta_0} \right) \tag{6.19}$$

where σ_0, E, g, $\dot{\epsilon}_0$, α and θ_0 are all parameters. The corresponding critical condition in the plane of ϵ and $\dot{\epsilon}$ for the occurrence of adiabatic shear banding is:

$$\left(1 + g\ln\frac{\dot{\epsilon}}{\dot{\epsilon}_0} \right) \left(\frac{\sigma_0}{E} + \epsilon \right) \left(a - \frac{\beta\alpha E}{\rho c\theta_0} \right) = 1 \tag{6.20}$$

where a is the constant of integration. The correlation of this equation with experimental results will be discussed in Section 6.4.

6.2. One-Dimensional Equations and Linear Instability Analysis

Although the simplified maximum shear stress criterion has been helpful in dealing with the problem of the occurrence of adiabatic shear bands, nevertheless the information provided by this criterion is restrictive. There are many reasons for this. Firstly, a number of assumptions are imposed such as adiabatic deformation and strain-rate-independent deformation. The assumptions made are obviously dependent on the expertise of the researchers and therefore it follows that the success of a particular model will also depend on the expertise and experience of the researchers. Secondly, adiabatic shear bands can only be dealt with using continuum mechanics. Therefore, the temporal and spacial variations can only be described adequately using field equations of a continuum. It is necessary to justify any empirical criterion proposed using continuum mechanics.

Here the governing differential equations will be presented and the homogeneous solutions and instability analysis related to the occurrence of adiabatic shear banding will be described. For simplicity, discussion will be confined to one-dimensional simple shear. Further discussions related to shear banding in multidimensional stress states will be left until Section 7.6.

As a starting point, we will state some fundamental physical laws.

(1) Adiabatic shear banding can be described using continuum mechanics (see Section 2.1) and the fundamental equations for conservation of mass, momentum and energy as well as the constitutive equation are used.

Thus:

$$v_{i,i} = 0$$
$$\rho \dot{v}_i = \sigma_{ij,j}$$
$$\rho c \dot{\theta} + \dot{e} = h_{i,i} + q + \sigma_{ij}\sigma_{i,j} \tag{6.21}$$

where v_i is the particle velocity, ρ is the density, σ_{ij} the Cauchy stress tensor, θ the temperature, c is the specific heat, h_i is the heat flux and e is the latent energy.

(2) Heat conduction is governed by Fourier's law:

$$h_i = -\lambda \theta_{,i} \tag{6.22}$$

where λ is the thermal conductivity, for simplicity taken to be a constant.

(3) The fraction of the plastic work converted into heat is a constant (Farren and Taylor, 1925; Taylor and Quinney, 1934) and for simplicity we will assume this constant is unity.

In the discussion of adiabatic shear bands above the following assumptions were adopted:

(1) Elastic deformation is ignored when compared to the large plastic deformations, so $\epsilon_{ij} = \epsilon_{ij}^p$.

(2) The material exhibits no strain-rate history effects, is plastically incompressible and isotropic, so:

$$\bar{\sigma} = f(\bar{\epsilon}, \dot{\bar{\epsilon}}, \theta)$$

where $\bar{\sigma}$ and $\bar{\epsilon}$ are the von Mises effective stress and strain, respectively. In addition, the external heat source is not taken into account, so $q = 0$. The basic equations concerning thermoplastic shearing are:

$$\left.\begin{array}{l} \rho \dot{v}_i = \sigma_{ij,i} \\ \rho c \dot{\theta} = \lambda \theta_{i,i} + \beta \sigma_{ij} v_{i,j} \end{array}\right\} \tag{6.23}$$

where $(1 - \beta)\sigma_{ij} v_{i,j} = \dot{e}$ according to Taylor and Quinney (1934).

Consider uniaxial motion along the x axis in the (x, y) plane (Fig. 6.3):

$$\left.\begin{array}{l} x_1 = x = u(x, y) + X \\ x_2 = y = Y \\ x_3 = z = Z \end{array}\right\} \tag{6.24}$$

where x_i are the Eulerian coordinates and X, Y, Z are the Lagrangian coordinates. In one-dimensional simple shear $u(x, y) = u(y)$ and the unique component of shear strain is $\gamma = \partial u / \partial y$. This mode of deformation only affects the (x, y) plane. Furthermore, the stress component σ_z should be zero due to incompressibility and the absence of any axial strains. The one-dimensional assumption also leads to a further simplification, i.e. the gradient in the x direction $\partial / \partial x \ll \partial / \partial y$

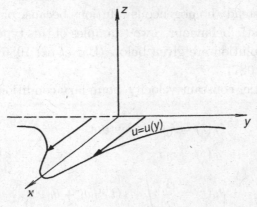

Fig. 6.3.   Schematic diagram showing one-dimensional simple shear.

or simply $\partial/\partial x \approx 0$. Other variables, such as the other normal stresses, σ_y and σ_x, do not appear in the model. This is in fact Prandtl's hypothesis for a boundary layer in fluid mechanics with a thin layer with a large gradient perpendicular to it. This is an analogue of adiabatic shear banding. The final governing equations become:

$$\rho\frac{\partial^2\gamma}{\partial t^2} = \frac{\partial^2\tau}{\partial y^2} \qquad (6.25\text{--}1)$$

$$\beta\tau\frac{\partial\gamma}{\partial t} = \rho c\frac{\partial\theta}{\partial t} - \lambda\frac{\partial^2\theta}{\partial y^2} \qquad (6.25\text{--}2)$$

$$\tau = f(\gamma,\dot{\gamma},\theta). \qquad (6.26)$$

Generally, under prescribed initial and boundary conditions the system of equations can be solved. Usually, the initial condition can be set since the shear stress or strain is zero and the temperature is homogeneous, i.e. the natural state. But in defect analysis, inhomogeneities of temperature or geometrical imperfections can be introduced. The mechanical boundary condition can either be stress or velocity controlled. The thermal boundary conditions can be idealized as adiabatic or isothermal.

After certain mathematical manipulations, homogeneous solutions to Eqs (6.25) can be obtained with certain combinations of boundary conditions and the natural initial conditions. The noticeable feature of the homogeneous solutions is their time-dependence, i.e. they are unsteady homogeneous solutions because of the nature of thermoviscoplastic behaviour. Two examples of this type of unsteady homogeneous solution are given below (Bai *et al.*, 1986; Fressengeas and Molinari, 1987).

(1) Adiabatic, constant velocity boundary condition:

$$\dot{\gamma}_h(t) = \dot{\gamma}_0 = \text{constant}$$

$$\gamma_h(t) = \dot{\gamma}_0 t$$

$$\theta_h(t) = \frac{1}{\rho c}\int_0^t \tau_h(t)\dot{\gamma}_0 dt + \theta_0$$

$$\tau_h(t) = f(\gamma_h(t),\dot{\gamma}_0,\theta_h(t)). \qquad (6.27)$$

(2) Adiabatic, constant stress boundary condition in the quasi-static case:

$$\tau_h(t) = \tau_0 = \text{constant}$$

$$\theta_h(t) = \frac{1}{\rho c}\tau_0\gamma_h(t) + \theta_0$$

$$\gamma_h(t) = \int_0^t \dot{\gamma}_h(t)dt$$

$$\dot{\gamma}_h(t) = g(\tau_0, \theta_h(t), \gamma_h(t)) \tag{6.28}$$

where $\dot{\gamma} = g(\tau, \theta, \gamma)$ is the constitutive equation.

If the constitutive equation has the following form $\tau = A\gamma^n\dot{\gamma}^m\theta^{-v}$ the above unsteady, homogeneous solutions have the following explicit forms:

(a)

$$\dot{\gamma}_h = \dot{\gamma}_0$$

$$\gamma_h = \dot{\gamma}_0 t$$

$$\theta_h = \left\{\frac{A(1+v)}{\rho c(1+n)}\dot{\gamma}_0^{m+n+1}t^{n+1} + \theta_0^{v+1}\right\}^{1/(v+1)} \tag{6.29}$$

(b)

$$\tau_h = \tau_0$$

$$\theta_h = \frac{\tau_0}{\rho c}\gamma_h + \theta_0$$

$$\dot{\gamma}_h = \left(\frac{\tau_0}{A}\right)^{1/m}\gamma_h^{-n/m}\left(\frac{\tau_0}{\rho c}\gamma_h + \theta_0\right)^{v/m}. \tag{6.30}$$

Now we come to the important question: when does the homogeneous solution become unstable and give way to inhomogeneous shear deformation? Clifton (1980), Clifton and co-workers (1984) and Bai (1981, 1982) examined the problem of instability using linear perturbation analysis. They linearized the governing equations for the undisturbed unsteady but homogeneous shear deformation and assumed that the rate of unsteady homogeneous deformation is much less than that of the rate of growth of disturbances. Therefore, the unsteady homogeneous solution can be chosen as a basis for calculation and the homogeneous variables can be treated as slow functions of time or simply as parameters. Both analyses include strain hardening, thermal softening, strain-rate sensitivity and heat conduction, and both investigate the growth of infinitesimal inhomogeneities from homogeneous simple shearing.

The infinitesimal perturbations were assumed to be:

$$(\gamma', \tau', \theta') \ll (\gamma_h, \tau_h, \theta_h).$$

Substitution of the perturbations into the governing equations (6.25) and only retaining first-order terms, i.e. $O(\gamma')$, etc., leads to the following linearized perturbated equations:

$$\rho \frac{\partial^2 \gamma'}{\partial t^2} = \frac{\partial^2 \tau'}{\partial y^2}$$

$$\beta \tau_h \frac{\partial \gamma'}{\partial t} + \beta \tau' \dot{\gamma}_h = \rho c \frac{\partial \theta'}{\partial t} - \lambda \frac{\partial^2 \theta'}{\partial y^2}. \tag{6.31}$$

We will now examine the equations for the following type of solution:

$$\left\{ \begin{array}{c} \gamma' \\ \tau' \\ \theta' \end{array} \right\} = \left\{ \begin{array}{c} \gamma_* \\ \tau_* \\ \theta_* \end{array} \right\} \exp(\alpha t + iky) \tag{6.32}$$

where the amplitude, γ_*, τ_*, θ_* and the values of the growth rate α and the wave number k are undetermined. Substituting Eq. (6.31) into Eqs (6.31), and recalling the differential of τ:

$$d\tau = Q d\gamma + R d\dot{\gamma} - P d\theta$$

where Q is the rate of strain hardening $(\partial \tau / \partial \gamma)$, R is the rate of strain-rate hardening $(\partial \tau / \partial \dot{\gamma})$ and P is the rate of thermal softening $(-\partial \tau / \partial \theta)$. We finally obtain a system of homogeneous algebraic equations for γ_* and θ_*:

$$\{\rho \alpha^2 + (Q + R\alpha)k^2\}\gamma_* - Pk^2 \theta_* = 0$$

$$\{\beta \tau_h \alpha + \beta \dot{\gamma}_h (Q + R\alpha)\}\gamma_* - (\beta \dot{\gamma}_h P + \rho c \alpha + \lambda k^2)\theta_* = 0. \tag{6.33}$$

To guarantee the existence of non-trivial solutions for γ_* and θ_* the determinant of the coefficients of Eqs (6.32) should be zero. This leads to the cubic characteristic equation:

$$\rho^2 c\alpha^3 + \rho\{\beta P\dot{\gamma}_h + (\lambda + cR)k^2\}\alpha^2$$
$$+(\lambda Rk^2 + \rho cQ - \beta\tau_h P)k^2\alpha + \lambda Qk^4 = 0. \qquad (6.34)$$

If the eigenvalue of α has a positive real part, this implies that instability is possible. Equation (6.34) can be non-dimensionalized as:

$$\bar{\alpha}^3 + \{C + (1 + A)\bar{k}^2\}\bar{\alpha}^2 + (A\bar{k}^2 + 1 - B)\bar{k}^2\bar{\alpha} \qquad (6.35)$$

where

$$\bar{\alpha} = \frac{\lambda\alpha}{cQ}, \quad \bar{k}^2 = \frac{\lambda^2 k^2}{\rho c^2 Q}, \quad A = \frac{cR}{\lambda}$$

$$B = \frac{\beta\tau_h P}{\rho cQ}, \quad C = \frac{\beta\lambda P\dot{\gamma}_h}{\rho c^2 Q} = B/Pr.$$

In fact A is Prandtl number Pd and B has been discussed in Section 2.4. The discussion about the cubic equation (6.35) cannot be explicit. Nevertheless, it can be deduced that for positive $\tilde{\alpha}$:

$$B > 1 + 2\sqrt{C}. \qquad (6.36)$$

Now as C can be evaluated for metals approximately as

$$C = \frac{\beta\lambda P\dot{\gamma}_h}{\rho c^2 Q} \approx \frac{10^2 \times 10^6}{10^3 \times 10^6 \times 10^8}\dot{\gamma}_h \approx 10^{-9}\dot{\gamma}_h \ll 1$$

then it follows that the criterion for instability simplifies to

$$B > 1, \qquad (6.37)$$

which is identical to the maximum shear stress criterion given in Eq. (6.5). Two special cases will be considered while omitting the algebra for clarity.

(1) The adiabatic case, $\lambda = 0$. The characteristic equation for this case is:

$$\rho^2 c\alpha^2 + \rho(\beta P\dot{\gamma}_h + cRk^2)\alpha - (\beta\tau_h P - \rho cQ)k^2 = 0. \qquad (6.38)$$

The criterion for instability is $B > 1$ and the maximum growth rate of the instability is

$$\alpha_m = \frac{\beta\tau_h P - \rho cQ}{\rho cR}, \qquad (6.39)$$

which occurs for disturbances with a short wavelength, $k \to \infty$.

(2) The non-strain hardening case, $Q = 0$. The characteristic equation for this case is

$$\rho^2 c\alpha^2 + \rho\{\beta P\dot{\gamma}_h + (\lambda + cR)k^2\}\alpha + (\lambda Rk^2 - \beta\tau_h P)k^2 = 0. \qquad (6.40)$$

The criterion for instability in this case is

$$\beta\tau_h P > \lambda Rk^2 \quad \text{or} \quad \bar{k}^2 < B/Pd. \qquad (6.41)$$

The maximum growth rate occurs at a disturbance with a certain wavelength such that $\alpha_m < \beta\tau_h P/\rho(\lambda + cR)$. Hence inertia, thermal conductivity and rate sensitivity all retard growth.

In both cases it is the inertia term that causes the characteristic equations to be non-linear. Therefore, for the bifurcation of solutions, the growth rates obtained in the linearized perturbation analysis will only be valid for a short time after instability and they cannot predict the whole evolution of the shear bands.

The above analysis provides us with two types of information about adiabatic shear banding. When the combination of variables in the homogeneous solution exceeds some critical condition, such as conditions (6.36), (6.37), (6.39) or (6.41), homogeneous deformation will have to give way to an inhomogeneous mode of deformation. The instability analysis has demonstrated that for most metals the adiabatic assumption is acceptable for the instability criterion because $C \ll 1$ in equation (6.36). In this case, the instability criterion, Eq. (6.37) or (6.39), is identical to the maximum shear stress criterion, Eq. (6.5). Secondly, inhomogeneities grow at different rates. In the adiabatic case, the inhomogeneity with the shortest wavelength will grow with the fastest rate. Thermal softening accelerates the growth of inhomogeneities whereas strain hardening and strain-rate hardening retard the growth.

Douglas and co-workers (1989) attempted to examine the stability of inhomogeneous shearing motions, as in the case of dynamic torsion of a thin-walled tubular specimen with thick flanges.

6.3. Localization Analysis

Since neither the principle of the maximum in the shear stress nor the perturbation technique describe shear localization itself, a new approach is needed. Although wave trapping, which is discussed in detail in Section 7.3 as a stage in the evolution of shear bands, provides a mechanism for localization beyond the peak in the stress, this is a special case occurring at the loading boundary. Suppose that inhomogeneities do occur but cannot be amplified by any mechanism, then no localization can occur in subsequent deformation. Therefore, a fundamental problem that we have is how to define localization and discover the fundamental mechanism for the occurrence of adiabatic shear bands.

One way of introducing localization is as follows. We define the inhomogeneities by $\Delta\gamma$ and $\Delta\dot\gamma$, where Δ denotes the spacial difference between disturbed and uniform deformation. For simplicity, let us suppose that the ratio of the inhomogeneities is constant with respect to time t:

$$\frac{\Delta\ln\dot\gamma}{\Delta\ln\gamma} = \chi = \text{constant}. \qquad (6.42)$$

This expression can be treated as the ordinary evolution equation of the inhomogeneity $\Delta\gamma$. After integration, the solution to the inhomogeneity $\Delta\gamma$ is:

$$\frac{\Delta\gamma}{\Delta\gamma_0} = \left(\frac{\gamma}{\gamma_0}\right)^\chi$$

where the subscript 0 denotes the initial value. The solution can be written as follows

$$\frac{\Delta\gamma}{\gamma} = \left(\frac{\gamma}{\gamma_0}\right)^{\chi-1}\frac{\Delta\gamma_0}{\gamma_0} = k(\gamma)\frac{\Delta\gamma_0}{\gamma_0}. \qquad (6.43)$$

The quantity $\Delta\gamma/\gamma$ represents localization better than $\Delta\gamma$. To show this, if both $\Delta\gamma$ and γ increase but the increase in $\Delta\gamma$ is less than that in γ, the inhomogeneity will become less important, although deformation is still concentrating. Therefore, in Eq. (6.43) $k(\gamma)$ effectively plays the role of the amplification factor in localization. $k(\gamma) > 1$ represents the process of localization, which requires $\chi > 1$ provided $\gamma > \gamma_0$. For a case study let us examine localization in a material deforming according to the constitutive equation $\tau = A\gamma^n\dot{\gamma}^m\theta^{-v}$. Under quasi-static conditions, a uniform shear stress exists along the whole gauge length of the specimen, despite disturbances in strain and temperature. It follows that the inhomogeneity in the stress is:

$$\frac{\Delta\tau}{\tau} = \frac{n\Delta\gamma}{\gamma} + \frac{m\Delta\dot{\gamma}}{\dot{\gamma}} - \frac{v\Delta\theta}{\theta} = 0. \qquad (6.44)$$

After introducing the critical condition for localization $\chi = \Delta\ln\dot{\gamma}/\Delta\ln\gamma = 1$ and the adiabatic assumption $\rho c\dot{\theta} = \tau\dot{\gamma}$ into Eq. (6.44), a critical shear strain can be expressed as:

$$\gamma_l = \frac{\rho c(m+n)\theta}{v}\frac{\int_0^t \frac{(\Delta\dot{\theta})}{\tau}dt}{\Delta\theta}. \qquad (6.45)$$

As a zero-order approximation, the shear stress τ is taken to be a constant from the occurrence of the inhomogeneity in the deformation field until the emergence of the shear localization, and so the critical expression becomes explicit:

$$\gamma_l = \frac{\rho c(m+n)}{v}\cdot\frac{\theta}{\tau}. \qquad (6.46)$$

It would be interesting to compare the critical strain for localization γ_l with its counterpart for instability, γ_i, Eq. (6.18). Both equations look alike apart from the strain-rate effect $d\ln\dot{\gamma}/d\ln\gamma$. This fact offers us a hint that for localization the effect of strain rate cannot be ignored, as shown by $\chi = \Delta\ln\dot{\gamma}/\Delta\ln\gamma = 1$. On the contrary, the omission of this effect in the maximum shear stress criterion for the occurrence of adiabatic shear banding still gives a reasonable prediction, Eq. (6.5).

Semiatin and co-workers (1984) extended the above approach to localization criteria under quasi-static conditions. They gave the variation of shear stress as:

$$\frac{\Delta\tau}{\tau} = \Delta\ln\tau = \frac{\partial\ln\tau}{\partial\gamma}\Delta\gamma + \frac{\partial\ln\tau}{\partial\dot\gamma}\Delta\dot\gamma + \frac{\partial\ln\tau}{\partial\theta}\Delta\theta = 0. \qquad (6.47)$$

Semiatin and co-workers prescribed that the spacial inhomogeneity of temperature $\Delta\theta$ can be described by the temperature increase at a fixed material element under the adiabatic assumption:

$$\Delta\theta = \beta\tau\Delta\gamma/\rho c. \qquad (6.48)$$

In addition, they assumed that the critical shear strain to instability γ_i in Eq. (6.6), is still valid, i.e. the strain-rate effect is negligible at instability. Substitution of Eqs (6.48) and (6.6) into Eq. (6.47) leads to an approximate law for the evolution of shear localization:

$$\left(\frac{\partial\ln\tau}{\partial\gamma} - \frac{n}{\gamma_i}\right)\Delta\gamma + \frac{\partial(\ln\tau)}{\partial\dot\gamma}\Delta\dot\gamma = 0$$

or

$$\chi(\gamma) = \frac{\Delta\ln\dot\gamma}{\Delta\ln\gamma} = \left(\frac{\gamma}{\gamma_i} - 1\right)n/m \qquad (6.49)$$

provided that $\tau = A\gamma^n\dot\gamma^m h(\theta)$. In the derivation of the evolution law (6.49), a further assumption was made, that is a constant rate of thermal softening during the course of localization. This is because it is only with this condition that we can introduce the material parameter γ_i, which is unchangeable during localization, into Eq. (6.49). Equation (6.49) indicates that the start of localization, $\chi = 1$, is not consistent with instability in this case. Furthermore, there is a gap between instability and localization. As mentioned previously, $\lambda(\gamma)$ represents the magnitude of the localization with increasing background or overall shear strain. When $0 < \chi < 1$, localization cannot occur, although inhomogeneities can occur in accordance with Eq. (6.42). Only when $\chi > 1$ or $\gamma \geqslant \gamma_l = \gamma_i(n + m)/n$ can inhomogeneities of shear deformation be amplified. Localization starts after this. Semiatin *et al.* (1984) suggested another variable $\alpha = \Delta\ln\dot{\bar\epsilon}/\Delta\bar\epsilon = \sqrt{3}\chi/\gamma$ as an indication of localization, where

$\bar{\epsilon} = \bar{\gamma}/\sqrt{3}$. By correlating some experimental results of Staker (1981), they pointed out that usually noticeable localization does not occur until $\alpha \approx 5$, which corresponds to $\gamma \approx \sqrt{3}\chi/5 \approx 0.35\chi$. From Table 6.3, we can see that noticeable localization appears much later than instability because the corresponding shear strain is $\gamma \sim 0.35$ if $\chi = 1$ and this is larger than the critical shear strain for instability, which ranges from 0.1 to 0.3.

Von Turkovich and Steinhurst (1986) emphasize the importance of real-life defects on instability but accept that it is possible to obtain unstable flow in a perfect defect-free material.

Molinari and Clifton (1983, 1987) dealt with shear localization from another point of view. Assuming the usual power-law constitutive equation $\tau = A\gamma^n\dot{\gamma}^m\theta^{-v}$ and assuming quasi-static deformation in simple shear with a uniform shear stress along the gauge length, they derived an equality between two different locations, A and B, along the gauge length,

$$(\tau/A)^{1/m} = \theta_A^{-v/m}\gamma_A^{n/m}\dot{\gamma}_A = \theta_B^{-v/m}\gamma_B^{n/m}\dot{\gamma}_B. \qquad (6.50)$$

With constant stress boundary conditions, integration gives

$$\gamma = \rho c(\theta - \theta_0)/\beta\tau$$

where θ_0 denotes the initial condition. The equality above for the two locations A and B can be rewritten as

$$\theta^{-v/m}\gamma^{n/m}\dot{\gamma} = \left(\frac{\rho c}{\beta\tau}\right)^{(n/m+1)} \theta^{-v/m}(\theta - \theta_0)^{n/m}\dot{\theta} \qquad (6.51)$$

where the factor $(\rho c/\beta\tau)^{(n/m+1)}$ remains unchanged during localization. Therefore, during time t a finite integration including the stress and proportional to the time t limits the corresponding integration with respect to the temperature θ,

$$\infty < \int_{\theta_0}^{\theta} \zeta^{-v/m}(\zeta - \theta_0)^{n/m}d\zeta. \qquad (6.52)$$

Assuming that the ratio of the temperature at element B where there is localization to that at element A where there is uniform deformation is $\theta_B/\theta_A \to \infty$ when severe localization occurs, we have to deal with the following integration:

$$\infty < \int_{\theta_0}^{\theta_B} \zeta^{-v/m}(\zeta - \theta_0)^{n/m}d\zeta \sim \int_0^\infty \zeta^{n-v/m}d\zeta \qquad (6.53)$$

provided θ_B/θ_A and $\theta_B/\theta_0 \to \infty$. The condition of integration is that the index of the integrand is less than -1, therefore localization requires

$$n + m - v < 0. \qquad (6.54)$$

The implication of this localization condition is that thermal softening outweighing the combined hardening effects of the strain and strain rate is the prerequisite for fully developed localization.

'So far the various descriptions of shear localization have been based on the behaviour of kinematic or geometric inhomogeneities on shear. As we have seen in the last section, localization can be determined by the development of these inhomogeneities against the background growth of the localized zone. Fressengeas and Molinari (1987) proposed a linear perturbation analysis to model localization.

The fundamental step is to examine the perturbations of relative measurements such as $\chi = \dot{\gamma}/\dot{\gamma}_h$, $\psi = \gamma/\gamma_h$ and $\varphi = \theta/\theta_h$ where the subscript h denotes unsteady but uniform solutions, for example, $\theta_h = \theta_h(t)$, etc. It is assumed that localization will occur provided the perturbations in these relative measurements grow rapidly enough. It is important to realize that the authors do not introduce any special inhomogeneities and this is therefore different from the analyses described above for localizations. Substitution of the variables χ, ψ and φ into the basic Eqs (6.25) for simple shear leads to:

$$\frac{\partial \chi}{\partial t} = \frac{\tau_h}{\rho \dot{\gamma}_h} \frac{\partial^2}{\partial y^2}(\varphi^{-v}\chi^m\psi^n)$$

$$\frac{\partial \varphi}{\partial t} = \kappa \frac{\partial^2 \varphi}{\partial y^2} + \frac{\beta\tau_h\dot{\gamma}_h}{\rho c\theta_h}(\varphi^{-v}\chi^{m+1}\psi^n - \varphi)$$

$$\frac{\partial \psi}{\partial t} = \frac{\dot{\gamma}_h}{\gamma_h}(\chi - \psi) \qquad (6.55)$$

if the constitutive equation $\tau = A\gamma^n\dot{\gamma}^m\theta^{-v}$ is assumed where the uniform time-dependent solutions θ_h, $\dot{\gamma}_h$ and τ_h have been considered. The next step is to apply the linear perturbation technique to Eqs (6.55), assuming that the perturbations are of the form $\chi = 1 + \chi_*$ $\exp(\alpha t + iky)$, etc., where $\chi = \varphi = \psi = 1$ correspond to uniform solutions.

We will only consider the quasi-static case here. After substitution of the perturbation expressions into Eqs (6.55) and retaining first-order variations in χ_*, ψ_* and φ_* we obtain the following characteristic equations of the linearized differential equations:

$$\begin{bmatrix} \dfrac{\dot{\gamma}_h}{\gamma_h}\dfrac{m+n}{m} & -\dfrac{\dot{\gamma}_h}{\gamma_h}\dfrac{v}{m} \\[2ex] \dfrac{\dot{\theta}_h}{\theta_h}\dfrac{m+v}{m} + \kappa k^2 & \dfrac{\dot{\theta}_h}{\theta_h}\dfrac{n}{m} \end{bmatrix} \begin{bmatrix} \psi_* \\[1ex] \varphi_* \end{bmatrix} = 0. \qquad (6.56)$$

Since the terms in the matrix are time dependent owing to the unsteady homogeneous solutions, the problem reduces to the linearized asymptotic stability of uniform but time-dependent solutions. A theorem described by Coddington and Levinson (1955) states that asymptotic stability is ensured by the long-term behaviour α_n where

$$\lim_{t \to \infty} \int_{t_0}^{t} Re(\alpha(Z))dt = -\infty \qquad (6.57)$$

and α_n are the eigenvalues of the non-integrable parts of the matrix of the linearized equations. But all terms in the characteristic matrix (6.55) are non-integrable, for example

$$\text{For } t \to \infty, \quad \frac{\dot{\gamma}_h}{\gamma_h} \approx \frac{\dot{\theta}_h}{\theta_h} \approx \frac{1}{t}$$

$$\int_0^\infty \frac{dt}{t} \approx \infty \quad \text{and} \quad \int_0^\infty \kappa k^2\, dt \approx \infty \qquad (6.58)$$

where $\dot{\gamma}_h/\gamma_h \approx \dot{\theta}_h/\theta_h \approx 1/t$ is derived for the case of an homogeneous solution under constant velocity boundary conditions; see Eq. (6.29).

Therefore, the characteristic equation that we are concerned with is

$$\alpha^2 + \alpha \left(\frac{\dot{\gamma}_h}{\gamma_h} \frac{m+n}{m} + \frac{\dot{\theta}_h}{\theta_h} \frac{m-n}{m} + k^2 \kappa \right)$$

$$+ \frac{\dot{\gamma}_h}{\gamma_h} \frac{\dot{\theta}_h}{\theta_h} \frac{m+n-v}{m} + \frac{\dot{\gamma}_h}{\gamma_h} \frac{m+n}{m} + k^2 \kappa = 0. \qquad (6.59)$$

Because of the asymptotic behaviour of the homogeneous solutions in Eqs (6.58), the characteristic eigenvalues should behave in the following way,

$$\alpha_1 \approx -\frac{1}{t} \quad \text{and} \quad \alpha_2 \approx -k^2 \kappa. \qquad (6.60)$$

When $t \to \infty$, these are all negative and non-integrable. According to Eq. (6.57), the mode of relative perturbation must be stable asymptotically, namely there is non-localization. However, if we let the coefficient of heat diffusion $\kappa = 0$, i.e. assume adiabatic deformation, the characteristic equation has the following asymptotic form:

$$\alpha^2 + \alpha A \frac{1}{t} + \frac{m+n-v}{m} B \left(\frac{1}{t} \right)^2 = 0$$

with $B > 0$. The asymptotic values of α should be $\alpha_1 \approx 1/t$ and $\alpha_2 \approx -1/t$, if the last term is negative, when

$$m + n - v < 0. \qquad (6.61)$$

This is the requirement for shear localization to occur and is identical to Eq. (6.54). However, linear perturbation carried out on the state variables using the same constitutive equation leads to the instability condition:

$$m + n - v > 0 \quad \text{and} \quad n - v < 0. \qquad (6.62)$$

Therefore, there is a gap between instability and localization. Table 6.2 summarizes the results and compares the instability and localization analyses for quasi-static deformation.

Table 6.2.   Comparison of linear instability and localization in relative perturbation analysis (quasi-static case).

		$m+n-v>0$ $n-v>0$	$m+n-v>0$ $n-v<0$	$m+n-v<0$
Constant velocity	$\kappa>0$	S	S	S
		NL	NL	L
	$\kappa=0$	S	U	U
		NL	NL	L
Constant stress	$\kappa>0$	S	S	U
	$\kappa=0$	S	U	U
	$\kappa\geq0$	NL	NL	L

S: stable, U: unstable, NL: non-localization, L: localization.

6.4. Experimental Verification

To examine the criteria and mechanisms that have been proposed, it is very important to design the tests carefully. The stress state needs to be well confined and well defined. Usually, therefore, the data for verification are taken from well-designed testing configurations such as: hydraulic torsion, impact torsion, split Hopkinson torsional bar technique and the contained exploding cylinder (CEC) technique. Nevertheless, comparison of the data from different tests should be made with the utmost care because of differences in the recognition of when shear bands occur and also differences in the material parameters used to characterize bands.

Table 6.3 gives the susceptibility of a number of materials to adiabatic shear. All these materials were tested under well-defined conditions. In this table the critical shear strain has been taken as the indication of the susceptibility to adiabatic shear. The experimental instability strains correspond reasonably closely with the theoretical calculation based on the maximum shear stress criterion, assuming adiabatic deformation and negligible strain-rate sensitivity. The theoretical shear strains are approximate because they are based on constant rates of strain hardening and thermal softening. Nevertheless, the theoretical results do provide validation for the maximum shear stress criterion and the discussion of it based on the perturbation method. This method shows that it is permissible, as a first-order approximation, to neglect strain-rate effects and assume

Table 6.3. Susceptibility to adiabatic shear localization in metals.

Material	$(-\partial\tau/\partial\theta)$ (kPa/K)	n	γ_{theo}	γ_{exp}	Constitutive relation (refer to Table 6.1)	Testing method	Ref.
OFHC copper		0.32	5.3	5.8	8	Hydraulic torsion	Lindholm and Johnson (1983)
Cartridge brass		0.34	3.7	3.0	8		
2024-T 351 Al		0.34	0.66	0.50	8		
7039 Al		0.41	0.77	0.55	8		
6061 T6 Al	496	0.075	0.438	0.35	3	Impact torsion	Culver (1972)
Nickel 200		0.32	0.03	0.18	8		Lindholm and Johnson (1983)
Armco IF iron		0.25	4.3	4.1	8		
Carpenter							
Electrical iron		0.43	4.4	5.8	8	Lathe torsion	Kobayashi (1987)
Low-carbon steel	423	0.04	0.36	0.28	3	Torsion	Lindholm and Johnson (1983)
1006 steel		0.36	3.3	3.5	8		
1018 CRS		0.05		0.16	7	Split Hopkinson bars	Costin et al. (1979)

(Continued)

Table 6.3.   (*Continued*)

Material	$(-\partial\tau/\partial\theta)$ (kPa/K)	n	γ_{theo}	γ_{exp}	Constitutive relation (refer to Table 6.1)	Testing method	Ref.
1020 steel	633	0.28	1.94	1.2	3		Culver (1972)
RHA steel		0.37	1.2	1.1	8		Lindholm and Johnson (1983)
AMS 6418 steel		0.18	0.16	0.20	8		
S-7 tool steel		0.18	0.16	0.50	8		
AISI 4340 steel tempering temperature							
466°C	1120	0.054	0.17	0.2–0.23	3	CEC	Staker (1981)
238–321°C	1610–1780	0.042–0.047	0.09	<0.11	3		
516°C	1020	0.061	0.22	0.22–0.26	3		
579°C	790	0.073	0.33	0.29–0.33	3		
α-Ti		0.17	0.326	1.15	3		Culver (1972)
CP-Ti	595	0.104	0.46	0.50	3		Kobayashi (1987)
Tungsten alloy (7% Ni, 3% Fe)		0.12	0.03	0.18	8		Lindholm and Johnson (1983)
Depleted uranium (0.75 Ti)		0.25	0.23	0.25	8		

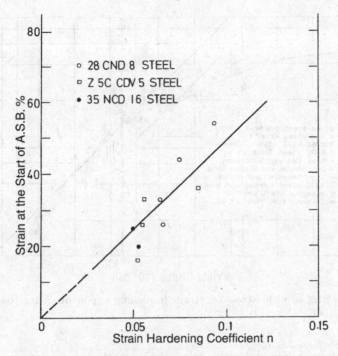

Fig. 6.4. Strain at start of adiabatic shearing versus strain hardening coefficient n in martensite steels. After Dormeval (1987), courtesy Elsevier Applied Science Publishers Ltd.

adiabatic deformation. Moreover, the data shows that copper has a good resistance to adiabatic shear although in the calculations heat conduction has not been taken into account. On the other hand, high-strength steels show a poor resistance to adiabatic shear, as shown in Fig. 6.4. Therefore, we cannot design a structure that is resistant to failure by taking into account its yield strength alone. We are left to ask why this type of behaviour occurs? A phenomenological explanation is that the rate of the strain hardening decreases with increasing strength (Fig. 6.5, after Rosenfield and Hahn (1966)) and it is therefore the rate of strain hardening which is important and not so much the strength in assessing the susceptibility to adiabatic shear banding.

Staker (1981) illustrated the importance of the strain hardening capacity of a material by carrying out tests on 4340 steel specimens that had been subjected to different tempering temperatures. In this

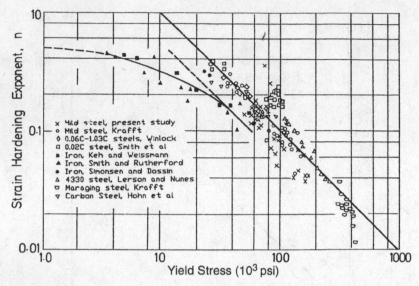

Fig. 6.5.   Effect of yield stress on strain hardening exponent. After Rosenfield and Hahn (1966).

case the thermophysical properties of the steel remain the same, but the mechanical properties change. Staker's original data are listed in Table 6.4 from which the reliability and scatter of the results can be seen. Figure 6.6 is Staker's correlation of his data showing a straight-line relation between the measured critical shear strain and the following combination of material variables: $n\rho c(-\partial\tau/\partial\theta)$. This is in good agreement with the predictions (see Table 6.4).

Lindholm and Johnson (1983) and Dormeval (1987) provide similar good correlations between their experimental data and the predicted critical shear strain based on the maximum shear stress criterion (see Fig. 6.7). It is important to notice that the agreement also holds for non-ferrous metals. More details of the data can be found in the appendices of this book.

We know that the rate of strain hardening is important in adiabatic shear banding and it is interesting to examine the effect of shear strain rate on the occurrence of bands. It is known that titanium alloys are highly strain-rate sensitive and also they are susceptible to shear banding and therefore we will consider experimental results for titanium alloys. Figure 6.8, after Wang and co-workers (1988), shows

Table 6.4. Experimental data for 4340 steel specimens, after Staker (1981).

Cylinder no.	Tempering temperature (°C)	n	$(\partial\tau/\partial\theta)$ (kPa/°C)	ϵ_A	ϵ_H	ϵ_R	γ	$\dfrac{-\rho c n}{\partial\tau/\partial\theta}$	Adiabatic shear band formation?
1	414	0.048	1360	0.12	0.24	-0.36	0.30	0.13	Yes
2	466	0.054	1120	0.12	0.17	-0.24	0.20	0.17	Yes
2	466	0.054	1120	0.12	0.27	-0.33	0.30	0.17	Yes
3	466	0.054	1120	0.15	0.19	-0.32	0.26	0.17	Yes
3	466	0.054	1120	0.15	0.18	-0.28	0.23	0.17	Yes
4	466	0.054	1120	0.13	0.14	-0.27	0.20	0.17	No
4	466	0.054	1120	0.13	0.24	-0.40	0.32	0.17	Yes
5	316	0.042	1610	0.10	—	-0.24	>0.17	0.09	Yes
6	274	0.043	1740	0.08	0.09	-0.17	0.13	0.09	Yes
6	274	0.043	1740	0.08	0.04	-0.17	0.11	0.09	Yes
7	238	0.047	1780	0.12	0.05	-0.17	0.15	0.09	Yes
8	321	0.042	1650	0.11	0.08	-0.19	0.15	0.09	Yes
8	321	0.042	1650	0.11	0.05	-0.16	0.11	0.09	Yes
9	429	0.049	1270	0.18	0.09	-0.19	0.14	0.14	No
9	429	0.049	1270	0.18	0.19	-0.35	0.27	0.14	Yes
10	516	0.061	1020	0.12	0.19	-0.32	0.26	0.22	Yes
10	516	0.061	1020	0.12	0.28	-0.34	0.31	0.22	Yes
11	516	0.061	1020	0.12	0.13	-0.24	0.19	0.22	No
11	516	0.061	1020	0.12	0.24	-0.35	0.29	0.22	Yes
12	516	0.061	1020	0.12	0.26	-0.29	0.28	0.22	Yes
12	516	0.061	1020	0.12	0.15	-0.29	0.22	0.22	No
13	649	0.092	620	0.21	0.30	-0.51	0.40	0.53	No
14	654	0.093	600	0.22	0.25	-0.46	0.35	0.56	No

(*Continued*)

Table 6.4.   (*Continued*)

Cylinder no.	Tempering temperature (°C)	n	$(\partial\tau/\partial\theta)$ (kPa/°C)	ϵ_A	ϵ_H	ϵ_R	γ	$\dfrac{-\rho c\, n}{\partial\tau/\partial\theta}$	Adiabatic shear band formation?
15	543	0.066	930	0.12	0.16	−0.28	0.22	0.26	No
15	543	0.066	930	0.12	0.25	−0.36	0.31	0.26	Yes
16	571	0.072	870	0.14	0.14	−0.26	0.20	0.30	No
16	571	0.072	870	0.14	0.24	−0.38	0.31	0.30	Yes
17	621	0.084	710	0.13	0.21	−0.33	0.27	0.43	No
17	621	0.084	710	0.13	0.30	−0.37	0.33	0.43	No
18	607	0.080	750	0.22	0.13	−0.24	0.19	0.39	No
18	607	0.080	750	0.22	0.23	−0.43	0.33	0.39	No
19	496	0.058	1110	0.11	0.20	−0.30	0.25	0.19	Yes
19	496	0.058	1110	0.11	0.17	−0.27	0.22	0.19	Yes
20	504	0.059	1060	0.11	0.15	−0.24	0.20	0.20	No
20	504	0.059	1060	0.11	0.25	−0.29	0.27	0.20	Yes
21	538	0.065	960	0.11	0.27	−0.38	0.32	0.25	Yes
21	538	0.065	960	0.11	0.19	−0.35	0.27	0.25	No
21	538	0.065	960	0.11	0.18	−0.28	0.23	0.25	No
22	568	0.071	860	0.11	0.27	−0.43	0.35	0.30	Yes
22	568	0.071	860	0.11	0.20	−0.38	0.29	0.30	Yes
23	610	0.081	730	0.09	0.31	−0.51	0.41	0.40	Yes
24	579	0.073	790	0.12	0.21	−0.30	0.26	0.33	No
24	579	0.073	790	0.12	0.27	−0.41	0.34	0.33	Yes
25	579	0.073	790	0.11	0.31	−0.35	0.33	0.33	Yes
25	579	0.073	790	0.11	0.24	−0.35	0.29	0.33	No
26	593	0.076	770	0.11	0.18	−0.30	0.24	0.35	No
26	593	0.076	770	0.11	0.29	−0.44	0.36	0.35	Yes

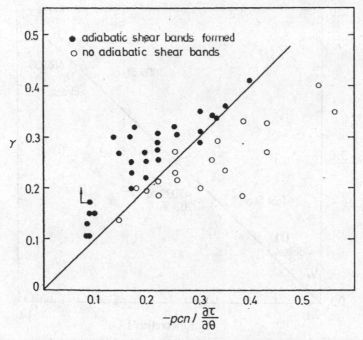

Fig. 6.6. Relation between measured true shear strain γ and a material parameter $-\rho cn(\partial t/\partial \theta)$ defining shear instability. After Staker (1981).

the susceptibility of titanium alloy TB2 (Ti-8Cr-5Mo-5V-3Al) as a function of strain and strain rate. The tests cover a wide range of environmental temperatures from $-110°$ to $20°$C. The diagram does show a rate dependence for the occurrence of adiabatic shear bands as predicted by Eq. (6.20). However, for the strain rates higher than 10^4/sec the rate dependence becomes weaker and appears to give way to a critical shear strain, which of course can be transformed into a critical strain criterion.

Before concluding our discussion of the experimental verification of the maximum shear stress criterion it is necessary to make two important observations. Firstly, all the data have come from metals, and secondly, the rates of strain hardening and thermal softening are not constant during deformation. Although all data come from metals, the differences in thermal conduction have been completely ignored. In fact there are big differences; for example, in titanium $\kappa \approx 0.08 \times 10^{-4}\,\mathrm{m}^2/\mathrm{sec}$ and for copper

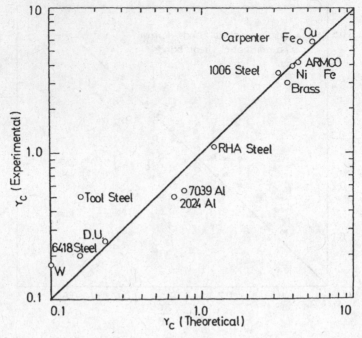

Fig. 6.7.   Correlation of experimental versus theoretical values of critical shear strain. After Lindholm and Johnson (1983).

$\kappa \approx 1.16 \times 10^{-4}\,\mathrm{m^2/sec}$. These differences may have some influence on the occurrence of adiabatic shear bands. Also the assumption of constant rates of strain hardening and thermal softening make the validation of the maximum shear stress criterion using experimental data not necessarily quantitatively correct. We should be very careful when using this experimental data in practical situations.

Careful examination of tested specimens, terminated at different stages of plastic deformation, can provide valuable information about what happens at the maximum shear stress. Let us study the deformation field within the gauge length. Figure 6.9 shows homogeneous shearing within the gauge length of a mild steel tube subjected to dynamic torsion by a split Hopkinson bar technique. Deformation within the gauge length is homogeneous shear, as described in Section 6.2.

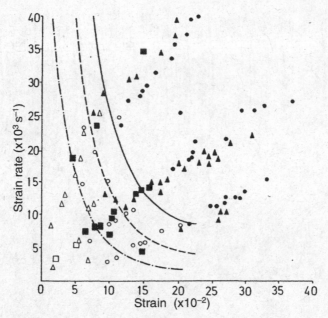

Fig. 6.8.   Dependence of occurrence of adiabatic shear bands on strain, strain rate and environmental temperature, where

   20°C ○ no shear band, ● occurrence of shear band
   −90°C △ no shear band, ▲ occurrence of shear band
   −110°C □ no shear band, ■ occurrence of shear band

theoretical curves: —— (−20°C) — — — — (−90°C) —·—·— (−110°C).
After Wang *et al.* (1988).

Now, what occurs in a specimen when the deformation has gone just beyond the shear stress peak? Huang '(1987) unloaded thin-walled tubular specimens tested on the Hopkinson bar just beyond the shear stress maximum. The material used was a hot-rolled low-carbon steel, with a critical shear strain of 0.82 ± 0.02. Figure 6.10 shows the deformation field at a nominal shear strain of between 0.85 and 0.9. Macroscopically, the shear deformation appears to be homogeneous. But careful examination with a microscope reveals a number of fine localized shear zones, about 30 μm wide, covering one or several grains with a spacing of 120 μm and distributed over the whole deformed region. This is quite different from the single fully developed shear band, about 400 μm wide, observed later in the process. However, the observation of a number of very narrow fine shear bands at instability is in qualitative agreement with the

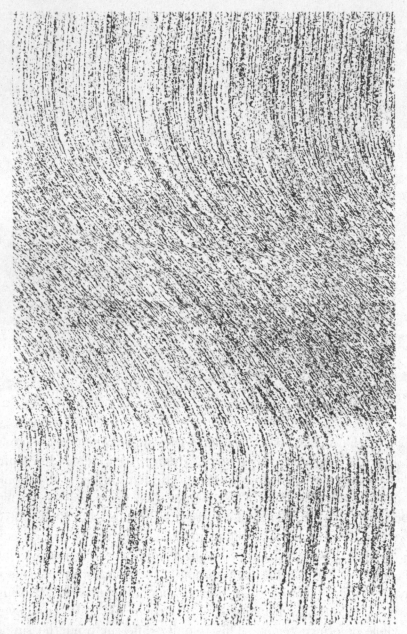

Fig. 6.9.　Homogeneous shearing within the gauge length of a thin-wall tube specimen subjected to dynamic torsion. After Xu *et al.* (1989).

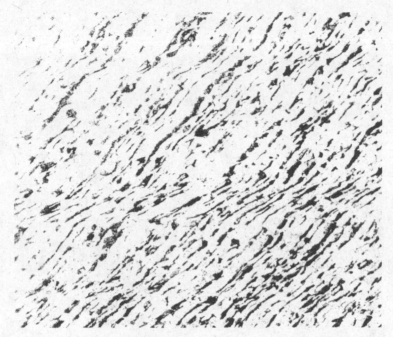

Fig. 6.10.   Fine localized shearing bands occurring just beyond instability. After Huang (1987).

prediction made using linear perturbation analysis. This analysis revealed that in thermoviscoplastic deformation, the mode that grows most rapidly are shorter wavelength perturbations. Hence, perturbation analysis is supported not only by its agreement with the maximum stress criterion and corresponding experimental data, but also by the observed mode of perturbated shear deformation. The transient process of adiabatic shear banding is further described in the next chapter.

Chapter 7

Formation and Evolution of Shear Bands

Much of the experimental data shows a reasonable consistency with the maximum shear stress criterion for the prediction of the onset of adiabatic shear bands. Also this type of criterion has been found to be helpful industrially. However, there is uncertainty amongst investigators as to how to identify the occurrence of adiabatic shear bands as well as the choice of the relevant material parameters such as the strain hardening index, the rate of thermal softening and so forth. According to the intrinsic nature of the bands, though, neither the empirical maximum shear stress criterion nor linear perturbation analysis describe the physical origin of shear bands.

The maximum shear stress criterion implies an unstable deformation regime beyond the strain at which the maximum shear stress occurs; it does not describe the nature or the mode of unstable deformation. Linear perturbation analyses can give more information than the simple maximum shear stress criterion; this information includes the threshold of the instability and the mode of unstable flow that develops most rapidly. Further, the perturbation technique has shown that the adiabatic assumption used in the maximum shear stress criterion is mostly valid and also the technique predicts that shorter wavelength disturbances, then fine bands, may grow quickly, which is supported by experiments as shown in Fig. 6.10.

Important questions that arise about the nature of adiabatic shear bands are: do these fine localized shearing zones represent the true occurrence of adiabatic shear bands or are these fine shearing areas what are usually observed and called adiabatic shear bands? Careful examination of experimental evidence shows that the formation of adiabatic shear bands is not simple. It is now accepted that the formation and evolution of adiabatic shear bands is a multi-stage process. The entire process is considered to occur as follows:

(i)   homogeneous shear deformation;
(ii) instability leading to inhomogeneous deformation;
(iii) the emergence of a localized shear zone;
(iv) formation of a fully developed localized shear zone;
(v)   shear fracture along the shear zones.

All of these features, together with when they occur in the overall process of shear localization, are crucially important in practice and therefore these aspects will be discussed in this chapter.

The whole process of the formation and evolution of shear bands is a challenging problem. Difficulties that may be encountered in their study are, for example: within a gauge length of several millimetres a narrow shear band of width between about 10 and 100 μm appears somewhere unexpectedly within hundreds of microseconds. With such a complex process, the problem is how to identify shear band formation. Experimentally, this identification requires the application of a number of different techniques; this is particularly so during the transient phase of up to about 1 msec.

The fact that it is necessary to use a number of different techniques, ranging from optical microscopy to transmission electron microscopy, implies the need for different scales of time and length for different stages in the evolution of shear bands. In fact, the choice of particular scales must depend on a full understanding of the relative significance of the various mechanisms and factors involved in shear band evolution. An important complication is the thermomechanical coupling of the non-linear governing equations and the constitutive equation.

It is now clear that for a full understanding of adiabatic shear banding there must be combined experimental, analytical and numerical studies because no individual method can be used to elucidate all the relevant information. However, this does not exclude any one of these techniques being used alone to obtain corroborative data on this multi-stage process.

7.1. Post-Instability Phenomena

One of the most appropriate machines for the study of adiabatic shear bands is the torsional split Hopkinson bar. Three test techniques are commonly used with the Hopkinson bar. These enable shear localization to be studied. The first and the most commonly used method is the measurement of the nominal (average) shear strain and shear stress with strain gauges mounted on the input and output bars. As well as providing the nominal stress and strain curve, other measurements can be made. For example, the nominal shear strain at maximum shear stress and the decrease in shear stress at relatively large strains can be monitored. However, a major disadvantage with this method is that it does not reveal any local changes but produces an overall representation of the stress and strain. Therefore, in many ways this technique can only provide necessary background information.

The second experimental technique is to terminate the torsion test at different amounts of plastic shear strain and to observe the change in microstructure with optical and/or scanning electron microscopy. Besides the microscopes, this technique does not require any special equipment but is clearly tedious and very time-consuming. Using this method, Huang (1987) examined the shear process in hot-rolled mild steel (20[#]).

The third method reveals, to some extent, the temporal progress of the deformation and therefore also reveals the multi-stage nature of the process. The way this is done is to make transient measurements of temperature and shear strain. However, the highlighting of the emergence of localized shear zones is still very difficult. Duffy (1984) and Marchand and Duffy (1988) adopted the third technique to study shear localization in 1018 cold-rolled steel, 1020 hot-rolled steel, HY 100 and 4340 steels. In this section a summary of the post-instability phenomena identified in these tests is given.

A number of fine shear bands with widths of about $30\,\mu$m and a spacing of $120\,\mu$m were observed at a nominal shear strain of between 0.85 and 0.9, which is just beyond the instability strain of

0.82 corresponding to the maximum shear stress. This compares with eventual shear band widths of $400 \, \mu m$ in hot-rolled $20^{\#}$ mild steel. With continued plastic deformation, the shear strain increases within a localized region. However, the shear strain outside the region shows hardly any increase, as observed by Huang (1987).

Giovanola (1987) studied the formation of adiabatic shear bands in 4340 steel using high-speed photography with an interframe time of $2.5 \, \mu sec$. From the time it took the incident torsional wave to reach the specimen, he determined that yield occurred at $25 \, \mu sec$. The first localization of shear strain occurs after $42.5 \, \mu sec$ within a width of about $60 \, \mu m$ in the centre of the specimen. At this point no significant stress drop was observed. Later in the process, between about 47.5 and $50 \, \mu sec$ in a narrow zone of about $20 \, \mu m$, secondary shear localization occurs together with a linear decrease of shear stress with strain until fracture finally occurs after about $60 \, \mu sec$. However, it is important to observe that the grid on the specimen had a line spacing of $100 \, \mu m \times 100 \, \mu m$, which is rather coarse compared to the shear band width and therefore the results can only be considered as qualitative information.

With improvements in the transient recording systems for temperature measurement, Marchand and Duffy (1988) achieved a simultaneous measurement of the nominal stress–strain curve and the temperature profiles at 12 stations within the gauge length as well as synchronized photographs at three locations along the circumference of the specimen.

The temperature measurement was performed by detecting infrared radiation emitted from a specimen's surface using 12 indium antimonide (InSb) cells. The response time of a cell was about $1 \, \mu sec$ and the spot width on the specimen was $35 \, \mu m$, with a spacing of $11 \, \mu m$. The entire length was $0.5 \, mm$. The local deformation was recorded at some specified nominal strain at three points $90°$ apart around the specimen circumference with three still cameras with short exposures of less than $2 \, \mu sec$. This equipment specification, although approximate, is a powerful way to examine shear localization.

Figure 7.1 shows a typical nominal shear stress–shear strain curve with the different stages of deformation for HY 100 steel. In stage 1,

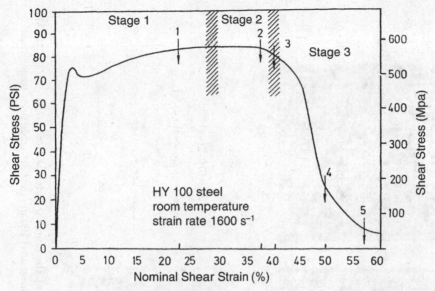

Fig. 7.1.   A typical stress–strain curve showing the strain values at which the photographs in Fig. 7.2 were taken. After Marchand and Duffy (1988).

$\gamma_I < 25\%$ up to about the maximum shear stress; the deformation appears to be homogeneous (see Fig. 7.2). This is in agreement with many other observations. Stage 2 of the process is characterized by inhomogeneous shear deformation (Fig. 7.2(2)), starting beyond γ_I and finishing at γ_{II}, which is somewhere between 35 and 50% and is where the shear stress begins to drop sharply. In this region the local strain may be well over 100% for hundreds of microns, indicating the non-uniform nature of plastic deformation in this stage. Although the deformation is inhomogeneous along the axial direction of the thin-walled tubular specimen, it remains homogeneous around the circumference of the specimen. The third stage of the process is that of a rapid, nominal stress drop with strain and at the same time the shear deformation begins to be confined to a narrow region, eventually reaching a width of about $20\,\mu m$ (Figs 7.2(3) and (4)). In this third stage, deformation appears to be confined to the shear band and fracture develops along the band.

Interestingly, the synchronized short exposure photographs taken at three points around the circumference show different local strains, ranging from 130 to 1900%, and also different band widths of between

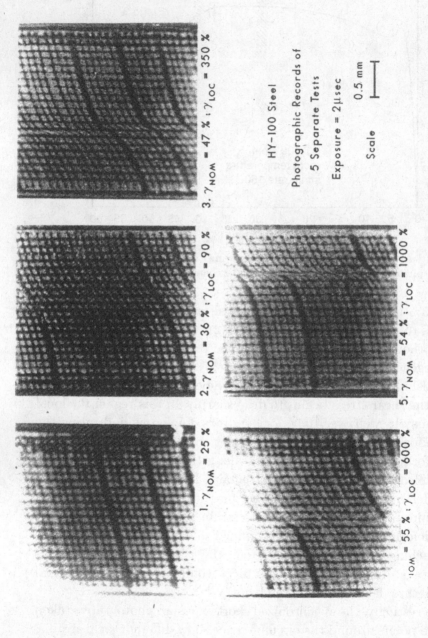

Fig. 7.2. Photographs of the grid patterns obtained in five separate tests at the nominal strain values shown in Fig. 7.1. The nominal strain rate is 1600/sec in each test. Each square in the grid pattern measured 100 microns on a side prior to deformation. After Marchand and Duffy (1988).

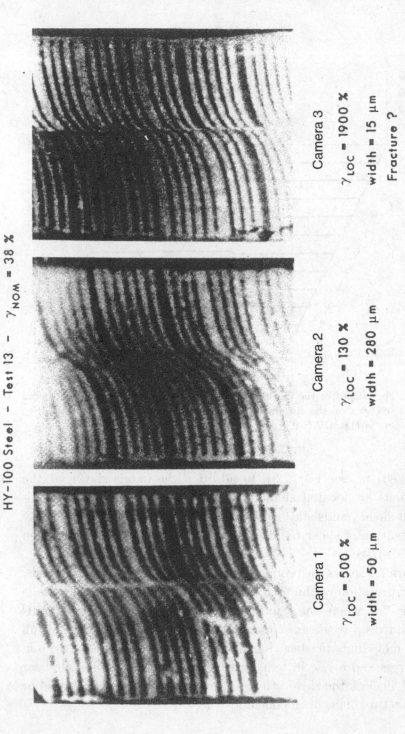

Fig. 7.3.   Simultaneous photographs taken at three locations on the same specimen at a nominal strain of 35%. The maximum in the stress–strain curve occurs at a nominal strain $\gamma = 21\%$. Note that the localized deformation is not the same at all locations. After Marchand and Duffy (1988).

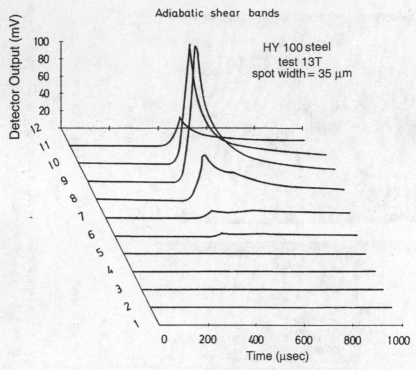

Fig. 7.4.   The output of the infrared detectors as a function of time and axial position. Each spot on the specimen is 35 μm wide and the space between two adjacent spots is 11 μm. After Marchand and Duffy (1988).

15 and 280 μm (see Fig. 7.3). In addition the bands shown at the three points are located at different axial positions. This may indicate that shear bands start separately. An estimation of the velocity of propagation of shear bands around the circumference is between 40 and 50 m/sec.

Figure 7.4 shows the temperature profiles detected by the 12 cells. The interpretation of the temperature profiles is more readily made from Fig. 7.5, where the rapid increase in temperature corresponds to a rapid drop in shear stress. It may be concluded that the full formation of adiabatic shear bands corresponds to an abrupt drop in shear stress and a rise in temperature. Finally, at a nominal shear strain of 40–55% the shear stress approaches zero and a fractured or partly fractured specimen concludes the test.

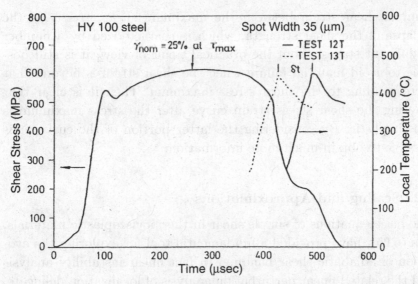

Fig. 7.5. Typical stress–time curve showing the temperature as a function of time in two separate tests. Note that the peak temperatures are separated by about 60 μsec. After Marchand and Duffy (1988).

In this section so far fracture or failure induced by adiabatic shear banding has not been mentioned. However, fracture associated with adiabatic shear banding plays a vital role in engineering. As most of the significant observations on fracture have been presented in Chapter 3, here we will concentrate on the interrelationships between shear localization and fracture. The time between the formation of adiabatic shear bands and eventual fracture is short, at least for the steels discussed in this section. For example, times of about 10 μsec for 4340 steel and 100 μsec or less for 20$^{\#}$ and HY 100 steels have been observed. It is reasonable to argue that the rapid and complete collapse of the shear strength is due to the formation of adiabatic shear bands or more realistically to the interactive formation of adiabatic shear bands and the accumulation of microdamage, in particular in materials with second-phase particles and or inclusions (as shown in Figs 3.9 and 3.10).

The attainment of the maximum shear stress corresponds to the end of homogeneous shear deformation along the gauge length and the appearance of inhomogeneities, but by no means does it mean the adiabatic shear bands are fully formed. There is often a significant nominal shear strain between the maximum shear stress and the collapse of the shear strength, which is characterized by a number of different stages. From the practical point of view, it is still possible to avoid material failure when the shear strain is beyond that corresponding to the shear stress maximum. Then it is clear that ignoring the shear stress–strain curve after the stress maximum is too simplistic. By considering the latter portion of the curve, it is possible to obtain much more information.

7.2. Scaling and Approximations

The basic equations of simple shear in thermoviscoplastic materials, Eqs (6.25), have provided a fundamental tool for exploring the evolution of adiabatic shear banding. In fact linear instability analysis and the related linear perturbation analysis of localization delineate, to some extent, the two key transition features: instability and localization. However, apart from a knowledge of the early growth rate of a disturbance offered by the linear instability discussion, nothing is revealed theoretically about the progress of adiabatic shear banding. Therefore, only a cross-check can be made for the occurrence of adiabatic shear bands, as has been shown in Section 6.4.

There are many obstacles preventing a full exploration of the essence of the evolution of adiabatic shear banding. It is possible, initially, that we have to deal with a new length scale, which defines the localization phenomenon. This new length scale is much less than the original gauge length. However, the value of the new scale cannot be prescribed beforehand. It is also necessary to study the problem by looking at timescales. As shown in the discussion of wave trapping in Section 7.3, the timescale t_w for a wave travelling through the gauge length $t_w \approx l/c \approx \mu\text{sec}$ must apparently be different from that determined from the nominal strain rate $t_{\dot{\gamma}}$, $t_{\dot{\gamma}} \approx 1/\dot{\gamma} \approx \text{msec}$. The difference can be as high as 10^3. This implies that to deal with the complete process from homogeneous shear to adiabatic shear banding, we must investigate the phenomena using different scales, otherwise we are in danger of missing banding.

The existence of different length and time scales is an indication that different mechanisms may underlie the whole process of adiabatic shear banding. It is instructive to list the possible variables and mechanisms that may dominate the process (Section 2.4), thus:

Variables: density, specific heat, heat conduction, strain rate, temperature, strain hardening, strain-rate hardening, thermal softening, viscosity, and a number of microscopic structural variables such as grain size, mobile dislocation density and so forth.

Mechanisms: wave trapping (see Section 7.3), hardening effects (see Section 6.3), effect of imperfections, discontinuous yielding and Lüders bands (see Section 2.1), viscous dissipation, heat diffusion and so forth.

The most logical step towards a correct and satisfactory solution to the problem is to evaluate the relative importance of a number of possible mechanisms and factors involved in the different stages of the evolution of adiabatic shear bands. Based on these evaluations, it should be possible to adopt an accurate simplified model together with appropriate assumptions for analytical and numerical studies. The significance of such an accurately posed problem together with scaled equations has been discussed by Burns (1986) and by Bai and co-workers (1986).

The scaling procedure is as follows. Firstly, we evaluate all the variables involved, i.e. note the appropriate quantity and its value which characterizes the phenomenon we are concerned with. Secondly, we determine scales in terms of these quantities and their combination. Thirdly, we non-dimensionalize all the terms in the equations under discussion. When doing this, the relative magnitude of each dimensionless term will be indicated by a dimensionless factor preceding the term. Such dimensionless equations are called well-scaled ones.

For adiabatic shearing, because a number of different quantities may have the same dimensions but very different magnitudes, we have to deal with different scales, in particular different time and length scales. As examples we list typical values within these scales.

Length scales: gauge length $L \approx$ mm, characteristic shear band width $\delta \approx (10-100\,\mu\text{m})$ and grain size $(1-10\,\mu\text{m})$.

Timescales: reciprocal of the strain rate $1/\dot{\gamma} \approx$ msec, characteristic time of wave propagation $L/\sqrt{(\partial\tau/\partial\gamma)/\rho} \approx \mu$sec, characteristic time for heat diffusion in the gauge length $L^2/\kappa \approx 10^1$ msec, characteristic time for heat diffusion within the shear band scale $\delta^2/\kappa \approx 10^{-1}$ msec. All of these figures are approximate and only used here by way of illustration. Importantly, though, we can obtain an impression that in one dimension there can exist various scales signifying various distinct physical processes.

It is probable that the following two timescales, which are closely related to localized shear zones, are crucially important in the discussion of the evolution of adiabatic shear banding. These two timescales are: $t_v \approx \rho\dot{\gamma}_*\delta^2/\tau_*$, the time representing rate-dependent diffusion and $t_h \approx \delta^2/\kappa$, the time representing thermal diffusion. The subscript $*$ denotes that the value characterizes the magnitude of the corresponding quantity within the localized shear band. The ratio of the two timescales is defined as the effective Prandtl number.

$$Pr = t_h/t_v = \left(\frac{\tau_*}{\rho\dot{\gamma}_*}\right) \Big/ \kappa = \left(\frac{c\tau_*}{\lambda\dot{\gamma}_*}\right). \qquad (7.1)$$

The effective Prandtl number represents the relative importance of the two dissipative mechanisms in the shear localization, characterized by its length scale δ. Therefore, the importance of the dimensionless number Pr in the evaluation of the mechanisms underlying shear banding is clear.

Suppose that the time and length scales are chosen as t_k and y_k respectively. The corresponding scaled equations of momentum and energy become (Bai, 1989):

$$a\frac{\partial\bar{\dot{\gamma}}}{\partial\bar{t}} = \frac{\partial^2\bar{\tau}}{\partial\bar{y}^2} \qquad (7.2)$$

$$\bar{\tau}\bar{\dot{\gamma}} = b\frac{\partial\bar{\theta}}{\partial\bar{t}} - d\frac{\partial^2\bar{\theta}}{\partial\bar{y}^2} \qquad (7.3)$$

where the quantities with a bar are well-scaled of order $O(1)$. The three dimensionless parameters a, b and d are

$$a = \rho\dot{\gamma}_*y_k^2/\tau_*t_k \qquad (7.4)$$

$$b = \rho c \theta_* / \beta \tau_* \dot{\gamma}_* t_k \qquad (7.5)$$

$$d = \lambda \theta_* / \beta \tau_* \dot{\gamma}_* y_k^2. \qquad (7.6)$$

In Eqs (7.4) to (7.6) the length and time scales have not been specified. Together with the three dimensionless parameters a, b and d, there are five undetermined quantities related by the three Eqs (7.4)–(7.6). Therefore, there are two quantities that can be specified arbitrarily. However, if we examine Eqs (7.2) and (7.3) in detail we find that the existing terms of order $O(1)$ are the stress differentiation in the momentum equation and the plastic work rate in the energy equation. These two terms are the cause of variations in momentum and energy in the process of shear deformation. The mechanisms that can dominate the process are those which aid the balance of these two terms. Therefore, the scaling has three choices:

$$(1) \ a, d = 1; \quad (2) \ b, d = 1; \quad (3) \ a, b = 1.$$

So the essence of the procedure of scaling is to discover the most probable mechanisms that predominate in shear banding.

Let us look at the three typical cases determined by the three choices given:

Case	a	b	d	$Pr \gg 1$	t_k	y_k^2
1	1	Pr	1	Isothermal	$\rho \lambda \theta_* / \tau_*^2$	$\lambda \theta_* / \beta \tau_* \dot{\gamma}_*$
2	$1/Pr$	1	1	Quasi-static	$\rho c \theta_* / \tau_* \dot{\gamma}_*$	$\lambda \theta_* / \beta \tau_* \dot{\gamma}_*$
3	1	1	$1/Pr$	Adiabatic	$\rho c \theta_* / \tau_* \dot{\gamma}_*$	$c \theta_* / \beta \dot{\gamma}_*^2$

It is clear that the dimensionless parameter Pr plays a significant role in scaling. For most metals the value of the effective Prandtl number Pr is much greater than unity; for example in metals: $c \approx 5 \times 10^2$ J/kg K, $\tau_* \approx 5 \times 10^8$ Pa, $\lambda = 10^2$ W/mK, $\dot{\gamma}_* \approx 10^5$/sec. Then $Pr \approx c \tau_* / \lambda \dot{\gamma}_* \approx (10^4 – 10^5)$. This large dimensionless number eventually allows us to make various approximations.

Case 2 is the most interesting mode. Since $b = d = 1$, all the energy terms are balanced and it represents a typical thermomechanical coupled process. Whereas if $a = 1/Pr \ll 1$, this leads to the conclusion that the inertia force can be treated as $O(1/Pr) \ll O(1)$ and then this can be ignored in a zero-order approximation. This is the quasi-static, heat-diffusion predominant stage. At this stage the relation between the time and length scales is governed by heat diffusion since $y_k^2/t_k \approx \kappa$.

Case 1 provides a different picture of shear deformation. Initially $t_{k1}/t_{k2} = 1/Pr \ll 1$ and $y_{k1} = y_{k2}$. Now $a = d = 1$ and $b = Pr$, which means that the plastic work rate and heat conduction, although both of order $O(1)$, can be taken to be negligible because of the high value of b. Therefore, the temperature increase is negligibly small. This is an isothermal and momentum-dominant process. The relation between the time and length scales in this case is $y_k^2/t_k = \tau_*/\rho\dot{\gamma}_*$, which indicates that rate-dependent diffusion, or say viscosity, is important in the localization and the deformation is quite unconnected with temperature.

If the timescale was even shorter, t_{k1}^s, the isothermal viscosity-dominated localized shear zone, $y_{k1} < y_{k1}^s$, would not be related to heat diffusion as well. Otherwise, if we enlarge the length scale from y_k to $(y_k^l)^2 = \xi\lambda\theta_*/\tau_*\dot{\gamma}_* \approx l^2$, $\xi > 1$, keeping the relation $(y_k^l)^2/t_k^l = \tau_*/\rho\dot{\gamma}_*$, then the timescale develops the same enlargement $t_k^l \approx \xi t_k$ and the three parameters a, b and d become $a = 1$, $b = Pr/\xi$ and $d = 1/\xi$. This maintains the dominant importance of momentum. However, the process cannot be treated as isothermal and must be treated as adiabatic because $d = 1/\xi = (\lambda\theta_*/\tau_*\dot{\gamma}_*)/l^2 \ll 1$ with a temperature increase if $b = Pr/\xi \leq O(1)$. In doing this, the length scale has changed from a localized area $\sqrt{\lambda\theta_*/\tau_*\gamma_*}$, which is defined as quasi-static, heat-diffusion-dominant approximation of case 2, to a large scale l. This indicates that if an object has a larger length scale, for example, the whole length rather than the localized shear zone, the adiabatic assumption will be successful.

Case 3 illustrates the above in a straightforward manner; since $a = 1$ and $b = 1$ are prescribed, then $d = 1/Pr$. Therefore, case 3 is the adiabatic mode with inertia and momentum in the zero-order approximation. Let us re-examine the time and length scales $t_{k3}/t_{k2} = 1$ and $(y_{k3}/y_{k2})^2 = Pr$. This is similar to the case 1 discussion where we consider a larger scale than the localized area. This larger scale can be the spacing of the shear bands or the macroscopic gauge length. By taking y_{k3} to be the spacing, then the ratio of band width to spacing should be in the ratio $(1/Pr)^{1/2} \approx (10^{-2}\text{--}10^{-3})$. Equivalently, if we were investigating an earlier time $t_k^1 = t_{k3}/\eta$ where $\eta > 1$, then the isothermal mode is approached following the same reasoning made for case 1.

Now, for simplicity, we shall summarize the assumptions made in these scaling approximations.

(1) If local deformation is examined in an area that is larger than the local region itself, then the adiabatic assumption is a good model with inertia and momentum relevant to the occurrence of instability throughout the gauge length.

(2) If deformation is concentrated into a relatively narrow region, two dissipative mechanisms may be in force, depending on the length and time scales of concern.

Heat diffusion predominates in quasi-static shear localization.

As discussed in case 1, the viscosity predominates in isothermal shear localization. The mechanism may be operative early before heat diffusion takes over or the localization may appear in a finer zone at an even earlier time.

7.3. Wave Trapping and Viscous Dissipation

As has already been discussed in the early stages, heat diffusion and the energy equation play an insignificant role in the formation of localized shear zones. Therefore, as a reasonable and acceptable approximation the effect of heat can be ignored, and hence the problem can be considered as quasi-isothermal.

In the last section it was explained how the mechanism predominating in this stage was the balance between the effect of inertia and the term $\partial^2 \tau / \partial y^2$ in the momentum equation. This early stage then is essentially isothermal with deformation governed by momentum.

If we examine carefully the momentum equation together with the thermoviscoplastic constitutive equation, we find that the strain-rate effect can lead to a dissipative mechanism, similar to the viscosity, as discussed in the last section. Moreover, strain hardening can provide another mechanism, that is wave propagation. Wave propagation plays a significant role in the dynamics of materials and cannot be ignored in the early stages of dynamic phenomena, namely before the quasi-static state has been reached by wave reflections. Importantly, if the stress–strain curve (e.g. $\tau = f(\gamma)$) is concave and has a maximum at some strain caused by heating or geometrical softening, what occurs?

Firstly, let us examine the momentum equation with a rate-independent constitutive equation:

$$\tau = f(\gamma). \tag{7.7}$$

Then

$$\frac{\partial^2 u}{\partial t^2} = a^2 \frac{\partial^2 u}{\partial y^2} \tag{7.8}$$

where u is the displacement and a is the wave velocity, $a = \sqrt{f'(\gamma)/\rho}$. The common concave stress–strain function without a stress peak will give a decreasing wave speed with increasing stress or strain. For a semi-infinite body loaded on its boundary surface with a monotonically increasing driving velocity, the wave pattern consists of continuous and divergent simple waves. However, as soon as the shear stress reaches its maximum, i.e. $d\tau/d\gamma = f'(\gamma_*) = 0$, the corresponding wave velocity is

$$a(\gamma_*) = \sqrt{\frac{f'(\gamma_*)}{\rho}} = 0. \tag{7.9}$$

Beyond the critical shear strain γ_* or the corresponding shear stress τ_* or the driving velocity v_*, all increments due to the driving velocity $v > v_*$ cannot propagate into the semi-infinite body and are trapped at the boundary surface. This phenomenon was termed 'wave trapping' and was first noticed by Erlich and co-workers (1980) (see Fig. 7.6).

The wave trapping mechanism can only provide a strong discontinuity, however, across which variables such as particle velocity suffer an abrupt jump. Such abrupt discontinuities are not observed in practice in adiabatic shear bands.

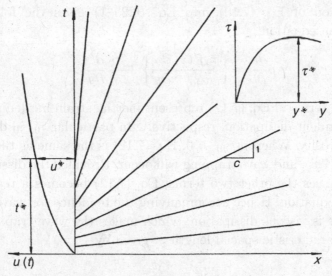

Fig. 7.6.   The (x, t) plane for one-dimensional wave propagation for $x > 0$ with boundary $x = 0$ subjected to a uniform tangential speed. The stress–strain curve has a positive slope for strain magnitudes less than γ_*, which is the boundary strain stress level at $t = t_*$. After Wu and Freund (1984).

Wu and Freund (1984) proposed that rate sensitivity was the dominant mechanism in wave trapping. The shear strain rate at the wave trapping position becomes infinite since the rate can be defined as $\bar{\dot{\gamma}} = \Delta v / \delta$, where Δv is the jump in particle velocity and δ is the thickness of the region affected by wave trapping. Because the wave trapped zone is a zero width discontinuity, $\delta \to 0$, and therefore $\bar{\dot{\gamma}} \to \infty$. Therefore, it follows that the effect of shear strain rate can no longer be ignored.

As illustrative models, Wu and Freund (1984) investigated both linear and logarithmic rate sensitivity

$$\frac{\tau}{\tau_0} = f\left(\frac{\gamma}{\gamma_0}\right) + \frac{\dot{\gamma}}{\dot{\gamma}_0}, \quad \dot{\gamma} \geq 0 \tag{7.10}$$

$$\frac{\tau}{\tau_0} = f\left(\frac{\gamma}{\gamma_0}\right) \{1 + p \ln(1 + \dot{\gamma}/\dot{\gamma}_0)\}, \quad \dot{\gamma} \geq 0 \tag{7.11}$$

where $f(z) = z^n/(1 + \alpha z^n)$ and p and α are constants. Here, for simplicity, the discussion is confined to linear rate sensitivity.

Substitution of Eq. (7.10) into Eq. (6.25–1) gives the following momentum equation:

$$\rho\frac{\partial v}{\partial t} = \left\{ \frac{\tau_0 f'(z)}{\gamma_0}\frac{\partial \gamma}{\partial y} \right\} + \frac{\tau_0}{\dot{\gamma}_0}\frac{\partial^2 v}{\partial y^2}. \tag{7.12}$$

The three terms in Eq. (7.12) represent inertia, strain hardening and rate-dependent dissipation, respectively. In particular $\tau_0/\dot{\gamma}_0$ denotes linear viscosity. When $\tau_0/\dot{\gamma}_0 \rightarrow 0$, Eq. (7.12) is the same as the wave equation (7.8), and wave trapping will occur. Provided the dissipative term balances the other two terms, Eq. (7.12) becomes a reaction-diffusion equation. Hence accompanying the tendency for wave trapping there is viscous dissipation, which makes the wave trap diffuse with a characteristic spacial length y, given by

$$y^2 \approx \frac{\tau_0 t}{\rho\dot{\gamma}_0}. \tag{7.13}$$

Therefore, according to this wave trapping-viscosity-dissipation mechanism of adiabatic bands, the observed shear band should be located near the loading surface. A typical distribution of shear strain in this type of shear band has been obtained by numerical simulation (see Fig. 7.7). A logarithmic rate sensitivity can lead to a similar reaction-diffusion equation to that in the linear case, Eq. (7.12), and the numerical simulation shows similarities.

Burns (1986) discussed wave trapping and dissipation due to strain-rate sensitivity in the evolution of adiabatic shear bands. He suggested that if the variation of the slope of the shear stress versus shear strain curve $f'(\gamma)$ is examined over the whole deformation range, beyond the yield point the increase in shear stress is less rapid than that in the elastic region. Thus, the slope of the stress–strain curve is large near the yield point and then decreases rapidly until the shear stress reaches a maximum at which $f'(\gamma) = 0$. Burns proposed that the inertial effects can dominate over viscous effects for a short time during which the process of shear deformation is governed by the wave equation. However, for slower timescales, the viscous effect becomes important and deformation is dissipation dominated.

It is clear from Eq. (7.12) that the magnitude of the term preceding $f'(\gamma)$ is determined by the slope. The time needed for the

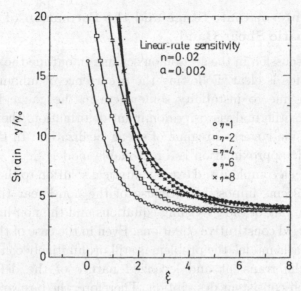

Fig. 7.7. Dimensionless total strain γ/γ_0 plotted against dimensionless positions ξ at different dimensionless times η for adiabatic deformation with linear strain-rate sensitivity. After Wu and Freund (1984).

development of wave trapping is determined by wave propagation, that is the inertia effect. As discussed in Section 7.2, this is the fastest timescale, microseconds for the whole gauge length. The timescale for viscosity can be estimated from Eq. (7.13) as

$$t \sim \frac{\rho \dot{\gamma}_0 l^2}{\tau_0} \sim \frac{10^4 \times 10^3 \times 10^{-6}}{10^8} \approx 10^{-1} \, \mu\text{sec}.$$

Therefore, the timescale for the viscosity effect is approximately the same as that for wave propagation.

Burns (1986) assumed that a small perturbation in strain can propagate as a wave near the yield point, but due to the dissipation caused by the viscosity, the higher frequency content of the perturbation is rapidly attenuated. Near the yield point then there is a combination of wave propagation and rate-induced viscosity. Burns suggested that this mechanism should be responsible for the nucleation of long-wavelength shear bands occurring in metals under dynamic loads. However, this suggestion still needs experimental verification.

7.4. The Intermediate Stage and the Formation of Adiabatic Shear Bands

From the discussion in the section on scaling, in outline the evolution of shear bands is clear. Following the occurrence of inhomogeneous deformation due to instability somewhere in the gauge length or early quasi-isothermal viscosity-dominant deformation, there will be a heat-diffusion-governed regime of shear localization. In this stage, a quasi-static approximation is a reasonable model.

To obtain a complete and exact analytical solution to the problem of banding seems impossible because of the non-linear thermomechanical coupling of the governing equations and the non-linearity of commonly used constitutive equations. Even in the case of the simplified quasi-static model, the problem is still prohibitively complicated. Moreover, the transient and localized nature of the deformation requires a self-consistent description. Therefore, the best approach is to combine experimental, analytical and numerical techniques. The major features observed in experiments on the post-instability phenomena have been summarized in Section 7.1. Some numerical analyses are introduced in Chapter 8. In this section a unified view of adiabatic shear bands is given, which results from the three different approaches.

A number of analytical procedures have been applied to adiabatic shear bands since the establishment of the governing equations [Eqs (6.25)]. For example, Bai and co-workers (1986, 1987) and Wright and Walter (1987) have all studied this problem. Here the description will follow that of Bai and co-workers (1986, 1987) and will be confined to one-dimensional simple shear for simplicity.

A simple constitutive equation is adopted here to help us reveal the mechanics and the evolution of adiabatic shear bands. From the observations of Campbell (1973) and others, material behaviour at high strain rates, for example $\dot{\gamma} > 10^3$/sec, is more strain rate than strain dependent and linear viscosity therefore appears to be a good approximation in this region. Also, for simplicity, linear thermal softening is assumed. When localization begins, the material within the localized band is assumed to behave in this simplified linearized manner.

As a first approximation, Bai and co-workers (1986) adopted the following linearized constitutive equation:

$$\frac{\tau}{\tau_*} = 1 + \frac{\dot{\gamma}}{\dot{\gamma}_*} - \frac{\theta}{\theta_*} \qquad (7.14)$$

where the subscript $*$ corresponds to properties within the shear band.

This simplification minimizes the non-linear thermomechanical coupling and the non-linearity of the governing equation to the lowest level, to facilitate the illustration of the underlying mechanism of adiabatic shear localization.

As observed by Marchand and Duffy (1987) and Huang (1987), when shear localization occurs, the material outside the localized shear zones appears to undergo no further plastic deformation. Therefore, it is reasonable to assume the material is rigid outside the shear zones. Correspondingly, the shear strain rate at the boundary of the shear zone should be zero. Therefore, in accordance with the definition of shear strain rate, the boundary velocity condition in simple shear can be formulated as

$$v = \int_0^\delta \dot{\gamma}\, dy$$

where the y axis origin is at the centre of the shear band and δ is the half-width of the zone. Substitution of the constitutive equation (7.14) into the boundary condition gives

$$v = \dot{\gamma}_* \int_0^\delta \left(\frac{\theta}{\theta_*} + \frac{\tau}{\tau_*} - 1 \right) dy$$

$$= \frac{\dot{\gamma}_*}{\theta_*} \left(\int_0^{\delta(t)} \theta\, dy - \theta_\delta(t)\delta(t) \right) \qquad (7.15)$$

where θ_δ is the current temperature at the boundary of the localized shear zone and both θ_δ and δ can vary with time. It is worth noting that $\dot{\gamma} = 0$ at $y = \delta$ has been used in the derivation of Eq. (7.15). Considering the representative case of a constant boundary velocity,

$v = $ constant, and differentiating Eq. (7.15):

$$\int_0^\delta \frac{\partial \theta}{\partial t} dy + \theta_\delta \frac{d\delta}{dt} = \frac{d\theta_\delta(t)}{dt} \delta(t) + \theta_\delta(t) \frac{d\delta}{dt} \qquad (7.16)$$

where

$$\frac{d\theta_\delta(t)}{dt} = \frac{d}{dt} \theta(t, \delta(t)) = \left. \frac{\partial \theta}{\partial t} \right|_\delta + \left. \frac{\partial \theta}{\partial y} \right|_\delta \frac{d\delta}{dt} \qquad (7.17)$$

$$\int_0^\delta \frac{\partial \theta}{\partial t} dy = \int_0^\delta \frac{1}{\rho c} (\beta \tau \dot{\gamma} + \lambda \partial^2 \theta / \partial y^2) dy$$

$$= \frac{1}{\rho c} \int_0^1 \beta \tau \dot{\gamma}\, dy + \frac{\lambda}{\rho c} \int_0^\delta \frac{\partial^2 \theta}{\partial y^2} dy$$

$$= \frac{\beta}{\rho c} \tau \int_0^\delta \dot{\gamma}\, dy + \frac{\lambda}{\rho c} \left. \frac{\partial \theta}{\partial y} \right|_\delta \qquad (7.18)$$

where the energy equation (6.25–2) and the quasi-static approximation $\tau = \tau(t)$ have been used. Substitution of Eqs (7.17) and (7.18) into Eq. (7.16) leads to (Bai *et al.*, 1986):

$$\frac{\beta \tau}{\rho c} \int_0^\delta \dot{\gamma}\, dy + \frac{\lambda}{\rho c} \left. \frac{\partial \theta}{\partial y} \right|_\delta = \left. \frac{\partial \theta}{\partial t} \right|_\delta \delta(t) + \left. \frac{\partial \theta}{\partial y} \right|_\delta \frac{d\delta}{dt} \delta(t)$$

or

$$\frac{d\delta}{dt} = \frac{\frac{\beta \tau \bar{\dot{\gamma}}}{\rho c} - \frac{\lambda}{\rho c} \left(\left. \frac{\partial^2 \theta}{\partial y^2} \right|_\delta - \overline{\frac{\partial^2 \theta}{\partial y^2}} \right)}{\left. \frac{\partial \theta}{\partial y} \right|_\delta} \qquad (7.19)$$

where

$$\bar{\dot{\gamma}} = \int_0^\delta \dot{\gamma}\, dy / \delta(t)$$

and

$$\overline{\frac{\partial^2 \theta}{\partial y^2}} = \left. \frac{\partial \theta}{\partial y} \right|_\delta \Big/ \delta(t) = \left(\int_0^\delta \frac{\partial^2 \theta}{\partial y^2} dy \right) \Big/ \delta(t).$$

In the derivation of Eq. (7.19), the energy equation at the boundary,

$$\left.\frac{\partial \theta}{\partial t}\right|_\delta = \frac{\lambda}{\rho c} \left.\frac{\partial^2 \theta}{\partial y^2}\right|_\delta ,$$

due to $\dot{\gamma}|_\delta = 0$ has been used.

Equation (7.19) is the equation describing the evolution of the band width in adiabatic shearing. The three terms on the right in the equation are as follows. In the denominator the temperature gradient at the boundary of the band δ always remains negative because the band acts as a heat source to the surroundings and is coupled to the other two terms. The first term in the numerator is the average plastic work rate within the shear band, which is always positive when plastic deformation occurs. The coupled effect of the positive plastic work rate and the negative temperature gradient at the band boundary provides a tendency for the deformed zone to shrink. On the other hand, the heat conduction of the second term in the numerator is

$$\left.\frac{\partial^2 \theta}{\partial y^2}\right|_\delta - \overline{\frac{\partial^2 \theta}{\partial y^2}} > 0$$

because in the centre of the heat source the temperature distribution is concave, i.e. $\partial^2 \theta / \partial y^2 < 0$, but near the boundary it is convex $\partial^2 \theta / \partial y^2 > 0$. Therefore, the general effect of heat conduction appearing in the numerator and denominator is positive and therefore diffusive.

Thus the two opposing effects in this coupled thermomechanical shear deformation are the plastic work rate, which tends to make deformation localized, and heat conduction, which is dissipative and tends to make the zone of deformation wider.

Although it can be seen that from these two opposing mechanisms adiabatic shear bands can form, the influence of individual material parameters on the details of the disturbances and timings of the process need separate calculations (see Chapter 8). Except for the average plastic work rate and a heat conduction factor, no other variables appear in the formulae for the evolution of adiabatic shear bands. Also the boundary condition, that is the driving velocity, does

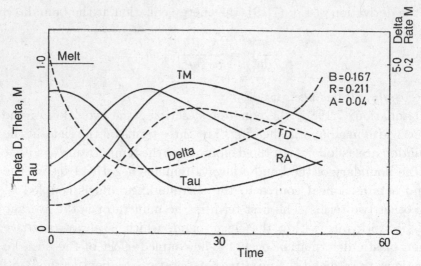

Fig. 7.8.   Variation of some characteristic parameters of shear band deformation with time, where TM and TD denote the maximum (at the centre) and minimum (at the boundary) temperature in the shear band. Rate M represents the strain rate at the centre. After Bai *et al.* (1987).

not appear explicitly in the evolution equation but it is included in the input plastic work rate. In addition, the initial condition can only be defined as $\delta = \delta_0$ at $t = t_0$. It appears that the details of the disturbance will only influence on a minor scale the shear band evolution. The variation in the two mutually opposing factors determines the evolution of adiabatic shear bands. An approximate model based on the aforementioned one is shown in Fig. 7.8. The calculation was started with a given wide disturbance. In the first phase, the shear stress decreases little, but the temperature and shear strain rate have a tendency to increase. After a certain time the shear stress begins to decrease; this is the second phase. At the same time the temperature and shear strain rate increase rapidly. Then, signifying the onset of the third phase, the strain rate reaches a maximum. For even longer times, for example more than 30 times the characteristic timescale of heat diffusion, the calculated results may not be accurate because of the assumed approximations. Also, other mechanisms of deformation may become important, such as the accumulation of microdamage.

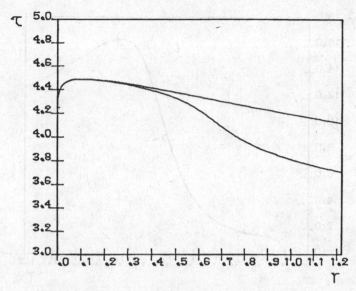

Fig. 7.9. Variation of shear stress with shear strain. The upper curve is the homogeneous solution and the lower curve is for a shear stress disturbed at its peak. After Xing (1988).

Here some reference is made to the results of numerical simulation techniques. However, a detailed description of these methods is left until Chapter 8. Figures 7.9 and 7.10 show the variations in shear stress and shear strain rate using a dimensionless time, which is different from that used in Fig. 7.8. In the numerical simulation, Xing (1988) used the quasi-static approximation. However, a fully non-linear thermoviscoplastic constitutive equation was used with the power form:

$$\tau = A\gamma^n \dot{\gamma}^m \theta^{-v}.$$

Figure 7.9 shows two stress curves. The upper one corresponds to the homogeneous solution for simple shear whilst the other is the result of disturbing the shear stress at its peak at about $t \approx 0.1$. For t between 0.1 and 0.42 (t_c), the disturbed shear stress exhibits little difference from the homogeneous one, but the local strain rate begins to increase as shown in Fig. 7.10. Between t_c and t_d the predictions are similar to those made by the simplified analytical study (Fig. 7.6); the shear stress decreases from the homogeneous solution. Accompanying this

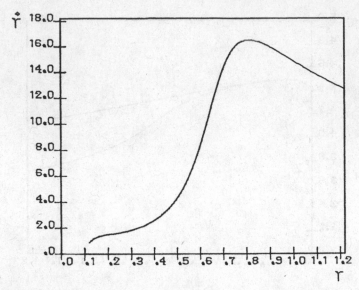

Fig. 7.10.   Variation of shear strain rate with strain. The shear strain rate reaches a maximum at t_d. After Xing (1988).

decrease in shear stress, the local shear strain rate increases rapidly to a maximum at t_d. Therefore, the times t_c and t_d signify two important characteristic phases in the evolution of adiabatic shear bands. Interestingly, the numerical simulation, which included non-linearity and strain hardening in the constitutive equation, shows no significant difference from the results obtained by using simplified analytical predictions. Examination of the three-dimensional results of the numerical simulation indicates that adiabatic shear bands form fully by time t_d where the local strain rate becomes saturated (see Fig. 7.11).

Although it is now possible to be convinced about the various mechanisms of adiabatic shear banding from the experimental observations described in Section 7.1, it is unfortunate that strain rate profiles are not available in the literature. Therefore, comparisons can only be made between variations in shear stress, temperature and band width. In the first stage, as reported by Giovanola (1987), localization occurs within a zone of $60\,\mu$m in width, with no obvious drop in shear stress. However, in the second stage, the shearing zone has shrunk to $20\,\mu$m, with a linear decrease in shear stress.

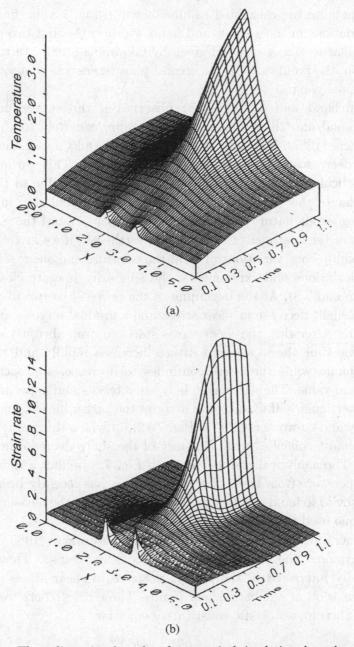

Fig. 7.11.   Three-dimensional results of a numerical simulation show shear band formation in a zone of (a) increased temperature and (b) increased strain rate. After Xing (1988).

Re-examining the calculated results shown in Fig. 7.8, we find that the variations in shear stress and band width in the first two phases are similar to the description given by Giovanola (1987). In the calculation, the bandwidth in the second phase contracts to about half its original value.

Marchand and Duffy (1988) observed a three-stage process. The second and third stage correspond approximately to those of Giovanola (1987). However, in Marchand and Duffy's case the band width decreases from hundreds to about $20\,\mu$m. This variation is not particularly near that calculated. This could be due to the initial value of the band width δ_0 selected in the calculation and the inhomogeneous nature of the shear localization around the circumference of the specimens in the test. Nevertheless, if we investigated the simultaneous variations of local temperature and shear stress in the calculations and experiments, the similarity is quite close (see Figs 7.5 and 7.8). At the beginning of the process the two diagrams show a slight decrease in shear stress and a gradual increase in temperature. After this, the shear stress starts to drop abruptly and at the same time the local temperature increases rapidly and reaches a maximum while the stress continues to decrease, approaching a minimum value. The agreement between analysis and experiment is good, particularly if we take into account the earlier measurements of Hartley and co-workers (1987) (Fig. 7.12), in which the temperature peak almost coincides with the start of the sharp decrease in shear stress. The analytical results shown in Fig. 7.8 predict a difference in temperature from those observed and this has recently been confirmed by Marchand and Duffy (1989) who felt that the measurement technique used by Marchand was more accurate.

Other authors, for example Wright and co-workers (1985, 1987), have explored the evolution of adiabatic shear bands. They were especially interested in the abrupt decrease in shear stress during the formation of adiabatic shear bands. These researchers used the simple thermoviscoplastic constitutive equation

$$\tau = (1 - a\theta)(1 - b\dot{\gamma})^m$$

where a and b are constants. They performed both numerical simulation and perturbation analysis and their results are shown in

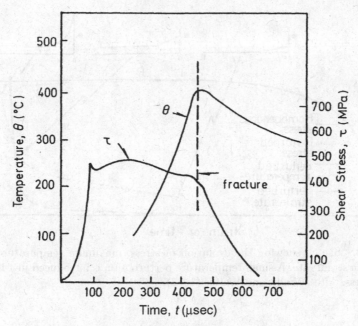

Fig. 7.12. Shear band temperature and stress as functions of time for 1018 cold-rolled steel. After Hartley (1987).

Fig. 7.13. The variations of shear stress, shear strain and temperature show the same trends given by the simplified analytical model (see Fig. 7.8).

From all the analytical models and experiments the same picture of adiabatic shear banding has emerged. The reason for this is in the physical nature of adiabatic shear bands and their band-like dissipative structure. However, there are some important details of shear localization, such as the timing of the decrease in shear stress, the inhomogeneous localization around the circumference of specimens and so forth, which have not been interpreted properly in the models discussed above.

7.5. Late Stage Behaviour and Post-Mortem Morphology

From the discussion in the last section, it is clear that the dominant dissipative mechanism of adiabatic shear banding leads to an

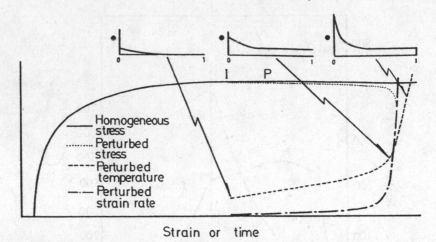

Fig. 7.13.   Sketch showing the evolution of stress, maximum temperature and maximum strain rate. A small temperature perturbation is introduced just before peak stress. After Wright and Walter (1987).

asymptotic quasi-steady structure of adiabatic shear bands. Using a constitutive equation of the general form

$$\dot{\gamma} = g(\tau, \theta) \qquad \cdot \qquad (7.20)$$

it has been shown by Bai and co-workers (1986) that this leads to a quasi-steady band-like solution to the basic equations (6.25).

As has been discussed in the last section, Eq. (7.20) is suitable for use in the high-temperature high strain rate regime within an adiabatic shear band. According to the linear perturbation analysis discussed in Section 6.2, assuming no strain hardening, the behaviour of the characteristic equation (6.40) will be decided by the term $(\lambda R k^2 - \tau P)$. With increasing deformation, the term can change sign from positive to negative due to the effects of thermal softening outweighing those of strain-rate hardening. According to bifurcation theory, this type of change in the linearized version indicates a transition from a stable mode to an unstable saddle. Further, in accordance with the principle of exchange of stability (see Nicolis and Prigogine, 1971), beyond this point the homogeneous solution is in a super-critical state, therefore it should be unstable. This indicates that there must be a new asymptotic stationary solution to the governing equations (6.25).

The asymptotic steady solution should satisfy the steady version of the basic equation (6.25):

$$\frac{d\tau}{dy} = 0 \tag{7.21}$$

$$\beta\tau\dot{\gamma} + \lambda\frac{\partial^2\theta}{\partial y^2} = 0, \quad \dot{\gamma} = g(\tau, \theta) \tag{7.22}$$

with an appropriate boundary condition, which may be a slow function of time making the corresponding solution quasi-steady. These equations can be solved easily (see Bai and co-workers, 1986, 1987). Firstly, Eq. (7.21) gives

$$\tau = \tau(t). \tag{7.23}$$

Supposing that

$$\xi = \frac{d\theta}{dy}$$

then

$$\frac{d^2\theta}{dy^2} = \frac{d\xi}{d\theta}\frac{d\theta}{dy} = \frac{1}{2}\frac{d\xi^2}{d\theta}.$$

Substitution of this differentiation and the constitutive equation (7.20) into Eq. (7.22) gives

$$\frac{1}{2}\frac{d\xi^2}{d\theta} = -\frac{\beta\tau}{\lambda}g(\tau, \theta) \tag{7.24}$$

where τ is independent of θ. Integration gives the temperature gradient

$$\xi = \frac{d\theta}{dy} = \sqrt{\frac{2\beta\tau}{\lambda}\int_\theta^{\theta_m} g(\tau, \eta)d\eta} \tag{7.25}$$

where the condition

$$\xi = \frac{d\theta}{dy} = 0 \quad \text{at } y = 0.$$

The centre of the shear band has been used. The second integration leads to the solution:

$$y = \sqrt{\frac{\lambda}{2\beta\tau}} \int_\theta^{\theta_m} \frac{d\xi}{\sqrt{\int_\xi^{\theta_m} g(\tau,\eta)d\eta}} \qquad (7.26)$$

where θ_m is the temperature at the centre of the shear band. The half-width of the asymptotic steady adiabatic shear band is

$$\delta = \sqrt{\frac{\lambda}{2\beta\tau}} \int_{\theta_\delta}^{\theta_m} \frac{d\xi}{\sqrt{\int_\xi^{\theta_m} g(\tau,\eta)d\eta}} \qquad (7.27)$$

where θ_δ is the temperature at the shear band boundary. This solution is suitable for the general form $g(\tau,\theta)$ of a thermoviscoplastic model and contains only two undetermined parameters, the shear stress τ and the temperature at the band centre θ_m. Both these parameters can be specified by the boundary conditions.

Although (7.27) is an exact and universal solution, it is still too complicated for engineers to use easily. An estimate of the shear band width based on the same asymptotic steady state balances the plastic work rate $\tau\dot{\gamma}$ with the heat conduction $\lambda(d^2\theta/dy^2)$; this results in Eq. (7.19). A very simple estimation of the shear band width has been given by Dodd and Bai (1985) and Bai and co-workers (1986). In this case

$$\delta = \sqrt{\frac{\lambda\theta_*}{\beta\tau_*\dot{\gamma}_*}} \qquad (7.28)$$

where the subscript $*$ denotes values within the shear band. The assumption of a monotonic temperature distribution within the band half-width was made in the derivation of Eq. (7.28).

Equation (7.28) indicates that when an adiabatic shear band develops it is a balance between opposing effects: plastic work and heat conduction. The plastic work-heat diffusion process governs the self-supporting band-like structure. Equation (7.28) can be rewritten in the following forms:

$$\delta = \sqrt{\kappa \frac{\gamma_*}{\dot{\gamma}_*} \frac{\rho c \theta_*}{\beta \tau_* \gamma_*}} \tag{7.29}$$

$$\delta = \sqrt{\frac{\lambda \theta_*}{\beta \eta} \frac{1}{\dot{\gamma}_*}} \tag{7.30}$$

where $\eta = \tau_*/\dot{\gamma}_*$ is the viscosity of the material at high rates. The most influential factor governing the shear band width is the shear strain rate within the band. Dodd and Bai (1985), Kobayashi (1987), Hartley and co-workers (1987) and Marchand and Duffy (1988) have separately compared estimations of δ from Eq. (7.28) with experimental observations. The comparisons are shown in Tables 7.1, 7.2 and 7.3. The agreement appears to be reasonably good.

Backman and co-workers (1986) carried out a series of penetration tests as mentioned in Section 3.2. They fired hard steel spheres of between 1/8 and 3/8 inch in diameter against aluminium alloy plates. The scaling factor was 3. They found that below the ballistic limit the band width ranged from 5 to 12 μm in the 1/8 inch system and 4 to 13 μm in the 3/8 inch system. The shear bands with well-defined boundaries had average shear band widths of between 5 and 7 μm, whereas the bands with less well-defined boundaries had average band widths of between 8 and 10 μm. The conclusion made by these researchers was that there was no scaling effect in relation to shear band width. All their results indicate that the shear band width is invariant with respect to linear measure. This is in

Table 7.1.   The comparison of the estimation of δ based on Eq. (7.28) and experimental observations.

Parameter	Units	Al	Cu	Mild steel
τ_f	MPa	300	200	300
θ_M	K	775–877	1355	1800
λ	$\mathrm{W\,m^{-1}\,K^{-1}}$	236	403	50
η	kPa S	2.1	3.6	2.8, 2.1
$2\delta_{\mathrm{pred}}$	mm	0.14	0.44	0.11
$2\delta_{\mathrm{exp}}$	mm		0.34	0.19

Dodd and Bai (1985).

Table 7.2.   The comparison of the estimation of δ based on Eq. (7.28) and experimental observations.

Parameter	Units	Low-carbon steel	Tool steel	CP-titanium
λ	W m^{-1} K^{-1}	48	40	17
$\Delta\theta_*$	K	201	241	265
τ_*	MPa	430	1169	435
$\dot{\gamma}_*$	1/sec	2226	1548	2283
$2\delta_{\mathrm{pred}}$	mm	0.1	0.073	0.067
$2\delta_{\mathrm{exp}}$	mm	0.13–0.17	0.1	0.06–0.10

Kobayashi (1987).
Noticeably, all the values $2\delta_{\mathrm{exp}}$ were obtained from thin-walled tube torsion tests, which offer quite long loading time. The values with subscript $*$ were taken from the tests and simulation in Kobayashi (1987).

Table 7.3.   The comparison of the estimation of δ based on Eq. (7.28) and experimental observations.

Parameter	Units	HY 100	4340	1018 CRS	1020 HRS
λ		54	54	54	54
θ_*	K	590	500 (assumed)	400	400
τ_*	MPa	270	400 (assumed)	400	400
$\dot{\gamma}_*$	1/sec	4×10^5	10^6	6100	6000
$2\delta_{\mathrm{pred}}$	μm	40	16	200	200
$2\delta_{\mathrm{exp}}$	μm	20	20	250	150
		Marchand and Duffy (1988)	Giovanola (1987)	Hartley *et al.* (1987)	Hartley *et al.* (1987)

agreement with the idea that the band width is mainly dependent on the behaviour of the material and the strain rate localized within the band [cf. Eqs (7.28) to (7.30)].

Wright (1987) also noticed that heat conduction is the significant factor for the removal of excess energy from a shear band to its surroundings if the process of localization continues for long enough. Therefore, it may provide a mechanism for the limitation of more severe localization and therefore the final stage of localization should be quasi-steady.

Wright assumed the following constitutive equation:

$$f(\tau, \dot{\gamma}, \theta) = K$$
$$\dot{K} = M(\tau, \theta, K)\dot{\gamma}. \tag{7.31}$$

However, it was found that no steady solution existed unless $\dot{K} = 0$. If Eq. (7.20) Is adopted as a constitutive equation, then the governing equations are the same as Eqs (7.20) to (7.22) except for $\lambda = \lambda(\theta)$. Therefore, Eq. (7.22) becomes:

$$\frac{d}{dy}\left(\lambda \frac{d\theta}{dy}\right) + \beta\tau g(\tau, \theta) = 0. \tag{7.32}$$

Suppose that $\xi = d\theta/dy$ and Eq. (7.32) is multiplied by a factor $2\lambda(\theta)$, then the first term becomes:

$$2\lambda\frac{d}{dy}\left(\lambda \frac{d\theta}{dy}\right) = 2\lambda\xi\frac{d}{d\theta}(\lambda\xi) = \frac{d}{d\theta}(\lambda\xi)^2. \tag{7.33}$$

Therefore, integration of Eq. (7.32) gives:

$$\left(\lambda\frac{d\theta}{dy}\right)^2 = 2\beta\tau \int_{\theta}^{\theta_m} \lambda(\eta)g(\tau, \eta)d\eta. \tag{7.34}$$

A second integration gives:

$$y = \frac{1}{\sqrt{2\beta\tau}} \int_{\theta}^{\theta_m} \frac{\lambda(\xi)d\xi}{\sqrt{\int_{\xi}^{\theta_m} \lambda(\eta)g(\tau, \eta)d\eta}}. \tag{7.35}$$

Wright proposed that the core of the adiabatic shear band should correspond to a central boundary layer.

Wright used a number of case studies to illustrate the late-stage steady morphology of adiabatic shear bands. He used four different constitutive equations calibrated with a set of typical data from a high-strength steel.

The calibration data used were:

$$\theta = 300\,\mathrm{K}, \quad \dot{\gamma} = 1000/\sec, \quad \tau = 500\,\mathrm{MPa},$$

$$m = \left(\frac{\partial \ln \tau}{\partial \ln \dot{\gamma}}\right) = 0.02, \quad P = -\left(\frac{\partial \ln \tau}{\partial \ln \theta}\right) = 0.2.$$

The constitutive equations were:

(1) An Arrhenius-type law:

$$\dot{\gamma} = \dot{\gamma}_0 \exp\left\{-\frac{v}{b\theta}(\tau_0 - \tau)\right\} \tag{7.36}$$

$\frac{v}{b} = 3 \times 10^{-5}\,\mathrm{K\,m^3/J}, \quad \tau_0 = 600\,\mathrm{MPa}, \quad \dot{\gamma}_0 = 2.2026 \times 10^7/\mathrm{sec}.$

(2) The Bodner–Partom–Merzer law:

$$\dot{\gamma} = 2\dot{\gamma}_0 \exp\left\{-\frac{1}{2}\left(\frac{\tau_0^2}{3\tau^2}\right)^{\frac{a}{\theta}+b}\right\} \tag{7.37}$$

$$\tau_0 = 1.05777\,\mathrm{GPa}, \quad a = 1653.75\,\mathrm{K}$$

$$\dot{\gamma}_0 = 4.6618 \times 10^4/\mathrm{sec}, \quad b = 0 \text{ (arbitrarily chosen)}.$$

(3) A conventional power law:

$$\dot{\gamma} = \dot{\gamma}_0 \left(\frac{\tau}{\tau_0}\right)^{1/n} \left(\frac{\theta}{\theta_0}\right)^{v/n} \tag{7.38}$$

$$\tau_0 = 500\,\mathrm{GPa}, \quad \theta_0 = 300\,\mathrm{K}, \quad \dot{\gamma}_0 = 1000/\mathrm{sec}.$$

(4) A Litonski-type law:

$$\dot{\gamma} = \frac{1}{b}\left[\left\{\frac{\tau}{\tau_0(1 - a(\theta - \theta_0))}\right\}^{1/n} - 1\right] \tag{7.39}$$

$$\theta_0 = 300\,\mathrm{K}, \quad b = 1000/\mathrm{sec} \text{ (arbitrarily chosen)}$$

$$n = m, \quad a = \frac{3}{2} \times 10^{-3}/\mathrm{K}, \quad \tau_0 = 379.29\,\mathrm{MPa}.$$

Typical distributions of plastic strain rate and temperature within the shear band in the steady state are shown in Fig. 7.14. The strain rate has a steeper profile than the temperature, which is similar to Xing's simulation (Fig. 7.11). In all four sets of calculations the shear band width and the driving shear stress are both much more strain-rate sensitive than temperature dependent (as shown in Fig. 7.15).

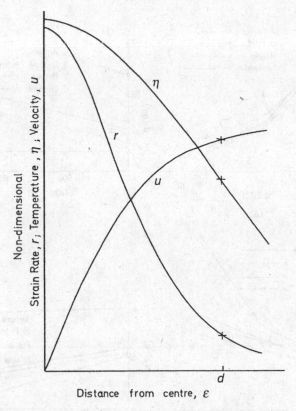

Fig. 7.14. Typical distributions of plastic strain rate, temperature and velocity versus distance from the centre of a shear band. After Wright (1987).

So far all observations considered were made assuming the asymptotic quasi-steady solution or the heat-diffusion-dominant solution. However, the adiabatic shear bands observed are not always fully developed. The development of a shear band can be interrupted and terminated at any stage of its evolution by unloading. Therefore, the features of adiabatic shear bands observed in the post-mortem state are not usually determined by the asymptotic quasi-steady solution. On the contrary, adiabatic shear bands can have some very different premature features.

Wu and Freund (1984) examined adiabatic shear banding dominated by wave trapping and viscous diffusion in the early stages.

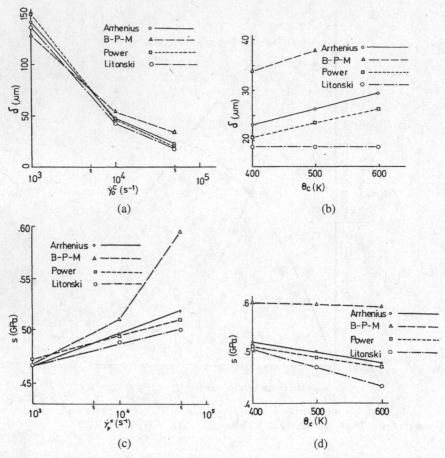

Fig. 7.15.   Width versus central plastic strain rate (a) and temperature (b) when central temperature is 400 K and plastic strain rate is 5×10^4/sec, respectively. Driving stress versus central plastic strain rate (c) and temperature (d), when central temperature is 400 K and plastic strain rate is 5×10^4/sec respectively. After Wright (1987).

Their numerical simulation gives the following:

$$\delta = 18.3 \frac{c_0}{\dot{\gamma}_0} \sqrt{\frac{(U + u^*)}{c_0}} \qquad (7.40)$$

where $c_0 = \sqrt{\tau_0/\rho}$, $u^* = \int_0^{\dot{\gamma}_*} c(\gamma)d\gamma$ and U is the driving velocity at the loading surface, for linear rate sensitivity, see Eq. (7.10), and

$$\delta = \frac{0.014c_0^2}{\dot{\gamma}_0(U + u^*)} \tag{7.41}$$

for logarithmic rate sensitivity, Eq. (7.11). The differences caused by the differences in rate sensitivity were attributed by Wu and Freund to the differences in the parameters preceding the diffusion term. They observed a thin layer (of about $20\,\mu$m in thickness) of intense shear adjacent to the surface in the material exhibiting low strain hardening, which supports the view of the importance of wave trapping and viscous dissipation in adiabatic shear banding.

Based on numerical simulations, Olson and co-workers (1981) and Merzer (1982) proposed that the width of an adiabatic shear band is controlled by heat conduction. Furthermore, they suggested that the band width was determined simply by the size of the heat-diffusion zone.

$$\delta \approx \sqrt{\kappa t} \approx \sqrt{\kappa \frac{\gamma_*}{\dot{\gamma}_*}}. \tag{7.42}$$

If the local strain and strain rate are 1–10 and 10^5–10^6/sec, respectively, then the band width should be about $10\,\mu$m for metals. It seems doubtful, however, whether adiabatic shear bands can be regarded simply as zones of heat diffusion, because heat diffusion itself is a continuing process without localization.

There are other discussions and evaluations of post-mortem morphology of adiabatic shear bands, such as that of Grady and Kipp (1987), which is concerned with the fine adiabatic shear bands that occur in planar impact. But approaches such as this are more concerned with specific problems in impact dynamics and discussion of such problems will be left until Chapter 9.

7.6. Adiabatic Shear Bands in Multidimensional Stress States

Hitherto, all the analytical investigations and proposed mechanisms for adiabatic shear banding have been concerned with simple shear. Previous discussions have all been based on the one-dimensional equations of momentum and energy, i.e. Eqs (6.25–1) and (6.25–2), and one-dimensional constitutive equations, Eq. (6.26). This one-dimensional model is excellent for exploring the fundamental mechanisms governing adiabatic shear banding. Practically in engineering, adiabatic shear bands are observed in multi-dimensional stress states. Therefore, from the practical point of view it is necessary to study the behaviour of shear bands in these more complex stress states. However, this is not a simple task, even in two dimensions, and consequently only a small number of researchers have studied this topic to date. Most of the research carried out so far in this area has utilized the effects of linear perturbation analysis on instability. Based on this approach, information about the orientation and growth rates of shear bands has been derived (see Anand *et al.*, 1987; Douglas and Chen, 1985). Other research has studied the late stage shear band width (see Dodd and Bai, 1989). In this section the basic concepts and techniques concerned with adiabatic shear banding will be introduced. To avoid unnecessary mathematical manipulation only the principles, starting points and results will be cited here, but their one-dimensional counterparts in Section 6.2 will be referred to in order to give a fuller understanding of the procedures.

Other research has considered the formation of adiabatic shear bands in multidimensional stress states such as that in planar plate impact and fragmentation. However, since this work is mainly concerned with specific applications, discussions of it will be found in Chapter 9.

The one-dimensional simple shear case from Section 6.2 is

$$
\begin{aligned}
x &= u(y) + X \\
y &= Y \\
z &= Z.
\end{aligned}
\tag{7.43}
$$

Douglas and Chen (1985) presented a linear perturbation analysis of anti-plane shear deformation (Fig. 7.16):

$$
\begin{aligned}
x &= u(y, z) + X \\
y &= Y \\
z &= Z.
\end{aligned}
\tag{7.44}
$$

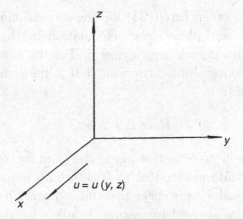

Fig. 7.16. Schematic diagram showing anti-plane shear deformation.

The basic equations of momentum and energy are

$$\rho \frac{\partial^2 \gamma_{xy}}{\partial t^2} = \frac{\partial^2 \tau_{yx}}{\partial y^2} + \frac{\partial^2 \tau_{zx}}{\partial z \partial y} \tag{7.45}$$

$$\beta(\tau_{yx}\dot{\gamma}_{yx} + \tau_{zx}\dot{\gamma}_{zx}) = \rho c \dot{\theta} - \lambda \left(\frac{\partial^2 \theta}{\partial y^2} + \frac{\partial^2 \theta}{\partial z} \right) \tag{7.46}$$

where $\gamma_{yx} = \partial u/\partial y$, $\gamma_{zx} = \partial u/\partial z$ and the coordinates y and z are interchangeable. The linear perturbation is similar to that given in Eqs (6.31)–(6.33). Using this approach, there is a supposition of the possible difference between the orientations of the homogeneous stresses and the perturbated shear strains:

$$\tau_{yx}^h = \tau_h \cos \beta_h \tag{7.47}$$

$$\gamma_{yx}^* = \gamma_* \cos \beta_* \tag{7.48}$$

$$\Delta\beta = \beta_h - \beta_*. \tag{7.49}$$

After similar manipulations to those carried out with Eqs (6.33) and (6.34), the corresponding characteristic equation is:

$$\rho^2 c \alpha^3 + \rho\{\beta P \dot{\gamma}_h \cos \Delta\beta + (\lambda + cR)k^2\}\alpha^2$$
$$+ (\lambda R k^2 + \rho c Q - \beta \tau_h P \cos \Delta\beta)k^2 \alpha + \lambda Q k^4 = 0. \tag{7.50}$$

The differences between Eq. (6.34) for one-dimensional simple shear and Eq. (7.50) for anti-plane shear deformation are the factor $\cos \Delta\beta$ and the homogeneous variables $\dot\gamma_h$ and τ_h. For the adiabatic approximation $\lambda = 0$ the instability criterion, that is the condition for positive real roots of α, is

$$\beta\tau_h P \cos \Delta\beta > \rho c Q. \tag{7.51}$$

Any instability will grow fastest along the path for which $\Delta\beta = 0$. Then criterion (7.51) is identical to (6.39). This demonstrates that the one-dimensional simple shear instability criterion can be generalized to anti-plane shear deformation. In other words, the inhomogeneity in shear deformation in the shear band plane, but perpendicular to the driving velocity, does not influence the occurrence of instability.

Anand and co-workers (1987) discussed more general forms of adiabatic shear bands in multidimensional stress states. The material was assumed to be isotropic, incompressible and thermoviscoplastic. Elastic effects were ignored. But these researchers included the effect of hydrostatic pressure on plastic flow. The constitutive equation adopted was:

$$\sigma_{ij} = -\bar{p}\delta_{ij} + 2\mu D_{ij} \tag{7.52}$$

$$\mu = \bar{\tau}/\bar{\dot\gamma} = \mu(\bar{\tau}, \bar{p}, \bar{\gamma}, \theta) \tag{7.53}$$

where the mean normal pressure $\bar{p} = -\sigma_{ii}/3$, the equivalent shear stress $\bar{\tau} = \sqrt{\frac{1}{2} s_{ij} s_{ij}}$, where $s_{ij} = \sigma_{ij} + \bar{p}\delta_{ij}$ is the stress deviator, $\bar{\dot\gamma} = \sqrt{\frac{2}{3} \dot\gamma_{ij} \dot\gamma_{ij}}$ is the equivalent deviatoric strain rate and the stretch tensor $D_{ij} = v_{ij} - W_{ij}$, where W_{ij} are the components of rigid rotation of the velocity gradient.

The basic equations can be written according to Eq. (6.23):

$$\rho\dot{v}_i = \sigma_{ij,j} \tag{7.54}$$

$$\beta\rho c\dot\theta = \lambda\theta_{i,i} + \sigma_{ij} D_{ij}. \tag{7.55}$$

Now consider it is possible to consider the linear perturbation for the multidimensional case. Anand *et al.* consider the following form of perturbation:

$$v'_i = v_{*i}\psi = v_{*i}\exp(\alpha t + ikx_in_i) \tag{7.56}$$

$$v_{*i}n_i = 0. \tag{7.57}$$

This means that the normal to the shear band perturbation has the orientation n_i in the current configuration. As we have already observed, Anand and co-workers also took into account the effects of hydrostatic pressure. Thus, besides strain hardening, strain-rate hardening and thermal softening, the pressure sensitivity appears in the perturbation:

$$P_H = \left(\frac{\partial \bar{\tau}}{\partial \bar{p}}\right)_{\bar{\gamma},\dot{\bar{\gamma}},\theta} > 0. \tag{7.58}$$

It is clear that a similar perturbation technique to that described in Section 6.2 can be used, with due allowance being made for mathematical manipulations required because of the three-dimensional shear deformation and perturbation. Anand *et al.* presented the result for plane motion, i.e. $v_z = 0$, with the trace of the shear band lying in the plane of motion. The characteristic equation is:

$$c_0\alpha^3 + c_1\alpha^2 + c_2\alpha + c_3 = 0 \tag{7.59}$$

where

$$c_0 = \rho^2 c(1 + \chi_2 P_H)$$

$$c_1 = \rho\left[\beta P\dot{\bar{\gamma}}_h + \left\{(1 + \chi_2 P_H)\lambda + \chi_1 cR\right.\right.$$

$$\left.\left. + c\left(\frac{\bar{\tau}_h}{\dot{\bar{\gamma}}_h}\right)(1 - \chi_1 + \chi_2 P_H)\right\}k^2\right]$$

$$c_2 = \left[\left\{\lambda R\chi_1 + \lambda\left(\frac{\bar{\tau}_h}{\dot{\bar{\gamma}}_h}\right)(1 - \chi_1 + \chi_2 P_H)\right\}k^2\right.$$

$$\left. + \rho cQ\chi_1 - \beta\{\bar{\tau}_h P(2\chi_1 - 1)\}\right]k^2$$

$$c_3 = \chi_1\lambda Qk^2 \tag{7.60}$$

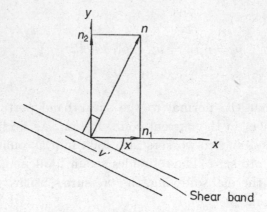

Fig. 7.17.   Schematic diagram showing the orientation of shear band perturbation.

and

$$\chi_1 = \sin^2 2\chi = 4n_1^2 n_2^2 \qquad (7.61)$$

$$\chi_2 = \cos^2 2\chi = n_2^2 - n_1^2. \qquad (7.62)$$

The unit vector n_i is the normal to the shear band perturbation and χ is the inclination of the trace of the shear band perturbation relative to the maximum principal stretching stress axis, x (Fig. 7.17).

If pressure sensitivity is neglected, i.e. $P_H = 0$, and the shear bands perturbed and developed at $\chi = \pm\pi/4$ with respect to the maximum principal stress direction are assumed *a priori*, Anand and co-workers found that Eqs (7.56) and (7.57) reduce to Eq. (6.34) for one-dimensional simple shear.

In the limiting case of adiabatic deformation $\lambda = 0$, the characteristic Eq. (7.59) reduces to a quadratic equation:

$$c_0 \alpha^2 + c_1 \alpha + c_2 = 0 \qquad (7.63)$$

where

$$c_0 = \rho^2 c(1 + \chi_2 P_H)$$

$$c_1 = \rho \left[\beta P \dot{\gamma}_h + \left\{ \chi_1 cR + c \left(\frac{\bar{\tau}_h}{\bar{\dot{\gamma}}_h} \right) (1 - \chi_1 + \chi_2 P_H) \right\} \right] k^2$$

$$c_2 = \{ \rho c Q \chi_1 - \beta \bar{\tau}_h P(2\chi_1 - 1) \} k^2. \qquad (7.64)$$

In a material insensitive to hydrostatic pressure, $P_H = 0$, and the coefficients $c_0, c_1 > 0$, then the condition for instability is

$$c_2 < 0. \tag{7.65}$$

According to the behaviour of the function $c_2(\chi_1)$, it is easy to obtain the condition for inequality (7.65) as:

$$\rho c Q < \beta \bar{\tau}_h P, \tag{7.66}$$

which is equivalent to criterion (6.39) for the multidimensional stress state. There is no restriction on the inclination of the shear band perturbation. Anand *et al.* found that as the wave number approaches zero, the shear band perturbation grows with the highest growth rate (similar to the rate in Section 6.2 for one-dimensional simple shear). Furthermore, if $\bar{\tau}_h / \bar{\dot{\gamma}}_h > R$ the shear band perturbation with $\chi = \pm \pi/4$ to the maximum stretching stress axis will grow at the fastest rate.

When Anand *et al.* considered pressure sensitivity, $P_H > 0$, they found that the shear band patterns were non-orthogonal; this is because the coefficient c_1 may be positive or negative. Examining the expression for c_1, the last term may not always be positive, depending on the function:

$$
\begin{aligned}
f_2 &= 1 - \chi_1 + \chi_2 P_H \\
&= \cos 2\chi (\cos 2\chi + P_H).
\end{aligned} \tag{7.67}
$$

For values of χ in the range $\pi/4 < \chi < \pi/4 + P_H/2$, the function f_2 is negative and reaches a maximum negative value $f_{2\max}^* = -P_H/2$ at $\chi^* = \pm(\pi/4 + P^0/4)$.

If $c_1 > 0$, then the reasoning follows the pressure-insensitive argument. If $c_1 < 0$ and $\rho c Q > \beta \bar{\tau}_h P$, instability can still occur because of the existence of the positive real part of the root in Eqs (7.63) and (7.64). In addition, the orientation of the shear band perturbation corresponding to the maximum growth rate is not necessarily $\chi = \pm \pi/4$. These two phenomena are interesting in adiabatic shear banding. Unfortunately, Anand and co-workers only discussed, as an example, quasi-static isothermal one-dimensional simple shear and found the shear band inclination for a nominal shear strain rate of

10^{-2}/sec. In addition, the conjecture that adiabatic shear bands may occur more readily in pressure-dependent materials contradicts the work of Dodd and Atkins (1983) and Seaman (1983).

The late stage band width in multidimensional stress states has been considered by Dodd and Bai (1989). It was assumed that once the shear bands have formed, the overall deformation in the vicinity of the shear bands will reduce to a plane strain stress state. This assumption will lead to the following form of the energy equation (6.21):

$$\rho c \left(\frac{\partial \theta}{\partial t} + u \frac{\partial \theta}{\partial x} + v \frac{\partial \theta}{\partial y} \right) = \beta \sigma_{ij} D_{ij} + \lambda \left(\frac{\partial^2 \theta}{\partial x^2} + \frac{\partial^2 \theta}{\partial y^2} \right) \qquad (7.68)$$

where $i, j = 1, 2$, provided that the characteristic length scale of the band width δ parallel to the y axis is much less than the band spacing l (parallel to the x axis). For a certain shear band, the scaled variables should be:

$$\bar{y} = y/\delta \approx O(1), \quad \bar{x} = x/l \approx O(1), \quad \bar{\theta} = \frac{\theta}{\Delta \theta} \approx O(1)$$

where $\Delta \theta$ represents the temperature difference between the shear band and remote material. Equation (7.68) becomes:

$$\frac{\delta^2}{\kappa} \left(\frac{1}{t_k} \frac{\partial \bar{\theta}}{\partial \bar{t}} + \frac{u}{l} \frac{\partial \bar{\theta}}{\partial \bar{x}} + \frac{v}{\delta} \frac{\partial \bar{\theta}}{\partial \bar{y}} \right) = \frac{\delta^2 \beta \sigma_{ij} D_{ij}}{\lambda \Delta \theta} + \frac{\partial^2 \bar{\theta}}{\partial \bar{y}^2} \qquad (7.69)$$

where the relation $\delta/l \ll 1$ has been assumed and t_k is the timescale corresponding to the temperature difference. Comparison of the three terms in the bracket gives:

$$\frac{u}{l} \approx \bar{\dot{\epsilon}}_x \approx \frac{v}{\delta} \approx \bar{\dot{\epsilon}}_y \ll \frac{1}{t_k} \approx \bar{\dot{\gamma}}.$$

$\bar{\dot{\epsilon}}_x$ denotes the average stretching rate in the shear plane over the spacing of the shear bands and $\bar{\dot{\epsilon}}_y$ is the average stretching strain rate across the shear band. Also, according to the one-dimensional analysis, the localized temperature initially increases slowly and then increases quickly. Hence, the time $t_k \gtrsim \delta^2/\kappa$. Therefore, Eq. (7.69) may be simplified:

$$\frac{1}{t_k}\frac{\delta^2}{\kappa}\frac{\partial\bar{\theta}}{\partial\bar{t}} = \frac{\delta^2\beta\sigma_{ij}D_{ij}}{\lambda\Delta\theta} + \frac{\partial^2\bar{\theta}}{\partial\bar{y}^2} \tag{7.70}$$

owing to the time for local heat diffusion $\delta^2/\kappa \ll t_k$.

The time for the whole process, Eq. (7.70), gives

$$\delta = \sqrt{\frac{\lambda\Delta\theta}{\beta\sigma_{ij}D_{ij}}}. \tag{7.71}$$

The band width is insensitive to the details of the combined stress state. Interestingly, the insensitivity of shear band width to the multidimensional stress state agrees with the observations on scaling by Backman and co-workers (1986) and the impact erosion work described by Timothy (1987).

Chapter 8

Numerical Studies of Adiabatic Shear Bands

Although discussions of adiabatic shear banding based on analytical investigations have revealed some of the fundamental mechanisms underlying the evolution of adiabatic shear bands, the results and conclusions are mostly qualitative in nature. One reason for this is that the assumptions made in theoretical analyses are oversimplified. Furthermore, any model with its corresponding equations is only suitable for a certain stage or aspect of the phenomenon. The transition from one stage to the next cannot be treated using the assumptions made for that stage only. For example, the transition from the stage dominated by wave trapping and viscous dissipation to the successive stage of quasi-static deformation coupled with heat diffusion should be understood in terms of the different mechanisms. From the engineering point of view, the timing of the full formation of adiabatic shear bands may be very important in the design of a structure subjected to short-duration dynamic loads.

Numerical simulation is a powerful tool, which is capable of unveiling the details of transient processes. If the numerical technique is well chosen and carefully designed, the calculation may cover several stages of adiabatic shear banding. However, it is outside the scope of this book to describe in detail the various numerical techniques; there are a number of books and papers on this matter, for example Douglas and Dupont (1970) on finite element methods. However, in this chapter the aims of numerical studies, a summary of available techniques and the difficulties to be encountered in the simulation of adiabatic shear banding will be emphasized. A number of important results obtained by numerical simulation are presented to give a better understanding of shear banding.

8.1. Objects, Problems and Techniques Involved in Numerical Simulations

Certain areas cannot be tackled well in analytical studies, such as the complicated geometrical configuration of disturbances or imperfections, multi-scaled transient processes and the non-linearity and coupling effects in the constitutive equations. Therefore, although numerical techniques can verify the experimental observations and theoretical predictions, the above-mentioned problem areas are of most interest in the application of numerical techniques.

Perhaps the first problems to be studied using numerical simulation techniques were the following.

The time from the first occurrence to the full formation of adiabatic shear bands, or the time from the maximum in the shear stress to the dramatic drop in shear stress, is of great interest in engineering. The important points are how long the transition takes and what parameters govern the timing.

The influence of disturbances, imperfections and the like on the occurrence and formation of adiabatic shear bands requires exploration. Whether the microstructure can influence the occurrence of shear bands and whether imperfections in the material can dominate the positions of shear bands are important questions. The areas in which numerical studies are concentrated are the effects of various inhomogeneities, imperfections, voids and perturbations on the various stages and transitions in the evolution of adiabatic shear bands.

The transitional processes, which have not been properly described by analytical studies, are another topic for numerical simulation. Although numerical simulation has advantages over analytical techniques for complex transient phenomena, the unusual features of adiabatic shear bands make the application of numerical techniques very complicated. An indication of this difficulty is the small amount of research into the transition from wave trapping to heat-diffusion-dominated shear localization.

Naturally the pattern or network of adiabatic shear bands in multidimensional configurations should be one of the central themes of the numerical techniques. This is because of the importance of the study of shear bands in multidimensional stress states in engineering and also the complexity of their analytical study.

There are some special problems we may encounter while conducting numerical simulations to study adiabatic shear bands. The three major problems appear to be: instability in the calculations, an unprescribed position for shear localization and adapting the mesh to the scale of localization. These three problems are all derived from the abrupt changes in variables in an extremely narrow region, even compared to the size of the mesh, during adiabatic shear banding. Therefore, the three problems are interrelated fairly closely. Moreover, in the study, the material instability may be unavoidably mixed with the instability of the discretized algorithm.

During the course of adiabatic shear localization, the region in which severe inhomogeneous shear deformation develops can become progressively narrower. Beyond some limit, the localized region may be even less than the size of the pre-engaged grid. Therefore, the object under calculation is beyond the spacial resolution provided by the technique. Olson and co-workers (1981) found that in their finite difference simulation of adiabatic shearing using the HEMP code, the intensification of shear strain within the localized area was significantly delayed when using a double-sized mesh with everything else kept constant. Clearly, this type of delay in the appearance of localized shear deformation is due to averaging of the variables over the whole size of a single grid. This will introduce possible misunderstandings and inaccuracies into the simulation of localization.

More seriously, computational instability can sometimes appear in calculations associated with shear banding. In early work on the simulation of banding, computational instability was not reported. This may be because these calculations did not extend to the fully developed shear bands which are accompanied by a sharp drop in shear stress, as described in Sections 7.1 and 7.4. In almost all of these computations, a sharp drop in the shear stress was not reported; for example see the work of Costin and co-workers (1979) and Clifton *et al.* (1984). Wright and Batra (1985a,b) did experience an instability in their calculations. They found that the computations in their finite element algorithm became unstable as soon as the abrupt drop in shear stress occurred. Thus, the validity of the results remained in doubt and the morphology of the fully developed adiabatic shear band could not be calculated. It is therefore clear that a stable algorithm is required throughout the successive stages of adiabatic shearing; this is especially so for the final phase, which includes the rapid drop in stress and rapid increases in shear strain rate and temperature. It appears that the material instability occurring at a maximum in the shear stress is not involved in computational instability. However, the difficulty is postponed to the later stage of the formation of adiabatic shear bands, which is accompanied by changes in all the mechanical and thermodynamic variables.

Finally, transitions from one mechanism to another can cause a number of problems. An example of such a transition is from wave trapping and viscous dissipation to coupled plastic work rate and heat diffusion. It is necessary to deal with different length and time scales for different mechanisms. Also the position of the localized shear zone may be located near the loading boundary, as predicted by wave trapping, or located in the centre of the torsional test piece, as determined by heat diffusion. A further possibility is that the localized zone with its developing localization may drift between these two extreme positions. In this case, the adaptive technique may be more helpful because it can follow and resolve localizing shear inhomogeneities.

Table 8.1 summarizes many of the numerical simulations, which have been carried out concerning the formation and evolution of shear bands. Mainly finite element techniques have been used because regular and irregular geometries can be encompassed readily using the method. If the intention is to study shear localization as well as the geometry of the specimen flanges in torsion tests, then this is more straightforward. Finite difference methods have also been used and some useful codes and procedures have been proposed (see Table 8.1). Also some semi-analytical and semi-numerical simulations have been developed. One technique is a type of numerical integration (see Wada and co-workers, 1978; Xing, 1988).

Table 8.1.   List of numerical studies on adiabatic shear bands.

Authors	Year	Method*	Conditions	Main points
Litonski	1977	FD	Adiabatic deformation	Geometrical imperfection
Costin *et al.*	1979		SHTB simulation, const. velocity	Geometrical imperfection leads to localization
Olson *et al.*	1981	FD (HEMP)	Adiabatic const. velocity	Wave trapping, mesh size effect
Merzer	1982		Quasi-static	Band width $\delta \sim \sqrt{\kappa/\dot{\gamma}_{av.}}$, geometrical imperfection exists
Johnson	1981	FE (EPIC-2)	SHTB simulation	Temperature gradient governs localization
Wu and Freund	1984	FE FD — MOL ID	Const. velocity	Wave trapping, viscous diffusion
Drew and Flaherty	1984	FE adaptive	Const. velocity, const. temperature	Wave trapping (early stage) central localization (late stage)
Wright and Batra	1985a,b	FE	Const. velocity adiabatic	Growth rate of localization depends on amplitude
Kuriyama and Meyers	1985a,b	FE	Quasi-static adiabatic	Notch effect on localization
Fressengeas and Molinari	1987	FD mixed explicit and implicit	Adiabatic boundary	Amplitude effect on localization
Grady and Kipp	1987	FD (WONDY)	Quasi-static approximate energy equation fixed band width	Three regimes of localization
Batra	1987a,b	FE	Const. velocity adiabatic boundary	Effect of viscosity, strain hardening and rate sensitivity on timing of stress collapse
Kobayashi	1987	FD	Quasi-static const. temperature	Void-induced softening effect on localization

(*Continued*)

Table 8.1.   (*Continued*)

Authors	Year	Method[*]	Conditions	Main points
Xing	1988		Quasi-static	Effect of multiple disturbances
Lemonds and Needleman	1986	FE	Plane strain compression	Coupling effects of rate, thermal softening, heat conduction and multiaxial constitutive response
Needleman	1989	FE	Plane strain compression	Dynamic, strain and rate hardening
Le Roy and Ortiz	1989	FE	Plane strain compression	Dynamic, rate-(in) dependent pressure sensitive

[*]FD, finite difference; FE, finite element; SHTB, split Hopkinson torsion bar

As illustration, here a brief introduction is given of the finite element techniques used by Johnson (1981), Wu and Freund (1984) and Wright and Walter (1987).

Johnson used an explicit finite element formulation. Both mechanical and thermal parts were combined to form a thermomechanical option in the EPIC-2 code. The capacitance and conductance matrices were not formed because, in Johnson's opinion, this allowed the non-linear solutions to be performed adequately. However, later we will notice that in his simulation, although the nominal shear strain attains a value of 7 in copper, the shear stress shows no apparent decrease.

Wu and Freund (1984) and later Wright and Walter (1987) and Batra (1987a,b) adopted the Crank–Nicolson–Galerkin procedure (see Douglas and Dupont, 1970) in the finite element algorithm. They applied Galerkin's principle for weak solutions of differential equations. Discretization in space was introduced with a linear variation of dependent variables within each element. This yields a system of non-linear ordinary differential equations with an independent variable time t for the nodal values of the dependent variables. Wu and Freund followed the iteration proposed by Douglas and Dupont. The technique can circumvent the need to solve the complete non-linear system at every time step. The procedure is implicit and unconditionally stable. In addition, the convergence of the scheme was achieved.

Wright and Walter (1987) followed the idea of weak form solutions, but divided the grid into equal segments on a logarithmic scale to give a large number of nodes at the centre of the shear band. In this case the system of ordinary differential equations can be formulated as

$$M\dot{u} = f(u) \tag{8.1}$$

where M is the combined mass/heat capacity matrix and u is the vector of nodal velocities and temperatures. They found that the matrix M is positive definite, banded and diagonally dominant. This clearly facilitates manipulations afterwards and reduces the storage requirements. These workers reported that the numerical analysis was stable through all the rapid transition regions.

Adaptive finite element methods are very helpful and powerful techniques for problems that are transient and of an unprescribed localized nature. Therefore, they are suitable for simulations of adiabatic shear banding. Unfortunately, so far only a small number of researchers appear to be working in this area and little has been published. More information on the evolution and morphology of adiabatic shear bands can be revealed using this technique, and further details of it will be given in Section 8.3.

8.2. One-Dimensional Simulation of Adiabatic Shear Banding

This section concentrates on the aspects and features of adiabatic shear banding that have not been explored very deeply by analytical means as described in Chapter 7. Nevertheless, it is difficult to describe the results of individual investigators comprehensively and comparably because each investigator has chosen their own set of parameters to study. As a compromise, descriptions are made under the heading of each author.

Litonski (1977) conducted an early numerical study on adiabatic shear deformation in a geometrically inhomogeneous tube of a thermoviscoplastic material, which was assumed to have the following constitutive equation:

$$\tau = A(1 - a\theta)(1 + b\dot\gamma)^m(\gamma + \gamma_0)^n. \tag{8.2}$$

The simulation demonstrate that when the torque M reached a maximum, the strain rate $\dot\gamma_B$ in the weaker part of the tube B increased, and $\dot\gamma_A$ in the stronger part A vanished. This implies that plastic flow starts concentrating in the weaker part of the tube when M falls. For rate-dependent materials the concentration of plastic flow and the change in temperature depend markedly on $\dot\gamma$ (see Fig. 8.1).

Costin and co-workers (1979) tried to combine Litonski's numerical simulation with their experimental results for 1018 cold-rolled and

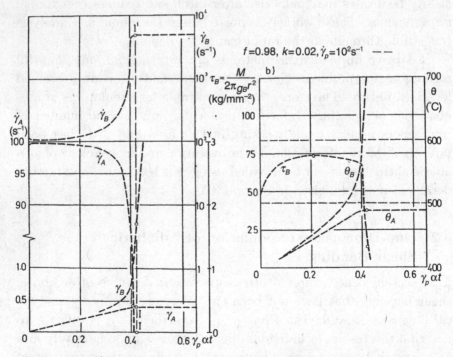

Fig. 8.1.   Computational results of strain, strain rate and temperature within the thinned section B and outside A, and the variation of shear stress. After Litonski (1977).

1020 hot-rolled steels. The constitutive equation assumed was:

$$\tau = A(1 - a\theta)(1 + b\dot{\gamma})^m \gamma^n. \tag{8.3}$$

Calculated results are shown in Fig. 8.2. These workers reported that the calculated nominal shear stress–strain curve in Fig. 8.2 showed

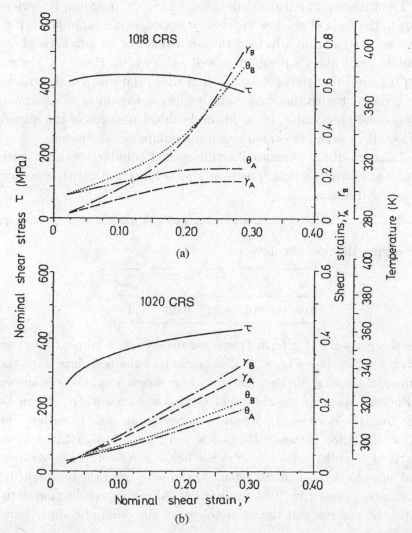

Fig. 8.2. Results of calculations showing strain and temperature within the thinned section B, compared to that in the remainder of the specimen A, dynamic loading, $\theta = 295\,\mathrm{K}$. After Costin (1979).

close agreement with their experimental results. In the case of the cold-rolled 1018 steel, the critical shear strain of 0.1 for the onset of softening is the same as that found from dynamic torsion tests. Furthermore, the calculated shear strains γ_A (uniform) and γ_B (localized) and temperatures θ_A and θ_B are also in agreement with experimental results.

From these numerical simulations, which are based on imperfections in the form of shallow grooves, it appears that an imperfection can have a significant effect on the occurrence of an adiabatic shear band, in particular its position as well as its width. However, Merzer (1982) argued that the assumption that the initial width of the groove was consistent with the shear band width, as Costin *et al.* assumed, seems to be improbable. In addition, both calculations of the imperfection effect were performed assuming adiabatic conditions.

Johnson (1981) simulated a thin-walled tubular specimen with flanges at either end acting as heat sinks. The constitutive equation he assumed was:

$$\tau = (a_1 + a\gamma^n)(1 + b\ln\dot{\gamma}/\dot{\gamma}_o)P. \tag{8.4}$$

For copper the relevant data are

a_1	a	n	b	$\dot{\gamma}_o/\sec$
0.069	0.016	0.32	0.027	1

The thermal softening term P was assumed to be a bilinear function of temperature (see Fig. 8.3). Comparison of the test data with the numerical solution for the nominal shear stress and strain is shown in Fig. 8.3. As with the results of Costin and co-workers, it can be seen that the decrease in stress is not very obvious. However, one major difference between Johnson's work and that of Litonski or Costin *et al.* is that in Johnson's case he did not assume any geometrical imperfections. The disturbance originates from the temperature distribution caused by heat conduction. Also heat conduction dominates the process and the morphology of the adiabatic shear band (see Fig. 8.4).

So far the numerical simulations have provided the general features of adiabatic shear bands and the effect of inhomogeneities on

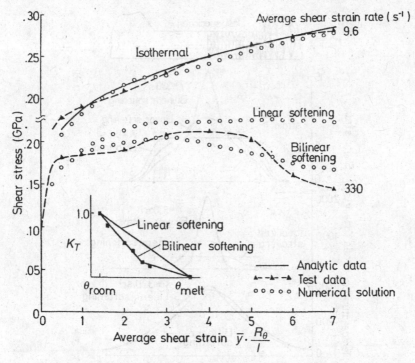

Fig. 8.3. Comparison of test data and numerical solution for strain relationship at $\dot{\gamma} = 9.6/\text{sec}$ and $\bar{\dot{\gamma}} = 330/\text{sec}$. After Johnson (1981).

the occurrence of bands. Although a preliminary picture of shear localization has been gained by this work, there are still some questions which remain unanswered. For example, what is the relationship between the width of the shear band and the size of imperfection, if any? Although some progress has been made on the role of inhomogeneities on the formation of shear bands, it is still not clear whether geometrical imperfections or disturbances in temperature are more important. An answer to this question together with advances made in theoretical investigations would lead to a more accurate and comprehensive understanding of adiabatic shear banding.

As mentioned in Section 7.3, Olson *et al.* (1981) and Wu and Freund (1984) began numerical simulations of the mechanism of shear localization at an early stage. Both groups of researchers used adiabatic stress–strain curves and studied the process of shear

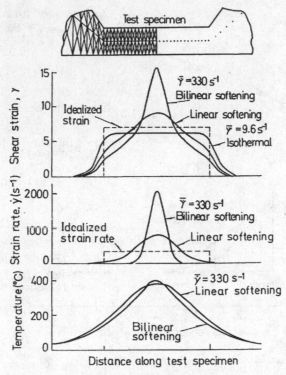

Fig. 8.4.   Strain, strain rate and temperature distributions in copper test specimen at $\bar{\gamma} = 7$. After Johnson (1981).

localization under constant boundary velocities. The numerical simulations showed that the strain localization occurs at the loading boundary in accordance with the concept of wave trapping. A typical distribution of shear strain near the surface is shown in Fig. 7.7. The strain distribution is the result of the coupling effect of wave trapping and viscosity. The shear strain rate approaches zero at some distance from the loading surface. This indicates that there is no further deformation outside the shear band. One of the calculated results of Olson *et al.* (1981) is particularly interesting. This is where a shear band occurred at the fixed end of the deforming body. The body itself was a 1 cm slab, which had a driving velocity of 10 m/sec imposed on it. The fixed end is equivalent to the centre of a test piece. Various aspects concerning wave trapping, viscosity effects and transitions to heat-diffusion-dominated localization will be discussed in Section 8.3, which is about adaptive techniques in numerical simulations.

At the other extreme, Merzer (1982) examined the intermediate and late stage behaviour of shear localization with an initial geometrical imperfection using numerical simulation. It was assumed that quasi-static equilibrium had been achieved, so the only equations required were the energy and constitutive equations. The material model used was the same as that used in the work of Bodner and Partom (1975):

$$\dot{\gamma} = \dot{\gamma}_e + \dot{\gamma}_p \tag{8.5}$$

$$\dot{\gamma}_p = 2D_o \exp\left\{-\frac{1}{2}\left(\frac{Z^2}{3\tau^2}\right)^n\right\} \tag{8.6}$$

$$Z = Z_1 - (Z_1 - Z_o)\exp(-mW_p). \tag{8.7}$$

In the particular case under consideration the data used were:

	$D_o\,(\text{sec}^{-1})$	$Z_o\,(\text{MPa})$	$Z_1\,(\text{MPa})$	$m\,(\text{MPa}^{-1})$	$n\,(\text{at }22°\text{C})$
A (normal rates)	10	1050	1240	0.085	2.4
B ($\bar{\dot{\gamma}} = 10^6/\text{sec}$)	10	1790	2110	0.085	1.43

The temperature dependence of the material was described by taking the index n to be a function of temperature:

$$n = (1407/\theta) = 2.37, \quad \text{normal rates}$$
$$= 422/\theta, \qquad\qquad \bar{\dot{\gamma}} \approx 10^6/\text{sec}.$$

To investigate the effects of an initial geometrical inhomogeneity, a narrow groove was introduced into the calculation. The results of simulation of the adiabatic shear banding process are shown in Fig. 8.5. Here, the letter E denotes the extent of the initial groove. It is clear that the eventual band width and its development are independent of the initial imperfection size. Merzer pointed out that the band width is dependent on heat conduction and strain rate. Figure 8.6 shows the results of simulation of the variation of the shear stress in the narrowing region of rapid straining with increasing nominal shear deformation. There is a clear, sharp decrease in shear

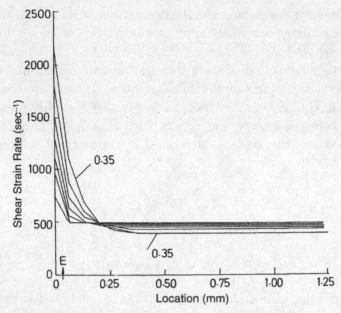

Fig. 8.5. Distributions of shear strain rates within the right-hand half of the specimen at different times for intervals of average strain of 0.05. E indicates the edge position of the initial groove. After Merzer (1982).

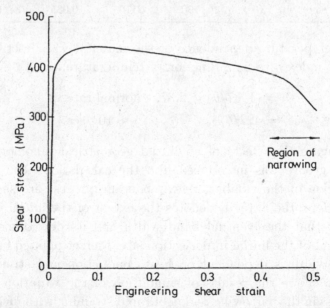

Fig. 8.6. Stress–strain curve with band collapse. Note the sharp stress collapse is associated with band narrowing in the region. After Merzer (1982).

stress accompanying the narrowing of the shear zone. This is in good agreement with theoretical predictions (Section 7.4).

The problem of the progress and timing of the occurrence through to the full formation of adiabatic shear bands has not been explained well in theoretical analyses to date. Only the predominant mechanisms have been described, but the quantitative effects of the various parameters on the progress remain unclear. From the results of numerical simulations described above, this would appear to be a powerful technique for identifying the influence of the individual material parameters on these aspects of adiabatic shear banding.

Batra (1987a,b) carried out numerical simulations with an improved version of the finite element analysis originally used by Wright and Batra (1985a,b). Here, the complete set of coupled thermomechanical non-linear equations was integrated numerically using the Crank–Nicolson–Galerkin method. Additionally, the mesh was finer near the centre of the slab. The boundary conditions were a constant driving velocity and adiabatic conditions. The constitutive equation used was a thermal, elastic, viscoplastic one, which includes strain hardening. The shear stress under dynamic plastic flow conditions is:

$$\tau = f(\tau, \theta, \dot{\gamma}_p)(1 - a\theta)(1 + b\dot{\gamma}_p)^m \qquad (8.8)$$

where f is the loading function:

$$f(\tau, \theta, \dot{\gamma}_p) = K \qquad (8.9)$$

where K is a measure of the strain hardening:

$$K = K_o(1 + \psi/\psi_o)^n \qquad (8.10)$$

where ψ is the plastic shear strain in an isothermal reference test. The relation between dynamic deformation is assumed to be

$$K\dot{\psi} = \tau\dot{\gamma}_p. \qquad (8.11)$$

Using this approach, the numerical simulation can examine the effects of strain-rate sensitivity b and m, strain hardening exponent n, thermal softening a and thermal conductivity λ on the timing of adiabatic shear banding from the maximum in the shear stress to its sudden decrease.

For reference, a set of typical values of material parameters were given for hard steel:

$$\rho = 7860 \, \text{kg/m}^3, \quad \lambda = 49.216 \, \text{W/m K}, \quad \mu = 80 \, \text{GPa},$$
$$K_\text{o} = 333 \, \text{MPa}, \quad a = 0.00552/\text{K}, \quad c = 473 \, \text{J/kg K},$$
$$m = 0.025, \quad n = 0.09, \quad \psi_\text{o} = 0.017, \quad b = 10^4 \, \text{sec}.$$

The calculations began with the following form of temperature disturbance:

$$\theta(y,0) = 0.1(1 - y^2)^9 e^{-5y^2}. \tag{8.12}$$

Figures 8.7–8.11 show the variations of the nominal stress–strain curves and the corresponding values of temperature and shear strain rate at the centre of the shear band. In these diagrams, P and F

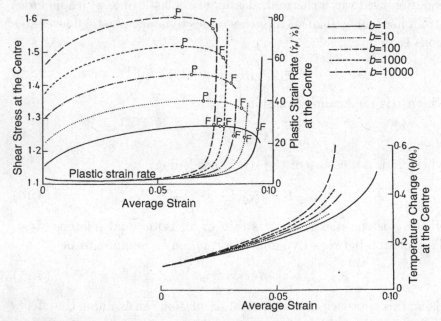

Fig. 8.7.   Effect of material rate parameter b on the evolution of the central stress, central plastic strain rate and central temperature. After Batra (1987b).

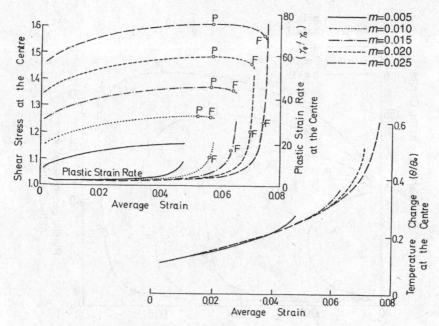

Fig. 8.8.   Stress–strain curve and the evolution of the central plastic strain rate and temperature for different values of rate sensitivity parameter m. After Batra (1987b).

indicate the peak of the nominal stress–strain curve and the point at which the shear strain rate at the centre starts increasing at an infinite rate, respectively.

The numerical simulations show that the existence of a peak in the shear stress acts as a necessary condition for the occurrence of adiabatic shear banding. More importantly, a higher shear strain rate sensitivity m and strain hardening index n can apparently retard the advance of shear localization towards its full formation (Figs 8.8 and 8.9). This is a significant result in practice.

Some features that derive from the simulations are worthy of closer study. Firstly, the parameter b is an indication of viscosity and coupled with the rate sensitivity parameter m, can lower the values of the critical strains for the peak stress and point F, the indication that an adiabatic shear band is fully formed. However, it seems that the value of $\gamma_F - \gamma_P$ is roughly constant, except for the unrealistic low value of b of unity, as shown in Fig. 8.7. Secondly, a very low

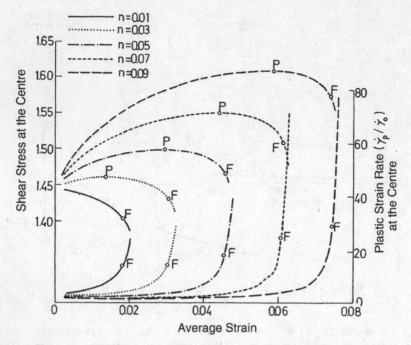

Fig. 8.9.   Stress–strain curve and evolution of the central plastic strain rate for different values of strain hardening parameter n. After Batra (1987b).

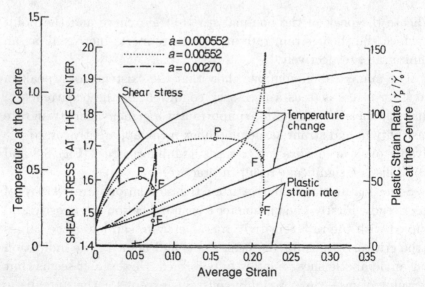

Fig. 8.10.   Effect of the value of the thermal softening coefficient a on the growth of the central stress, plastic strain rate and temperature. After Batra (1987b).

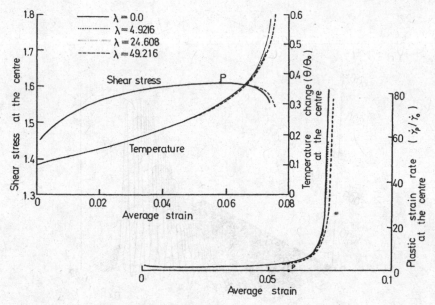

Fig. 8.11.   Effect of the thermal conductivity λ on the growth of the central shear stress, temperature and the plastic strain rate. After Batra (1987b).

thermal softening $a = 0.00055$, which is one order of magnitude less than the typical case, gives no peak in the stress–strain curve until a shear strain of 35% is reached. Also at $a = 0.00552$, the localized temperature at the centre shows a considerable increase (Fig. 8.10). However, halving the thermal softening gives a lower peak in shear stress and a smaller, shear strain for this peak. This contradicts the usual experimental observations. There is no simple answer to these contradictory results. Finally, the conclusions of the numerical simulations were that the thermal conductivity ranging from zero to 49.2 had little effect on the critical shear strain corresponding to the maximum shear stress. This is in agreement with most of the results that we have already mentioned. However, surprisingly this parameter also has little effect on the lapse between points P and F and even the width of the band.

Wright and Walter (1989) studied the effect of elasticity on adiabatic shear banding at the stage of intense localization. Elasticity could well help to provide an elaborate structure for the late stage

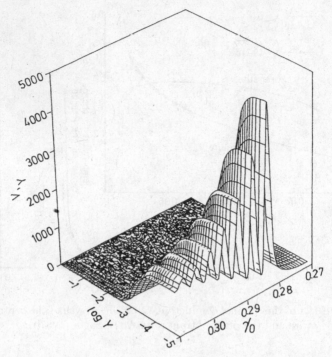

Fig. 8.12.   Velocity gradient for elastic perfectly plastic case with $\dot{\gamma}_o = 750/\text{sec}$. After Wright and Walter (1989), reproduced by permission of the Institute of Physics.

morphology. Figure 8.12 shows the results of their numerical simulations. It is interesting to note that severe oscillations appear when an intense shear band forms. The authors attribute this oscillatory behaviour to elastic unloading.

As well as the effects of material parameters on the evolution of adiabatic shear bands, which are very clear from the earlier numerical simulations of Costin *et al.* (1979) and Johnson (1981), the effects of various types of inhomogeneities, both geometrical and physical, appear to be very important. Despite the complicated effects of these inhomogeneities, numerical simulations appear to be the only means to deal with them. The following three groups of researchers concentrated on this problem and attempted to delineate the effect of the nature of the disturbance on the timing between the occurrence of adiabatic shear bands and the sudden collapse in shear stress: Wright and Batra (1985a,b), Wright and Walter (1987) and Xing (1988). The effect of voids on the formation of adiabatic shear bands was incorporated in the constitutive equation by Kobayashi and Dodd (1989).

Wright and Batra (1985a,b) demonstrated that the amplitudes of disturbances have a decisive effect on the time between the maximum shear stress and its sudden collapse. Their simulations were made on a complete system of momentum and energy equations with a constant driving velocity and adiabatic boundary conditions using the finite element method. They used two symmetrical humps of temperature in an initial disturbance. The humps were similar but had different amplitudes and widths. The larger amplitude hump was five times greater and broader than the smaller. The curves of nominal shear stress and strain obtained using the simulations are shown in Fig. 8.13. Both the homogeneous and disturbed curves of stress versus strain are shown and the letters I and P indicate where the disturbance was added and the location of the peak in the homogeneous curve, respectively. The shear stress peak acts as the threshold for the growth of disturbances. However, it seems that for a small temperature perturbation, the shear stress follows closely the homogeneous curve, even to strains beyond the critical shear strain at

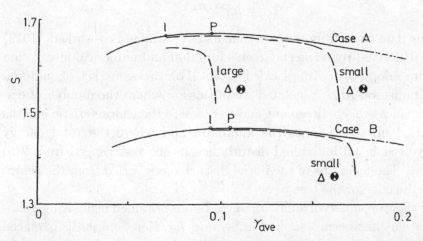

Fig. 8.13. Stress versus average strain (or time). Homogeneous and perturbed responses. After Wright and Batra (1985a,b).

which the maximum stress occurs (marked P). Thereafter, the curve drops suddenly. On the other hand, a large temperature perturbation has a much greater influence on the process; the shear stress deviates significantly from that expected for homogeneous response and collapses catastrophically at about the critical shear strain (marked P). The conclusion is that the amplitude of the temperature disturbances governs the strain between the shear stress peak and the stress collapse.

Kwon and Batra (1988) examined the effects of multiple disturbances in temperature on the evolution of shear bands using a similar model material and numerical technique as Batra (1987b). They reported that deformation localized at points where the perturbed temperature has relative minima when the average shear strain rate is $\dot{\gamma}_0 \approx 500$/sec but at points of relative maxima of perturbation when $\dot{\gamma}_0 > 1000$/sec (see Fig. 8.14). The explanation given for this difference was based on the different length and time scales.

Xing (1988) investigated the effects of amplitudes, energy and the distribution of temperature and strain rate on the evolution of adiabatic shear bands in more detail. He adopted the quasi-static assumption for the calculation but the boundary conditions were as before: constant velocity and adiabatic conditions. The constitutive equation used was

$$\tau = \mu \dot{\gamma}^m \gamma^n \theta^{-\nu}.$$

The data came from the experiments of Costin and co-workers (1979) on 1018 cold-rolled steel. Green's function and numerical integration were adopted in Xing's calculations. For the same set of material parameters, Xing calculated various cases: where the number of disturbances was between one and five, where the characteristic lengths varied up to 20 times the minimum and where the total energy involved in an individual disturbance in one test ranged from 2 to 20%. The locations of the initial disturbances varied from the centre to the edge of the test piece.

The evolution of adiabatic shear bands obtained using Xing's simulations has been described in Section 7.4. Here, emphasis is placed on the progress of disturbances with finite amplitudes. Figure 8.15 shows the typical evolution of two disturbances into a single shear

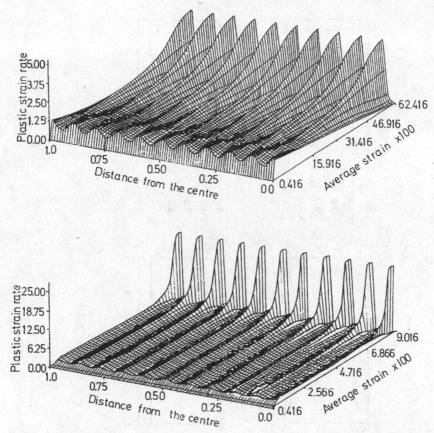

Fig. 8.14. Evolution of plastic strain rate at $\dot{\gamma}_0 = 500/\text{sec}$ (a) and $\dot{\gamma}_0 = 50000/\text{sec}$ (b). After Kwon and Batra (1988).

band. The two disturbances become flat and then merge; this is in fact delocalization. Comparing Figs 7.9 and 7.10 it can be seen that the stress in this stage is close to that for the homogeneous case. Later, a hump emerges from the centre of the plateau and becomes very sharp and narrow. This corresponds to the apparent divergence of the stress from the homogeneous distribution. The three-dimensional diagram shown in Fig. 7.11 demonstrates the temporal and spacial evolutions of temperature and strain rate; with the stress variation shown in Fig. 7.9, the relation between the variations becomes clearer. In the simulations, Xing reported that compared to other aspects of the disturbances, such as energy, width and so forth,

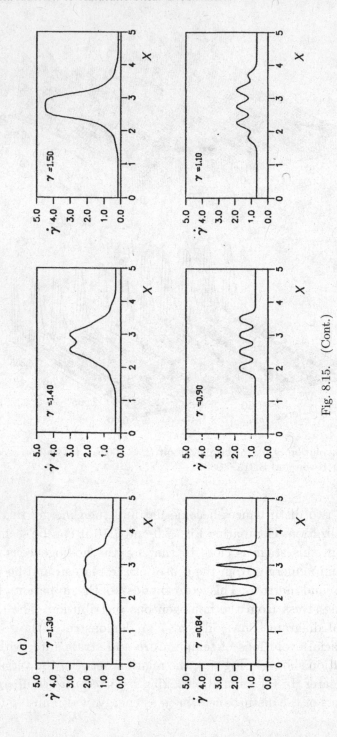

Fig. 8.15.   (Cont.)

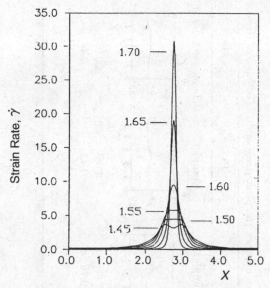

Fig. 8.15. The evolution of two disturbances into one fully developed adiabatic shear band. After Xing (1988).

the distribution of finite amplitude disturbances appears to play the most significant role in the formation of adiabatic shear bands. Generally, the quicker the merging of the disturbances, the shorter the time between the stress peak and the collapse of stress (see Table 8.2 and Fig. 8.16).

Kobayashi and Dodd (1989) examined the effect of imperfections by using a different approach. They studied the coupling of the nucleation and growth of voids with adiabatic shear. A finite difference technique was used in the study. The constitutive equation used was

$$\tau = A\dot{\gamma}^m \gamma^n (1 - c\theta). \tag{8.13}$$

The authors considered that in metals such as steels, second-phase particles and inclusions are potential positions for void nucleation either by particle cracking or by particle–matrix interface decohesion. Voids once formed weaken the material. Thus in the simulation it was assumed that at the instability shear strain γ_i voids nucleate at some point P in the gauge section. Then the local stress, according

Table 8.2.   Comparison of various disturbances and their corresponding times between the peak stress and the collapse of stress.

Work No	03	15	18	19	24
Disturbance distribution	Double	Triple	Complex	Complex	Long wave single
Scheme of disturbance distribution of temp.					
Disturbing energy	5%	7.5%	7.0%	3.6%	20.0%
t_c (time between the peak stress and the collapse of stress)	0.43	0.30	0.26	0.42	0.48

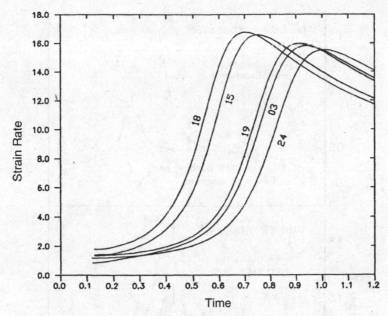

Fig. 8.16. Effect of the distribution of disturbances on the timing from stress peak to collapse. After Xing (1988).

to quasi-static equilibrium, should be

$$\tau_p = \frac{\tau_S}{S_p} = \tau \Big/ \left\{ 1 - \frac{a}{2\pi r_m} \langle \gamma - \gamma_i \rangle \right\} \qquad (8.14)$$

where $\langle \varphi \rangle = 0$ or φ depending on whether $\varphi < 0$ or $\varphi > 0$, r_m is the average radius of the tube cross-section for a thin-walled specimen and a is an amplification factor for void growth. For steel and commercial purity titanium a was taken to be between 3 and 6. Numerical simulations demonstrate that the shear bands can occur non-centrally (see Fig. 8.17). However, the band width appears to be thermal diffusion dominated (see Table 7.3). These results are similar to those obtained by Merzer (1982). However, the role of microstructural inhomogeneities in guiding the location of adiabatic shear band is revealed clearly in the later work.

Thus far the reader may have obtained the impression that our knowledge about the timing of the occurrence to the full formation of adiabatic shear bands and the effects of the various material parameters, inhomogeneities, imperfections and disturbances on the

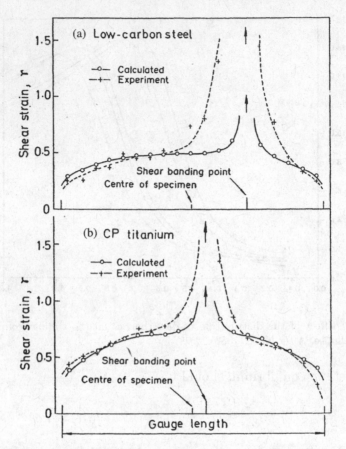

Fig. 8.17.   Comparison of calculated shear strain profiles along the gauge length
with experimental data. After Kobayashi (1987).

occurrence and timing has been widened. For example, which material parameters or which features of disturbances may retard or
accelerate the shear stress collapse and the full formation of the
shear band, as well as their relative significance in the process. But
actually some results are still contradictory. For example, does the
stress collapse and the rapid increase in local shear strain rate occur
simultaneously, or does the shear strain rate approach infinity or
become saturated at some stage? Above all, a full understanding of
the various aspects of adiabatic shear banding still has many questions to be answered.

8.3. Simulation with Adaptive Finite Element Methods

It has been shown that adiabatic shear banding is a type of transient and localized structure consisting of a number of stages. The unprescribed position of the shear band and the transition from one stage to another add to the difficulties to be solved in the numerical simulation.

Adaptive finite element techniques can automatically locate the localizing zones and adaptively concentrate elements at these zones to minimize the discretization error due to the large gradients in the narrow regions. Therefore, it is quite clear that this technique seems to be very helpful and powerful in the study of adiabatic shear banding. Generally, adaptive techniques can be divided into two groups; these are the moving grid method and the local refinement method. Both groups are rezoning techniques accompanying successive steps in time. In the moving grid technique a fixed number of computational elements are used, but the elements are moved to follow and resolve the localizing inhomogeneities. However, in the local refinement method, fine uniform grids are added in the localizing area when the discretization error in the original coarse grid becomes too large owing to the increasing gradients of the variables. Drew and Flaherty (1984) studied adiabatic shear bands using the moving grid technique. Batra (1987b) worked on the central part of a shear band using the adaptive method.

Drew and Flaherty (1984) used a moving finite element–Galerkin method to discretize the governing equations of one-dimensional simple shear, such as Eq. (6.25), by taking a grid of trapezoidal space-time elements, because the moving mesh technique is easier to implement than local refinement. The trapezoidal grid was moved to minimize the discretization error per time step approximately. The basic idea is asymptotically equivalent to distributing the local discretization error equally. Namely, for each time step, a mesh is selected so that the error is equal in every element. The details of the procedure to select the mesh are described by Drew and Flaherty (1984).

By using this moving mesh finite element method Drew and Flaherty calculated the following one-dimensional model of simple shear in a temperature-dependent, elastic–viscous material. The constitutive equation is as follows:

$$\tau = G(\theta)\gamma + \mu\dot{\gamma} \tag{8.15}$$

$$G(\theta) = \frac{G(0)}{2}\left\{(1 + G_\infty) - (1 - G_\infty)\tanh\left(\frac{\theta - \theta_m}{\Delta\theta}\right)\right\}. \tag{8.16}$$

Hence thermal softening occurs when $\theta \approx \theta_m$; when $\theta \to \infty$ the modulus $G \to G(0) \cdot G_\infty$, where $G(0)$ is the modulus at ambient temperature.

The boundary was considered as being at constant ambient temperature. One side of the slab was supposed to be fixed and on the other side a velocity-governing boundary condition was imposed:

$$v(t) = v_0 \tanh(t/t_0). \tag{8.17}$$

The front of an elastic wave should coincide with a velocity discontinuity, provided that the driving velocity $v(t) = v_0$, i.e. $t_0 \to 0$. Undoubtedly, viscous effects will be very important there. Viscous dissipation at the elastic wavefront converts the mechanical work into heat and leads to a temperature increase. In particular, the wave emanating from the driving boundary has its maximum strength there as well. Therefore, like wave trapping (Section 7.3), a localized temperature together with the shear deformation should be anticipated near the driving boundary.

Figure 8.18 shows the evolution and distribution of the temperature and velocity in a case study with a maximum driving velocity $v_0 = 0.5C_e = 0.5\sqrt{G(0)/\rho}$, i.e. half of the initial elastic shear wave velocity. This is a high velocity for materials and the computation lasts for a short time — the dimensionless time $t = L/C_e t_* \leq 0.25$.

In another example a lower driving velocity $v_0 = 0.25C_e$ was used. The computation time was taken to be ≤ 3.0 the effect of the reflective wave can be demonstrated. The temperature profiles are shown in Fig. 8.19(a). Although there is a bump near the loading boundary $X = 1$ at an early time $t = 0.5$, the temperature bump becomes smaller at $t = 1$ and 1.5. Subsequently, when the reflective wave returns to the area near the loading boundary, a rapid local rise in temperature occurs at $t = 2.5$. Because the temperature bump was

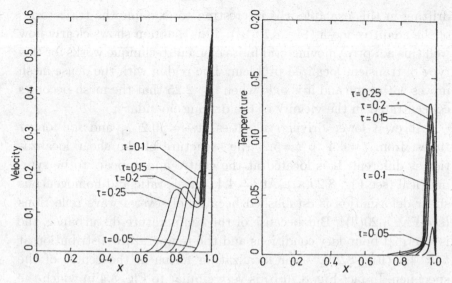

Fig. 8.18. The evolution and distribution of velocity (a) and temperature (b) in the case of high driving velocity $v_0 = 0.5\ C_e$ showing the trapping. After Drew and Flaherty (1984).

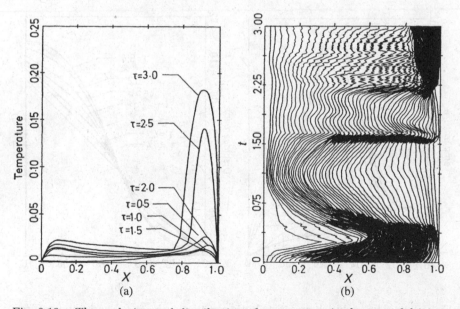

Fig. 8.19. The evolution and distribution of temperature in the case of driving velocity $v_0 = 0.25\ C_e$ (a) and the corresponding adaptive mesh trajectories (b). After Drew and Flaherty (1984).

drifting in this example it is interesting to examine the trajectories of the adaptive mesh (Fig. 8.19(b)). This diagram shows clearly how well this adaptive moving grid finite element technique works for this type of transient localized problem. The region with the dense mesh moves with time and it is only when $t \geq 2.25$ that the mesh becomes concentrated in the vicinity of the driving boundary. ·

At even lower driving velocities, $v_0 = 0.2C_e$, and for longer times, from $t = 4.4$ to $t = 5.2$, the structure of the shear localization is different. It is located at the centre and appears to be symmetrical (see Fig. 8.20(a)). At $t = 4.4$, quasi-static and homogeneous shear deformation is established because of several wave reflections (see Fig. 8.20(b)). But because of the temperature disturbance, the isothermal boundary conditions and the temperature distribution at $t = 4.4$ (Fig. 8.20), shear localization forms in the centre of the specimen. In fact Fig. 8.20(b) is very similar to Fig. 8.4 in which the dominant mechanism is heat diffusion as opposed to viscosity.

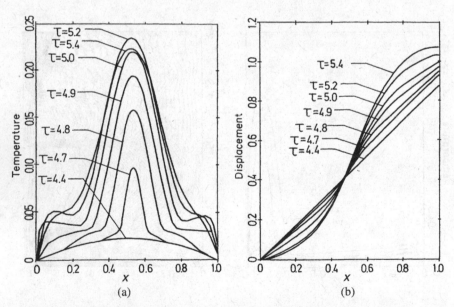

Fig. 8.20. The evolution and distribution of displacement (a) and temperature (b) in the case of driving velocity $v_0 = 0.2\,C_e$ showing a symmetric shear band. After Drew and Flaherty (1984).

The most interesting feature of the results of this adaptive finite element technique is that the two mechanisms predominate at different stages. In the early stages, wave and viscosity-diffusion governs the localization. But in the later stage of one model this gives way to the coupling of the deformation work and heat diffusion. Thus the relation between of these mechanisms becomes clearer. These simulations justify the concept that adiabatic shear banding is a process with many stages, from early wave trapping and viscous dissipation to coupling of plastic work and heat dissipation as proposed in the theoretical analysis given in Chapter 7.

8.4. Adiabatic Shear Bands in the Plane Strain Stress State

As mentioned previously, adiabatic shear bands occur in multi-dimensional stress states in practical engineering problems. In addition, the complicated geometrical configuration and constitutive equations in multidimensional objects tend to prevent analytical approaches going too deeply into the details of the features of the bands. That is, except for some characteristic aspects such as the instability condition, the growth rate of the disturbances as well as the width and spacing of the bands (see Sections 7.2 and 7.6). Therefore, in this case, the advantages and potential capacity of numerical simulation appear to be considerable. It is desirable that the exploration of the features and evolution of adiabatic shear bands in multi-dimensional stress states using numerical simulation will provide some evidence to show which mechanisms and controlling factors derived for the one-dimensional case still work for multidimensional situations. If this is possible, it would benefit greatly the applications of the concepts and theories developed for the one-dimensional case to realistic engineering practice.

Plane strain compression of a rectangular block is a typical configuration in mechanical analysis. It is also an important and representative testing method in the study of adiabatic shear banding (see Section 4.3). Therefore, the numerical simulation of the dynamics of adiabatic shear banding under dynamic plane strain compressive loading is, of course, instructive and significant. Moreover, an enormous amount of research has been carried out into shear localization in static plane strain compression, both analytically and numerically. Therefore, the results form a background to which a comparison of the simulation of dynamic compression can be made, increasing our understanding of the similarities and differences between static shear localization and adiabatic shear banding. Finally, it is hoped that it will be possible to extrapolate the techniques and description of plane strain compression to multidimensional configurations encountered in industry and that it may be possible to use the data for computer-aided design (CAD) and computer-aided manufacturing (CAM) systems.

The normal approach to the problem of shear localization in static plane strain compression is an analysis based on a material instability or imperfection. Numerical simulations in quasi-static isothermal plane strain compression concentrate on the appearance and development of shear bands from an internal imperfection. So far, then, numerical simulations of adiabatic shear bands formed in dynamic plane strain compression are within the framework of static, isothermal imperfection-induced shear localization. As is well known in static isothermal deformation, a yield surface vertex together with discrete crystallographic slip are the key factors dominating shear localization. Therefore, the aim of a simulation of adiabatic shear bands is to understand the interrelationship between multi-axial hardening and those factors dominating adiabatic shear bands in simple shear.

The results of several simulations on this subject have been published (see Table 8.1). The main points concerning the numerical simulations are listed in Table 8.3.

The finite element technique used in these simulations (Lemonds and Needleman, 1986a,b; Needleman, 1989; Le Roy and Ortiz, 1989) basically used four-noded quadrilateral elements consisting of crossed constant strain triangular sub-elements. This type of mesh has two advantages. One is its ability to represent the volume-preserving plastic strain rate without mesh locking. The other advantage is that the mesh, consisting of crossed triangles, can resolve narrow shear bands in four directions, i.e. parallel to either of the sides or the diagonals of

Table 8.3.   Finite element simulation of shear localization for 'dynamic' plane strain compression.

	Material model	Imperfection	Approximations	Remarks
Lemonds and Needleman (1986a,b)	Elastoviscoplastic solid with thermal softening	Periodic array of inclusions	Quasi-static with heat conduction	1. $\dot{\epsilon} = 500$/sec thermal softening is primary destabilizing factor 2. Shear band development is sensitive to the curvature of flow potential
Needleman (1989)	Elastoviscoplastic solid with strain softening no heat effect	Internal inhomogeneity	Dynamic no heat effect	1. In softening $\bar{\epsilon} > \bar{\epsilon}_m$ and further strain is in an ever narrowing band 2. Inertia retards shear band localization
Le Roy and Ortiz (1989)	Rate independent and rate dependent [pressure sensitive (Drucker–Prager hardening law due to friction)]	Internal inhomogeneity	Dynamic	1. Inertia and rate sensitivity delay shear localization 2. Inertia and rate dependence affect band width and spatial distribution

the element. Therefore, it is possible to reproduce localization oblique to the quadrilateral element.

The constitutive model used in the work of Lemonds and Needleman (1986a,b) and Needleman (1989) can be summarized as the following elastoviscoplastic one. The flow potential surface is $\sigma_f = \left(\frac{3}{2}\bar{s}_{ij}\bar{s}_{ij}\right)^{1/2}$, where $\bar{s}_{ij} = s_{ij} - \alpha_{ij}$ and s_{ij} is the Kirchhoff stress deviator:

$$\sigma_f = \left(\frac{\dot{\bar{\epsilon}}}{\bar{\epsilon}_0}\right)^m g(\theta, \bar{\epsilon}; \lambda_0) \tag{8.18}$$

where $\dot{\bar{\epsilon}}$ is the effective strain rate, $\dot{\bar{\epsilon}} = \left(\frac{2}{3}\dot{\epsilon}_{ij}\dot{\epsilon}_{ij}\right)^{1/2}$, and g is a hardening function. The hardening function can be chosen in several ways:

$$g(\theta, \bar{\epsilon}; \lambda_0) = \sigma_0 \left\{\lambda_0 \left(1 + \frac{\bar{\epsilon}}{\epsilon_0}\right)^n + (1 - \lambda_0)\right\}\{1 - \beta(\theta - \theta_0)\} \tag{8.19}$$

where $\bar{\epsilon}$ is the effective strain. The first bracket corresponds to strain hardening, which is isotropic if $\lambda_0 = 1$ and kinematic if $\lambda_0 = 0$. Further:

$$\dot{\alpha}_{ij} = \frac{2}{3}(1 - \lambda_0)bD_{ij}^p \tag{8.20}$$

$$b = \left(\frac{\dot{\bar{\epsilon}}}{\dot{\epsilon}_0}\right) \frac{dg(\theta, \bar{\epsilon}; 1)}{d\bar{\epsilon}} \tag{8.21}$$

where D_{ij} is the rate of total deformation given by

$$D_{ij} = D_{ij}^e + D_{ij}^p. \tag{8.22}$$

The hardening function can include strain softening to simulate thermal softening within the isothermal assumption (Needleman, 1989):

$$g(\bar{\epsilon}) = \sigma_0 \left(1 + \frac{\bar{\epsilon}}{\epsilon_0}\right)^n \bigg/ \left(1 + \frac{\bar{\epsilon}}{\epsilon_1}\right)^2. \tag{8.23}$$

Then, the normal von Mises viscoplastic relation can be written as:

$$D_{ij}^p = \frac{3}{2}\left(\frac{\dot{\bar{\epsilon}}}{\sigma_f}\right)\bar{S}_{ij}. \tag{8.24}$$

The data used in the simulation (Lemonds and Needleman, 1986a,b) are representative of structural steel: $\sigma_0 = 1250$ MPa, $E = 200$ GPa, $\nu = 0.3$, $\epsilon_0 = 0.003$, $n = 0.08$, $m = 0.01$, $\dot{\epsilon}_0 = 10^{-3}$/sec, $\rho = 7833$ kg/m^3, $c = 465$ J/kg K, $\lambda = 54$ W/m K, $\theta_0 = 293$ K and $\beta = 0.0016$/K. In Needleman's later simulation (1989), similar data were used except that $\epsilon_1 = 200\,\epsilon_0$ or $100\,\epsilon_0$.

Le Roy and Ortiz (1989) adopted a rate-independent and rate-dependent pressure-sensitive model with a Drucker–Prager type of monotonic hardening law. The dilatancy angle was null and the plastic response was non-associative in the calculation.

Let us examine the results obtained by these numerical simulations of plain strain compression. Firstly, Table 8.4 summarizes the main features of the quasi-static simulation made by Lemonds and Needleman (1986a,b). The calculations were performed on a square cell $2h_0 \times 2h_0$ with one inhomogeneity at the centre of the cell.

The first columns in Table 8.4 list the working conditions. $\dot{\epsilon}_n$ is the imposed nominal strain rate, $\lambda_0 = 1$ or 0 denotes isotropic or kinematic strain hardening and $\xi = \lambda/\rho c\dot{\epsilon}_n h_0^2$ is a dimensionless measure of the heat conduction. Case I corresponds to the higher

Table 8.4. Summary of two-dimensional FE simulation of adiabatic shear bands.

	Conditions			(U/h_0) σ_{max}	$(U/h_0)_{sb}$ discernible shear band	Band width	Remarks
Case	$\dot{\epsilon}_n$ (1/sec)	λ_0	ξ				
I	500	1	0	0.14	0.16	Mesh size controlled	
II	500	1	2.965×10^{-2}	0.18	0.20	$\sim \sqrt{\lambda/(\rho c\dot{\epsilon}_n)}$ $\sim 225\,\mu$m	
III	500	0	2.965×10^{-2}	0.19	0.20	Narrow well-defined	Abrupt drop in load
IV	50	0	0.2965		0.31	Broad	
	50	1	0.2965		0.31	Broad	
V	5	0	2.965		0.40		
	5	1	2.965	No sign of band until $U/h_0 = 0.6$			

In the table $\xi = \lambda/(\rho c\dot{\epsilon}_n h_0^2)$; U is the displacement at $x^2 = h_0$.

strain rate and adiabatic case. Case V corresponds to the low strain rate state with an apparent effect of heat conduction. Therefore, it is quite clear from the variations of the boundary displacement at the stress maximum $(U/h_0) < \sigma_{max}$ and the further boundary displacement at a discernible shear band $(U/h_0)_{sb}$ that the high imposed strain rate and the low heat conduction are the substantial factors for the occurrence, development and eventual morphology of a shear band. Compared to the above effects, the particular combination of isotropic and kinematic hardening has a secondary effect in high strain rate loading. As the strain rate decreases, the precise formulation of the flow function begins to play a greater role, as seen in case V. Moreover, a material that hardens kinematically is more prone to shear localization. This is in agreement with available static analyses of shear localization, i.e. shear localization is sensitive to the curvature of the yield surface. These results demonstrate that there is a gradual transition from static-isothermal shear localization dominated mainly by the curvature of the yield surface to adiabatic shear banding. In addition, the computations show a dramatic increase in shear band width with decreasing strain rate. This is consistent with the conclusion drawn from one-dimensional shear banding (see Section 7.5). Figures 8.21 and 8.22 show the deformed finite element meshes and contours of maximum principal logarithmic strain and temperature in case III.

Needleman (1989) simulated dynamic plane strain compression without a heat effect. Instead, a strain softening parameter was included in the computation, i.e. the parameter ϵ_1 in Eq. (8.23). When ϵ_1 decreases, the hardening function g decreases. As an illustration, Fig. 8.23 shows contours of constant effective plastic strain. The first impression that we get might be that the general features of shear band development under dynamic loading are similar to those occurring under quasi-static loading. However, the simulations show that a fingerlike contour of plastic strain emanates from the inhomogeneity at $45°$ to the compression axis, regardless of whether the material hardens or softens. However, for the softening model, beyond the critical strain where the hardening function reaches a maximum, further straining takes place in an ever narrowing band. It is also found that inertia retards shear band development. These two points are in agreement with linear perturbation in simple shear

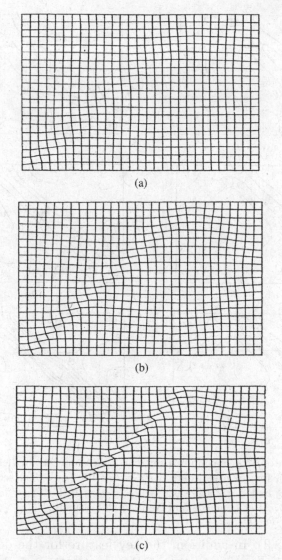

Fig. 8.21. Deformed finite element meshes for the case $\dot{\epsilon}_n/\dot{\epsilon}_0 = 5 \times 10^5$ ($\dot{\epsilon}_n = 500$/sec) with kinematic hardening and $\xi = \lambda/(\rho c \dot{\epsilon}_n h_0^2) = 2.965 \times 10^{-2}$: (a) $U/h_0 = 0.19$, (b) $U/h_0 = 0.21$ and (c) $U/h_0 = 0.23$. After Lemonds and Needleman (1986a,b).

(Section 6.2). Since the computation time is very short (see Fig. 8.23), only between 3.24 and 3.76, the consistency is reasonable. Needleman (1989) also noticed that an inhomogeneity is not necessary for initiating shear localization in the dynamic case owing to the wave effect

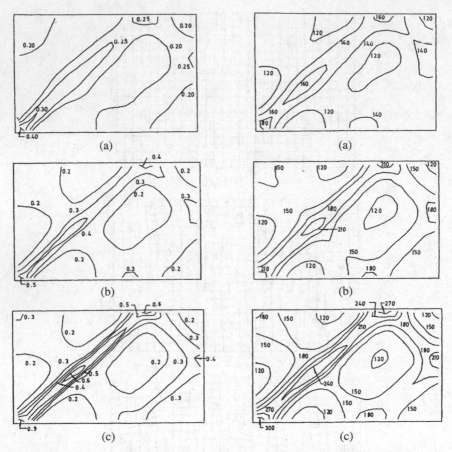

Fig. 8.22. Contours of maximum principal logarithmic strain (left) and temperature (right), for the case where $\dot{\epsilon}_n/\dot{\epsilon}_0 = 5 \times 10^5$ ($\dot{\epsilon}_n = 500/\text{sec}$) with kinematic hardening and $\xi = 2.965 \times 10^{-2}$: (a) $U/h_0 = 0.19$, (b) $U/h_0 = 0.21$ and (c) $U/h_0 = 0.23$. After Lemonds and Needleman (1986a,b).

(see Section 8.3). In addition, the key feature for the development of shear bands is the reduction of the propagation of plastic strain contours in the direction normal to the band. This agrees with the conclusion made by Hill (1962) and Rice (1977) about the coincidence of band-like localization and acceleration waves with vanishing speed (see Dodd and Bai, 1987). Finally, the main strain gradients occur across the band and large strain gradients do not develop along the formed shear band. This has been mentioned and applied in the shear band analysis (see Section 7.6).

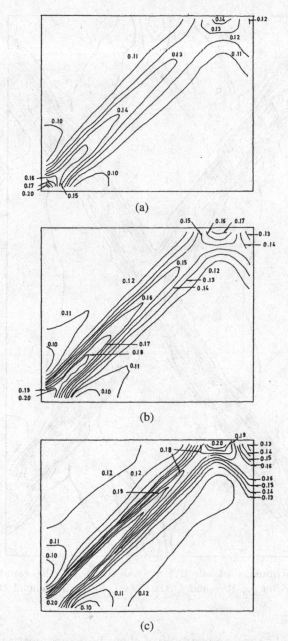

Fig. 8.23. Contours of constant effective plastic strain $\bar{\epsilon}$ for a softening solid $\epsilon_1 = 200 \times \epsilon_0$ in dynamic plane strain compression, driving velocity is $30\,\text{m/sec}$: (a) $t = 3.24\,\mu\text{sec}$, (b) $t = 3.5\,\mu\text{sec}$, and (c) $t = 3.76\,\mu\text{sec}$. After Needleman (1989).

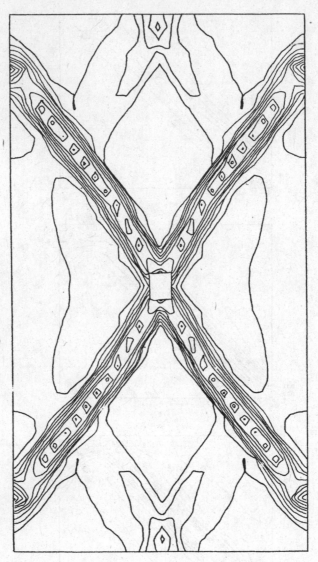

Fig. 8.24. Distribution of effective plastic strain for the rate-independent dynamic test. After Le Roy and Ortiz (1989), by permission of the Institute of Physics.

Figure 8.24 shows an example of shear band geometry taken from Le Roy and Ortiz (1989) for the dynamic rate-independent pressure-sensitive case. They found that the geometry of the shear bands, their width and spatial distribution were quite sensitive to rate dependence and inertia. It is interesting to note that the shear bands were non-orthogonal in the pressure-sensitive model. Compare this result with the analysis made by Anand *et al.* (1987) see Section 7.6.

From the numerical simulations of plane strain compression, it seems that the essential features, mechanisms and important factors in the two-dimensional cases are essentially the same as those given for the one-dimensional analysis. These corroborations potentially benefit our quantitative understanding of adiabatic shear bands in various configurations found in practice.

Further Reading

Batra, R. C. (2012) 'Analysis of adiabatic shear bands by numerical methods'. In *Adiabatic Shear Localization: Frontiers and Advances* (eds. Dodd, B. and Bai, Y.) pp. 173–214, Elsevier, London.

Wright, T. W. (2002) *The Physics and Mathematics of Adiabatic Shear Bands*, Cambridge University Press, Cambridge.

Wright, T. W. (2012) 'Theory of adiabatic shear bands'. In *Adiabatic Shear Localization: Frontiers and Advances* (eds. Dodd, B. and Bai, Y.) pp. 215–246, Elsevier, London.

Chapter 9

Selected Topics in Impact Dynamics

In the first three chapters of this book a number of phenomena and subjects concerned with adiabatic shear banding were introduced. Some characteristic features of adiabatic shear bands occurring in some typical loading conditions as well as the relation between adiabatic shear bands and fractures were discussed in detail in Chapters 2 and 3, respectively. Two important general conclusions can be drawn from these earlier parts of the book. Firstly, adiabatic shear banding profoundly affects the deformation process. In particular, it has a very significant effect on the variation of the supporting load, fracture and failure of materials, since cracks and voids either show distinct tendencies to follow the same patterns as the adiabatic shear bands or have pronounced links with the formation of shear bands. Secondly, it is very clear that adiabatic shear banding is a common and important mode of deformation and failure under dynamic loading. In the design and application of materials and structures subjected to impulsive, high-rate loadings, adiabatic shear banding is an important consideration.

The mechanisms and laws that govern the occurrence and formulation of adiabatic shear bands are understood to a degree. However, this knowledge cannot be considered in any way as equivalent to design standards. Each process, such as fragmentation, requires individual investigation, based upon a fundamental understanding of the essential underlying mechanisms and laws available. Similarly the data concerning the various aspects of adiabatic shear bands obtained by using various specially designed testing techniques and the constitutive equations outlined in Chapter 5 should be used properly in the study of selected topics in impact dynamics. However, the selection and choice of testing techniques and constitutive equations requires experience.

In this chapter, selected topics are chosen as case studies to illustrate how a specific engineering problem can be examined or quantified if possible, based on the physical ideas, concepts, mechanisms and rules of adiabatic shear banding described in previous chapters. The advances and understanding of the various topics are uneven. In some cases a large amount of data has been amassed and this has allowed a model to be proposed to quantify the phenomena. Even though in some cases the model is unsatisfactory at the moment, descriptions are helpful in understanding the interaction of shear bands with process variables and can introduce novel features.

9.1. Planar Impact

Planar impact is a fundamental technique in impact dynamics. It uses a light gas gun or explosive lens. Due to the large aspect ratios of the diameter to thickness of both the flyer and target plates, the stress state is assumed to be uniaxial, as shown in Fig. 9.1. Under the high pressure that occurs during an impact, often as high as 10 GPa, extreme compression is achieved in the direction of impact. This compressive deformation is normally attributed to uniform shearing in the material element (Fig. 9.1). However, a large quantity of convincing experimental evidence has been accumulated, suggesting that heterogeneous shear deformation occurs. This shear deformation takes the form of a network of very fine shear bands, which forms during the passage of the shock wave induced by the planar impact in the target material. Asay and Chhabildas (1981) first reported this phenomenon in 6061-T6 aluminium (see Fig. 9.2). Later Rohde *et al.* (1984) published similar results.

The characteristic feature of heterogeneous shear deformation in shock compression is a periodic pattern of fine bands in all grains. These bands have resolved spacings of about 2 to 10 μm, and a very fine width of between 0.1 and 0.3 μm (Fig. 9.2).

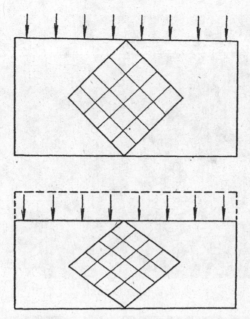

Fig. 9.1.   Schematic diagram showing the shear deformation occurring under uniaxial conditions.

Usually, the term adiabatic shear band is associated with macroscopically narrow shear bands due to thermal softening. As introduced in Chapter 2, the normal observations made in engineering practice show that adiabatic shear bands are non-crystallographic in nature. Furthermore, generally speaking, these shear bands usually cover tens or hundreds of grains. They then have a thickness of 10^{-4} to 10^{-5} m. In this sense adiabatic shear bands are different from these shear bands, like slip bands or a glide packet, both of which are essentially isothermal and take place along active crystallographic planes within a single grain.

However, we should note that the width of an adiabatic shear band is relatively sensitive to local strain rate (see Section 7.5). In the shock front created by a planar impact, the loading strain rate is given approximately by

$$\dot{\gamma} \approx \frac{u}{w} \approx \frac{u}{U t_r}$$

where u is the particle velocity behind the shock front, t_r is the rise time of the shock front and w is the width of the shock front. For a

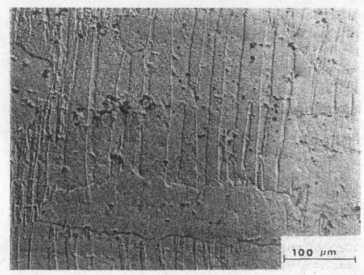

14.7 GPa shock

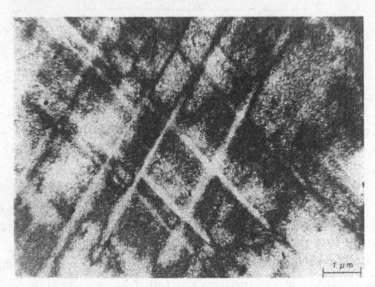

9.0 GPa shock

Fig. 9.2.   Optical and TEM metallographs of shock-deformed aluminium. After Grady *et al.* (1983), published by permission of Plenum Publishing Corp.

10 GPa shock wave in aluminium, $\dot{\gamma} \approx 500/6000 \times 10^{-9} - 10^8/\text{sec}$, where $t_r \approx 1\,\text{nsec}$. Consequently, it may be supposed that adiabatic shear banding may occur in an intergranular as well as in a transgranular manner. Based on this idea, Grady *et al.* (1983), Grady and Kipp (1985, 1987) and Swegle and Grady (1985) proposed that the mechanisms of rate-dependent dissipation and heat diffusion were predominant in adiabatic shear banding. These researchers proposed an approximate solution to the heterogeneous shear deformation appearing after a planar impact.

Before discussing the approximations made by Grady *et al.*, let us make some preliminary estimates of the network of the fine adiabatic shear bands under the unusual loading conditions obtained in planar impact, according to the fundamental concepts of shear bands that have already been described.

Estimates of the width of a shear band can be made for several stages (see formulae (7.28), (7.30) and (7.42)), provided that the heterogeneous shear deformation is due to the mechanism of thermal diffusion in adiabatic shear banding. For a planar impact with aluminium, the following data were used to make the estimation:

$$c \approx 10^3\,\text{J/kg K}, \quad \lambda = 200\,\text{W/m K}, \quad \kappa \approx 10^{-4}\,\text{m}^2/\text{sec}$$
$$\theta_* \approx 10^3\,\text{K}, \quad \tau_* \approx 500\,\text{MPa}, \quad \dot{\gamma}_* \approx 10^8/\text{sec}$$
$$\eta \approx 2 \times 10^3\,\text{Pa} \quad (\text{see Campbell, 1973})$$
$$t_r \approx 10^{-9}\,\text{sec} \quad (\text{see Barker, 1968}).$$

Then the estimates of the shear band width are

$$\delta \approx \sqrt{\frac{\lambda\theta_*}{\tau_*\dot{\gamma}_*}} \approx 2\,\mu\text{m} \tag{7.28}$$

$$\delta \approx \frac{1}{\dot{\gamma}_*}\sqrt{\frac{\lambda\theta_*}{\eta}} \approx 0.1\,\mu\text{m} \tag{7.30}$$

$$\delta \approx \sqrt{\kappa t} \approx 0.3\,\mu\text{m}. \tag{7.42}$$

Compared with the observed values of between 0.1 and 0.3 μm these seem fairly reasonable.

The spacing between adiabatic shear bands requires some thought. Firstly, the spacing is assumed to be much larger than the band width, as observation suggests. Due to the fact that the network of adiabatic shear bands, including the band width, the spacing between the bands and their morphology should form simultaneously, then the timescales for their formation should be the same. According to the discussion on scaling in adiabatic shear banding, the model for the spacing between the shear bands should be case 3 in Section 7.2, namely the adiabatic and momentum-dominated mechanism. The characteristic length scale is $(c\theta_*/\rho\dot{\gamma}_*^2)^{1/2}$, corresponding to the same timescale; $t_{k3} = t_{k2}$ in shear band formation should represent the spacing l.

$$l \approx (c\theta_*/\rho\dot{\gamma}_*^2)^{1/2}. \tag{9.1}$$

The same set of data for aluminium gives

$$l \approx \left(\frac{c\theta_*}{\dot{\gamma}_*^2}\right)^{1/2} \approx 10\,\mu\text{m}.$$

The agreement between this value and the observed spacing of between 2 and $10\,\mu$m is quite good. Both agreements in the band width and spacing appear to support the idea that the heterogeneous shear deformation occurring in the target material due to the shock wave in planar impact can be attributed to adiabatic shear banding, although the bands are extraordinarily fine. However, it should be pointed out that the above estimations were made using the one-dimensional model of adiabatic shear banding, which is clearly a different stress state to that found in uniaxial planar impact. Moreover, all the data used in the estimations are also very approximate, necessarily so as they are not the true values of the parameters at such ultra-high strain rates. Finally, at strain rates of $\dot{\gamma} \approx 10^8$/sec, the range of the effective Prandtl number Pr may not be in $O(10^4-10^5)$, in which case the assumptions made in Section 7.2 should be re-examined.

Now let us study the approach followed by Grady and his co-workers. Although 6061-T6 aluminium is a precipitation hardened alloy, there are indications that thermal trapping within the shear band prevails, at least for higher amplitude shock waves (Grady and

Asay, 1982). Therefore, the observed heterogeneous shear deformation was mainly attributed to adiabatic shear banding, rather than a pre-existing metallurgical microstructure. Further, for a steady shock wave compression, there must be a viscous stress component that emerges to diffuse the discontinuous shock. Also, due to the uniaxial strain configuration, both the pressure and the shear stress were established by the shock wave. Because compressive deformation can be regarded as volumetric strain under uniaxial conditions, then the shear in the shock wave is more important in plastic flow. An approximate measure of the maximum shear can be provided by the offset between the Rayleigh and the hydrostatic Hugoniot. For a 10 GPa shock in aluminium, the maximum shear stress is approximately 0.5 GPa.

Against this background, Grady and Kipp (1987) proposed that thermal trapping exists in shear bands, and a momentum diffusion, due to the shear stress relaxation within the band, would be transferred from the band to the surroundings. Then the momentum diffusion to adjacent shear bands forms the network of heterogeneous shear deformation in planar impact.

It has been assumed that the shear stress within the shear band decreases below its maximum. However, at this stage and in the timescale for the formed shear band the relaxation of the stress cannot be explained by wave propagation. Therefore, it is further assumed that this could be accomplished by slower momentum diffusion. In this case there should be an unloading interface $\xi(t)$ emanating from the band into its vicinity, which was originally deformed uniformly and plastically, $\dot{\gamma} = \dot{\gamma}_0$, $\tau = \tau_y$. The following discussion is concentrated on the material outside the shear band, not the material of the band itself. Therefore, the band width can be ignored and the shear band taken to be a discontinuity at $y = 0$. In addition, for simplicity, the unloading interface was supposed to be rigid-plastic. Therefore, from conservation of momentum in the region of $y = 0$ to $y = \infty$, the stress relaxation of the band and the momentum diffusion require

$$\frac{d}{dt}\left(\rho\dot{\gamma}_0\xi\xi + \int_\xi^\infty \rho\dot{\gamma}_0 y\,dy\right) = \tau_y - \tau(\psi). \tag{9.2}$$

The right-hand side of this equation is the difference in the stresses at $y = \infty$ and $y = 0$, where $\tau(\psi)$ is the related stress in the shear band. $\int_\xi^\infty \rho\dot{\gamma}_0 y \, dy$ is the momentum per unit area in the uniformly and plastically deformed material outside the unloading from $\xi(t)$, whereas $\rho\dot{\gamma}_0\xi\xi = \rho v \xi$, denotes the momentum per unit area in the unloaded and rigid body between the shear band and the front $\xi(t)$, where $v = \dot{\gamma}_0\xi$ is the current velocity of the rigid body. $\psi = \int_0^t v \, dt = \int_0^t \dot{\gamma}_0\xi \, dt$ is the displacement at the boundary of the shear band due to the relaxation. Supposing the relaxation function $\tau(\psi)$ is known, for simplicity let $\tau(\psi) = \tau_y(1 - \tau_y(\psi/w))$, where w is the energy dissipated in the shear band. We can simplify equation (9.2) to

$$\begin{cases} \xi\dot{\xi} = \psi\tau_y^2/\rho\dot{\gamma}_0 w \\ \dot{\psi} = \dot{\gamma}_0\xi \end{cases} \tag{9.3}$$

with the boundary condition $\psi = 0, \xi = \dot{\xi} = 0$, when $t = 0$. Denoting $\zeta = \dot{\xi}$, then $\ddot{\xi} = \zeta(d\zeta/d\xi)$, we can transform Eq. (9.3) to

$$\zeta^2 + \frac{\xi}{2}\frac{d^2\zeta}{d\xi} = \xi\frac{\tau_y^2}{\rho w}. \tag{9.4}$$

Then, the solutions to Eqs (9.4) and (9.3) under the prescribed boundary conditions are

$$\zeta^2 = \frac{2}{3}\frac{\tau_y^2}{\rho w}\xi \tag{9.5}$$

$$\xi = \frac{1}{6}\frac{\tau_y^2}{\rho w}t^2 \tag{9.6}$$

$$\psi = \frac{1}{18}\dot{\gamma}_0\tau_y^2 t^3/\rho w. \tag{9.7}$$

If $\psi = \psi_c = w/\tau_y$, the shear stress in the shear band is relaxed to zero, according to the assumed linear law of relaxation. Hence a limiting time t_c for relaxation and the limiting distance ξ_c of influence induced by the relaxation from expressions (9.6) and (9.7) are given by

$$t_c = \left(\frac{18\rho w^2}{\dot{\gamma}_0\tau_y^3}\right)^{1/3} \tag{9.8}$$

$$l = 2\xi_c = \left(12 \times \frac{w}{\rho\dot{\gamma}_0^2}\right)^{1/3} \tag{9.9}$$

where l is assumed to be the band spacing, being twice ξ_c, to provide a minimum separation for shear bands.

Now it is necessary to determine the parameter w, the energy dissipated in the shear band. So we have to consider the shear band from the point of view of the surroundings. Grady and Kipp (1987) proposed the mechanism for momentum diffusion between the shear band and the surroundings and they also assumed thermal trapping dominated the behaviour of the shear band. Recalling the discussion of scaling (Section 7.2), the two propositions, in fact, correspond to cases 3 and 2 respectively. Therefore, naturally, they adopted Eq. (6.25–2), i.e. the quasi-static heat diffusion model, to describe the shear band itself as

$$\bar{\theta} \approx \frac{1}{\rho c}\overline{\tau\dot{\gamma}_l} + \kappa\overline{\frac{\partial^2\theta}{\partial y^2}} \tag{9.10}$$

where $\dot{\gamma}_l$ denotes the local shear strain rate with the shear band and the bar implies the mean value of the quantity at time t_c. The mean values could be expressed as

$$\overline{\tau\dot{\gamma}_l}t_c = w/2\delta \tag{9.11}$$

$$\overline{\frac{\partial^2\theta}{\partial y^2}} \approx -\frac{\bar{\theta}t_c}{2(2\delta)^2} \tag{9.12}$$

where δ is the half-width of the shear band.

Then, if linear thermal softening is assumed,

$$\tau = \tau_y\{-\alpha(\theta - \theta_0)\} \tag{9.13}$$

at a limiting time t_c, the temperature increment

$$\bar{\theta}t_c \approx \frac{1}{\alpha} \tag{9.14}$$

to make the shear stress vanish. Combining (9.11), (9.12) and (9.14) the parameter w would be

$$w = \frac{\rho c(2\delta)}{\alpha}\left\{1 + \frac{1}{2}\frac{\kappa t_c}{(2\delta)^2}\right\}. \tag{9.15}$$

When coupling Eq. (9.15), which describes the interior of a shear band, to Eqs (9.8) and (9.9), which describe conditions outside the band, there are three equations but four unknowns: t, w, l and 2δ.

Grady and Kipp (1987) found that combination of Eqs (9.8) and (9.15) leads to the relation $t_c = t_c(2\delta)$, and there is a minimum time t_{c0}:

$$\frac{dt_c}{d(2\delta)} = 0. \tag{9.16}$$

Thus, the four equations (9.8), (9.9), (9.15) and (9.16) lead to the following solutions:

$$t_{c0} = \left(\frac{36\rho^3 c^2 \kappa}{\tau_y^3 \alpha^2 \dot{\gamma}_0}\right)^{1/2} \tag{9.17}$$

$$2\delta_0 = \sqrt{\frac{\kappa}{2} t_{c0}} \tag{9.18}$$

$$l_0 = \sqrt{\frac{2\tau_y}{\rho \dot{\gamma}_0} t_{c0}} \tag{9.19}$$

$$w_0 = 2\rho c(2\delta_0)/\alpha. \tag{9.20}$$

It is helpful to notice that the ratio of the band width to the spacing is

$$\frac{2\delta_0}{l_0} = \frac{1}{2}\left(\frac{\tau_y/\rho\dot{\gamma}_0}{\kappa}\right)^{-1/2} \sim Pr^{-1/2}.$$

This is exactly the same expression as for y_{k2}/y_{k3} in the discussion of scaling in Section 7.2.

During the course of stress relaxation in shear bands, the specific work done by an external stress is

$$\epsilon_0 = \frac{1}{\rho}\tau_y \dot{\gamma}_0 t_{c0} = \left(\frac{36\rho c^2 \kappa \dot{\gamma}_0}{\tau_y \alpha^2}\right)^{1/2}. \tag{9.21}$$

Comparison of expressions (9.19) and (9.21) leads to another more explicit formula for the spacing:

$$l_0 = \sqrt{\frac{2\epsilon_0}{\dot{\gamma}_0^2}}. \tag{9.22}$$

In fact, (9.22) is similar to the estimation of (9.1).

By taking the following data for aluminium under a 10 GPa shock wave loading:

$$\dot{\gamma}_0 \approx 10^8/\text{sec}, \qquad \tau_y \approx 500\,\text{GPa}, \qquad \alpha = 0.003/\text{K},$$
$$\rho = 2700\,\text{kg/m}^3, \quad \kappa = 9.7 \times 10^{-5}\,\text{m}^2/\text{sec}, \quad c = 900\,\text{J/kg K},$$

Grady and Kipp (1987) calculated the variables concerned with shear bands formed in planar impact: $2\delta_0 \approx 1\,\mu\text{m}$, $l_0 = 9\,\mu\text{m}$, $t_{c0} \approx 22\,\text{nsec}$. They thought that these values were somewhat larger than the observed results and there was still some discrepancy. However, the calculation provided some support for the view that adiabatic shear banding dominates the heterogeneous shear deformation induced by the shock wave.

This example clearly shows that adiabatic shear banding is a common phenomenon in impact dynamics, even under very different stress states and different loading rates. However, a proper identification of the relevant mechanisms underlying adiabatic shear banding is of great importance for successful analysis. Moreover, the fundamental concepts and mechanisms proposed and elucidated in Chapter 7 are fairly universal. Even very simple scaling rules can produce accurate estimations to actual problems in engineering practice.

9.2. Fragmentation

This section is concerned mainly with fragmentation in the military context, that is the fragmentation of rounds and similar devices. Fragmentation resulting from other phenomena, such as penetration, will be mentioned in later sections to complete the subject.

A fragmentation round consists of a hollow cylinder filled with explosive. Accompanying the detonation of the explosive inside the cylinder, shock loading acts on the wall of the cylinder and adiabatic shear bands and cracks form along and on the wall. When the shear bands and cracks intersect, fragmentation occurs. Figure 9.3 shows a schematic cross-section of a typical fragment formed from a cylinder wall. Usually cracks nucleate near the outer surface of the cylinder and grow inwards radially, whereas shear bands form near the inner surface and grow on 45° planes outwards.

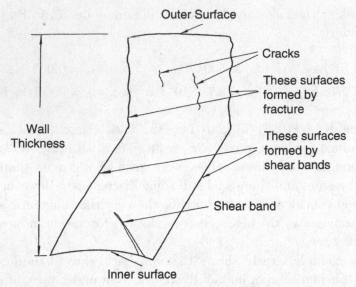

Fig. 9.3.   Cross-section of a typical fragment from an exploding cylinder. After Seaman (1983).

Stelly *et al.* (1981) and Curran *et al.* (1987) systematically studied the phenomenon by carefully designing explosion tests, a recovery device and afterwards direct observations and modelling. The former paid attention to the metallurgical aspects of fragmentation, whilst the latter attempted to develop a microstatistical model to stimulate the microscopic evolution of adiabatic shear bands to form the observed fragments.

Some of the observations made by Stelly *et al.* (1981) are introduced firstly to obtain a practical feeling for the relation between the fragments and adiabatic shear bands.

There are two experimental arrangements: spherical and cylindrical shells. From the records obtained by a fast image camera, the average strain rates in the two arrangements are $5 \times 10^3 - 5 \times 10^4$/sec and $10^5 - 5 \times 10^5$/sec, respectively. Two types of rupture are observed from recovered fragments. The first is fairly ductile, with a chisel edge, which occurs in copper and aluminium alloys and materials of higher elongation. The second type is relatively brittle, occurring in materials with lower elongations and is closely associated with adiabatic shear bands occurring typically in uranium and titanium alloys. In other materials such as steels, there is a mixed rupture mode. Typical shear bands in uranium and titanium are between 10 and 30 μm in width across grains. Rupture surfaces follow the trajectories of shear bands and form a regular network. More importantly, the greater the number of shear bands, the lower the material ductility.

The research group at Stanford Research Institute designed and developed a contained fragmenting cylinder to quantify shear banding in fragmentation. The reason for using this device was to make the sample undergo uniform amounts of stress, strain and shear banding. In fact, it is not in the fragmenting cylinder but within the central part of the cylinder that the states can be considered enough to draw comparable information. As a case study, the following is taken from this group's model of fragmentation (Erlich *et al.*, 1980; Seaman and Dein, 1982; Seaman, 1983; Seaman *et al.*, 1983).

The approach adopted by Seaman *et al.* is essentially a microstatistical study of a great number of shear bands. The model treats the statistical distribution of the adiabatic shear bands as a probability density for a material element and then correlates the relevant quantities to the fragmentation.

Firstly, the number of shear bands on the inner surface of the tested cylinder is counted. The cylinder was split into several short segments along its axis. In addition, within each segment plastic strains were assumed to undergo the same history and the damage was assumed to be homogeneous. Figure 9.4(b) shows a typical result of counting the surface density of shear bands greater in length than R. Then the surface density is converted into a volume-free density. The steps of the procedure are shown in Fig. 9.4. The volumetric distributions can be fitted into an exponential formulation and the two constants N_{g0} and R_n determined:

$$N_g = N_{g0} \exp(R/R_n) \qquad (9.23)$$

where N_g is the number of shear bands greater than or equal to R in unit volume. To model the fragmentation, we must combine the statistical distribution and macroscopic dynamics of deformation.

The plastic strains sustained by each axial element can be calculated in accordance with an elastic–plastic model. The correlation of

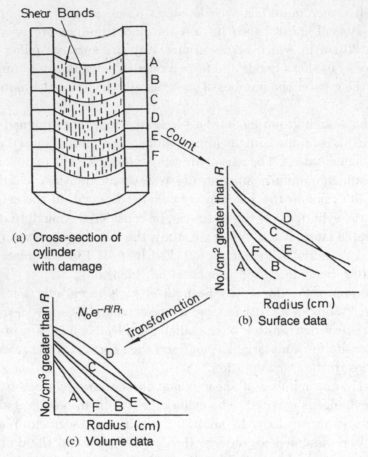

Fig. 9.4. Steps in obtaining cumulative shear band distributions from contained fragmenting cylinder data. After Curran *et al.* (1987).

the mechanical variable — the plastic strain — with the statistical microscopic variables of the shear bands — the density N_g and size R — can be given.

Figures 9.5 and 9.6 show these two correlations on a specified shear plane (45° between the radial and circumferential directions, termed mode 1). The correlations show extensive scatter. However, considering the microscopic statistics of shear bands in materials, it may be necessary to deal with this scatter. Using the plastic strains, it is possible to establish the statistical dynamics for the initiation, rate of nucleation, growth and coalescence of adiabatic shear bands.

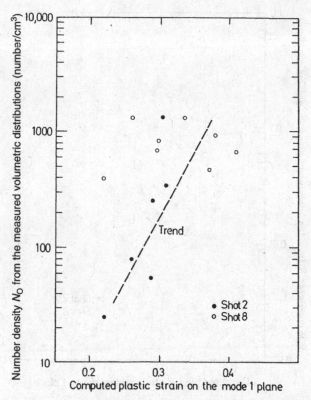

Fig. 9.5. Measured number of shear bands versus computed plastic strain for determining an initial estimate of the nucleation rate parameter. After Seaman (1983).

Based on the concept that adiabatic shear bands occur at a critical shear strain (see Chapter 6), it is supposed that in this multi-dimensional stress state, the initiation of adiabatic shear bands occurs in any orientation in a material element, only after the plastic shear strain on that plane with orientation s, γ_s^p, exceeds a critical value γ_c. In this model of fragmentation, γ_c should be specified by other testing procedures and then provided as an input parameter.

The nucleation rate of shear bands is determined by the total number of potential nucleations per unit volume N_0 the probability of shear banding f and the time required for the shear band to appear:

$$\dot{N} = N_0 f/t \qquad (9.24)$$

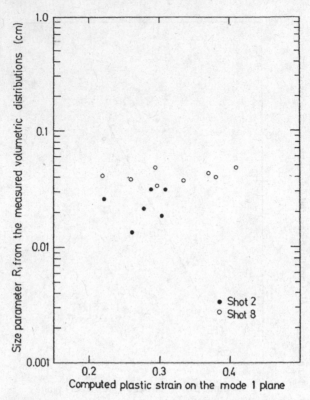

Fig. 9.6.   Measured size parameter for shear bands versus computed plastic strain for determining an initial estimate of the growth rate parameter. After Seaman (1983).

provided one specified orientation s is taken into account. Furthermore, the probability f is presumed to be a function of the plastic shear strain and strain rate:

$$f = f\gamma_s^p f\dot{\gamma}_s^p = H(\gamma_s^p - \gamma_0)\exp(-\dot{\gamma}_0/\dot{\gamma}_s^p)\qquad(9.25)$$

where H is the Heaviside function, γ_s^p and $\dot{\gamma}_s^p$ are the in-plane plastic strain and strain rate respectively and γ_0 and $\dot{\gamma}_0$ are two parameters. According to the definition of H,

$$H = H(\zeta) = \begin{cases} 1 & \zeta \geq 0 \\ 0 & \zeta < 0 \end{cases}.\qquad(9.26)$$

γ_0 defines the threshold for the nucleation of adiabatic shear bands whilst $\dot{\gamma}_0$ is a scale for the strain rate. If the strain rate $\dot{\gamma}_s^p$ is much less than the rate scale, it is impossible for shear bands to nucleate. The time t can be estimated as $t \approx pE_m/\sigma_y\dot{\gamma}_s^p$, where E_m is the specific melt energy. Therefore, the time t corresponds to the time necessary for plastic work to heat the material to its melting point. Finally, the nucleation rate can be estimated as

$$\dot{N} = N_0 \frac{\sigma_y}{\rho E_m} \dot{\gamma}_s^p H(\gamma_s^p - \gamma_0) \exp(1 - \dot{\gamma}_0/\dot{\gamma}_s^p). \tag{9.27}$$

A growth rule for shear bands is required. The rule suggested is similar to viscous growth of a void in a fluid:

$$\frac{dR}{dt} = cR_i \frac{d\gamma_s^p}{dt}. \tag{9.28}$$

The coalescence of shear bands is a very difficult problem to solve because of the interaction of the various shear bands. However, it is the most significant part of the whole model, linking adiabatic shear bands to eventual fragmentation. In order to propose a model, several simplifications were made. It is reasonable to assume that the shear bands are randomly oriented, thus the fragment size distribution must reflect the shear band size distribution, as sketched in Fig. 9.5.

In this model it is assumed that the number of fragments is related to the number of shear bands by a factor ξ:

$$N_i^f = \rho \xi N_i \tag{9.29}$$

where the subscript i denotes the ith size group. Because each crack can form two sides for two fragments and each fragment should have at least four sides, $\rho\xi$ should be less than a half; a value of $\frac{1}{3}$ or $\frac{1}{4}$ for ξ was suggested. Similarly, the size of fragment was also assumed to be related to the band size by a factor η:

$$R_i^f = \eta R_i$$

where η should be equal approximately to unity. Over all orientations and size groups, the volume fraction of fragments v can be estimated as

$$v \approx \sum_i N_i^f \frac{4\pi}{3} (R_i^f)^3$$

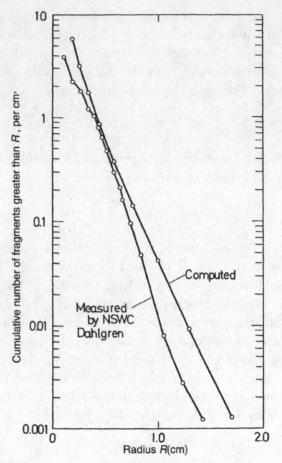

Fig. 9.7. Comparison of computed and observed fragment size distribution for sample fragmenting projectile. After Seaman (1983).

$$\approx \frac{4\pi}{3}\xi\eta^3 \sum_i N_i R_i^3$$

$$\approx \frac{4\pi}{3}\xi\eta^3 \sum_i \sum_s N_{is} R_{is}^3. \qquad (9.30)$$

Clearly, $v = 1$ indicates complete fragmentation.

The above formulations allow us to calculate fragmentations according to the statistical evolution and distribution of shear bands. However, it is necessary to calibrate this model to determine all the parameters involved in the formulae.

Calculations have been made for the two-dimensional case with the effect of normal stress on shear stress neglected. Figure 9.7 shows the comparison of calculated fragment size distribution with experimental measurements. In the two-dimensional case, only mode 1 shear bands were activated. In the computation, it was assumed that fragmentation occurs only by shear banding. Cracking initiated at the outer surface of the cylinder was completely ignored, despite the agreement between the two distributions being quite good. In addition, the total computed equivalent plastic strain imposed up to complete fragmentation was 60–80%: 40% of this was homogeneous deformation; the remaining 20–40% was accommodated by shear banding. Experimental measurements show that about 35% of the strain is homogeneous and is retained within the fragment.

This model of adiabatic shear banding and fragmentation could be improved because many simplifying assumptions were made. However, as an operational approach to engineering practice based on adiabatic shear banding, this is a good model.

This research group later modified their shear band model (Curran and Seaman, 1985) and a comprehensive review is available (see Curran *et al.*, 1987).

In engineering practice, shear-controlled fragmentation of warheads was investigated by Pearson and Finnegan (1981). A diamond grid system with a V-shaped groove was machined and formed onto the inner surface of a steel sample as a mechanical stress raiser. The orthogonal grid was designed to follow the trajectories of maximum shear. Controllable fragmentations were obtained because the shear fractures originated at the roots of the grooves. It was found that the size of the fragments was mainly dependent on the grid spacing as well as the shell wall thickness. This is an excellent example of the application of controllable local shear deformation and fracture.

9.3. Penetration

The aim of penetration mechanics is to design a target with the minimum weight and cost that can protect against a projectile impact and further to design a projectile that uses the minimum energy and material for target destruction. To achieve this goal, the major task is to clarify the mechanisms involved in the whole process of penetration, cratering, penetration and perforation and to understand the relevant material parameters governing these mechanisms.

Regardless of what failure mechanisms occur in projectiles or targets in the penetration process, the mechanical failure is induced by the kinetic energy of the impinging projectile. Therefore, generally, the ratio between the kinetic energy of the projectile and the characteristic strength σ_y of the target $D = \rho v^2 / \sigma_y$, is called the damage number and is used to identify the various deformation and failure modes in impact dynamics. Johnson (1972) outlined the range of the damage number D and the corresponding modes of deformation and failure (see Table 9.1). Ballistic penetration usually occurs when D is between about 1 and 10^2.

Depending on the specific penetration configuration, that is the plate material, thickness and so forth, and the geometry of the nose of the cylinder, there can be various modes of failure. These are spalling, enlargement and hole deepening by pushing of material aside, plugging, petalling, bulging, cracking, shear banding, as well as shattering. Perhaps, of the mechanisms mentioned above, hole enlargement and plugging are the most significant in penetration and eventual perforation. Above all, each failure mode is effective in a specific stage of the whole process of penetration. More details concerning ballistic materials can be obtained from the monograph by Laible (1980).

Table 9.1. Damage number D and deformation and failure modes.

ft/sec	$\rho v^2 / \sigma_y$	Regime
2.5	10^{-5}	{Quasi-static / Elastic
25	10^{-3}	Plastic behaviour starts
250	10^{-1}	Slow bullet speeds
2,500	10^1	Extensive plastic deformation — ordinary bullet speeds
25,000	10^3	Hypervelocity impact
—	—	Laser, electron beam

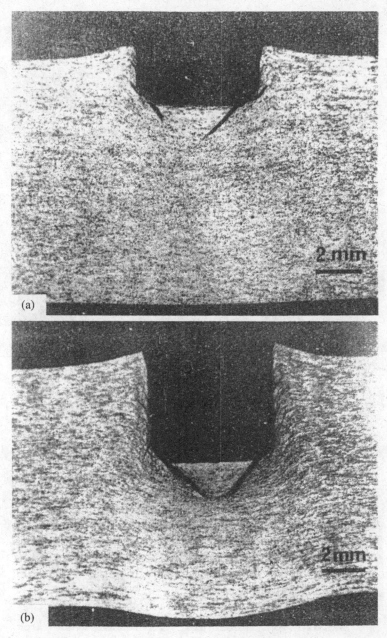

Fig. 9.8.   (a) and (b).

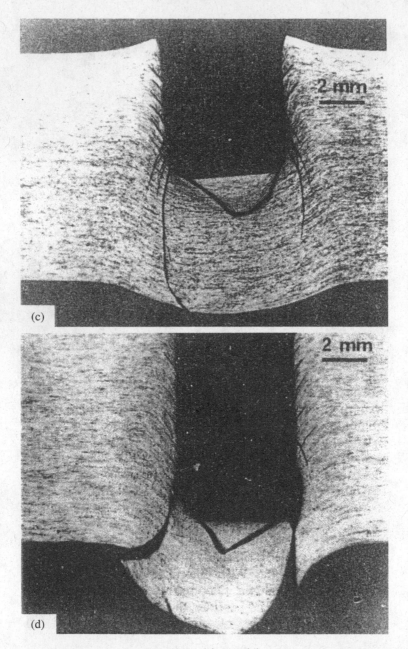

Fig. 9.8. (c) and (d).

Sectioned aluminium targets impacted by projectiles with a 1 mm corner radius. The cases are for different impact velocities: (a) 196 m/sec, (b) 257 m/sec, (c) 315 m/sec, (d) 353 m/sec. After Woodward *et al.* (1984), by permission of the Institute of Physics.

Plugging is a special mode of failure and is assumed to be dominated by adiabatic shear banding. This failure mode usually governs the perforation of medium-thickness plates, i.e. those for which the ratio of the projectile radius to the plate thickness is about unity. This deformation mode also dominates the final stage of perforation (see Fig. 9.8). Plugging occurs especially when blunt or round-nosed projectiles are used.

Apart from plate thickness and the geometry of projectile nose, which may favour plugging, a higher material strength can also favour plugging failure. This is because an increase in material strength, like an increase in plate thickness, increases the stiffness of the plate and then prevents the plate from bulging and petalling, as happens for a thin plate, and plugging is then the preferred failure mode. A material with a higher strength is prone to plugging more readily than materials of low strength. Therefore, there is a limit above which plugging becomes the dominant failure mechanism.

The preference for plugging failure implies that less energy is needed for this type of failure mode for a certain combination of geometry and material parameters of the projectile and target. Experience gained in penetration tests show that little plastic deformation occurs in the rest of the target and there is almost no lateral compression in the target material when plugging is the active penetration mechanism.

Penetration in fairly thick plates consists of two stages. Firstly, the target material is pushed away from the advancing projectile to form a penetration hole. When the thickness of the remaining target ahead of the projectile approaches the projectile radius, penetration changes from pushing material aside to plugging, which is a less energy-consuming failure mode.

Plugging and adiabatic shear banding are apparently important. From an analysis of the material parameters dominating adiabatic shear banding (see Sections 2.4 and 6.4), the penetration of high-strength targets through plugging may be attributed to their susceptibility to adiabatic shearing.

To identify the plugging mechanism and the response of material structures to impact, Rogers (1983) developed flat, stepped-nosed projectiles. A projectile was a cylinder with a blunt tip but connected

Table 9.2.   Comparison of shear deformations in various materials under different impact velocities.

Material	Condition	Impact velocity	Deformation	Reference
1018	Annealed	100 m/sec	Rather general plastic deformation	
1018	Cold-rolled (67%)	94 m/sec	Deformed shear band	Rogers (1983)
		100 m/sec	Transformed shear band	
4340	Annealed (Rc20)	800 m/sec	No shear band at all	Mescall and
4340	Rc52	800 m/sec	Transformed shear band	Papirno (1974)

to a relatively massive cylinder. The shoulders of the massive cylinder can stop further penetration of the blunt projectile when they impact the target. The tests carried out by Rogers (1983) and the ballistic observations of Mescall and Papirno (1974) were made using different steels, 1018 and 4340 steels, respectively, for annealed and cold worked states. Also different impact velocities were used ranging from 100 to 800 m/sec. Despite these differences, the results are quite similar (see Table 9.2). Therefore, the conclusion can be drawn from the stepped projectile tests that plugging does occur due to the initiation and development of localized adiabatic shear bands. The high-strength metal favours plugging results from its higher susceptibility to adiabatic shear banding.

Woodward *et al.* (1984) carefully examined the role of adiabatic shear bands in the formation of plugs. Two metals were tested, 9.5-mm-thick 7039-T6 aluminium plate and 6-mm-thick titanium–6% aluminium–4% vanadium plate, with the same flat-nosed steel projectile with a diameter of 4.76 mm. The radius of the corner end of the projectile ranged from 0.25 mm to 1.5 mm. Figure 9.9 demonstrates the dependence of plug thickness on the impact velocity for the two target plates. The corner radius appears to have a minor effect on the general features of plugging failure in the aluminium plate (Fig. 9.9(a)). However, the different radii produce a large scatter in the titanium alloy plate (Fig. 9.9(b)).

In the titanium plate, the thicknesses of the plugs were only about 10% less than the original plate thickness (Fig. 9.9(b)). This indicates that the compression before the full formation of the plug in

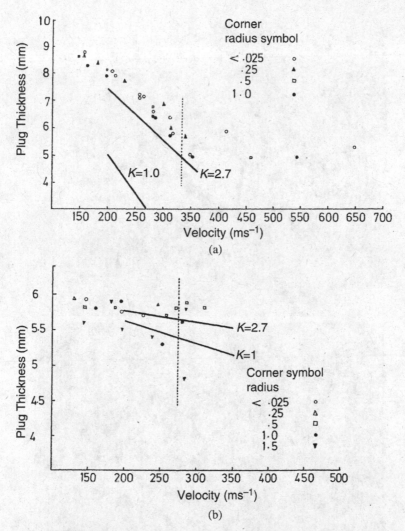

Fig. 9.9.   Plot of plug thickness as a function of velocity for (a) aluminium and (b) titanium alloy plates. After Woodward *et al.* (1984) by permission of Institute of Physics.

the titanium alloy was comparatively small. This point is confirmed because impact velocities from 150 to 270 m/sec produced a range of effects from slight indentation to complete perforation. Therefore, plugging failure dominated by adiabatic shear bands is decisive in the penetration and perforation of a titanium alloy plate. More careful examination of the adiabatic shear bands (see Fig. 9.10) shows

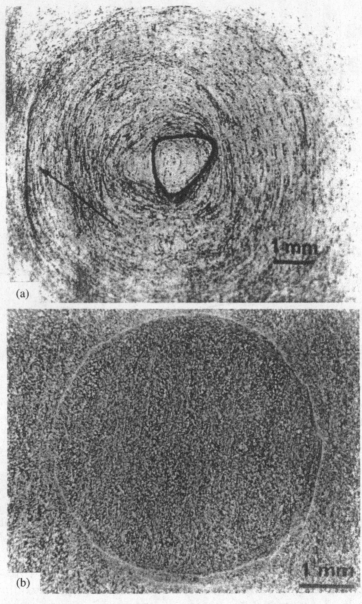

Fig. 9.10.   (a) A section cut just above the bottom of the dead zone of an aluminium plate impacted at a velocity of 313 m/sec showing that the shear band, associated with plugging, does not encircle the plug. (b) A section cut through a titanium alloy target impacted at a velocity of 213 m/sec showing that the plug is almost completely surrounded by shear bands, which have nucleated separately and joined at several places. After Woodward *et al.* (1984) by permission of the Institute of Physics.

that during the formation of the plug, the adiabatic shear bands not only propagate towards the rear of the plate (see Fig. 9.8), but also propagate circumferentially as Marchand and Duffy (1988) reported from torsion tests (Section 7.1). This indicates that in forming a plug, several shear bands initiate and then join circumferentially, forming an almost symmetrical circle (see Fig. 9.10(a)). This leads to the eventual separation of the plug from its surroundings.

The modes of deformation and failure during penetration and perforation of an aluminium plate are more complicated. The series of photographs of sectioned aluminium targets subjected to different impact velocities (Fig. 9.8) illustrates the rapid decrease in plug thickness with increasing impact velocity (Fig. 9.9(a)) and reveals clearly the extensive compression and lateral movement of material, before the plug associated with the primary adiabatic shear bands is eventually formed. Before the formation of the plug, some severe shear bands, inclined to the impact direction, form dead metal attached to the head of the projectile, similar to the theoretical slip-line field solution for indentation. These shear bands will be relevant to later fragmentation but not to plugging. Finally, unlike the titanium alloy plate, the adiabatic shear bands do not encircle the plug (Fig. 9.10(a)). The eventual separation of the plug from the target is accompanied by a ductile tearing mode of failure at the rear surface of the target. All of these features make penetration of an aluminium alloy more complicated. Even so, adiabatic shear banding is a major mechanism governing penetration in metals that are less sensitive to adiabatic shear banding.

There have been a number of attempts to use the above ideas in design calculations against penetration. For example, there are a number of empirical formulae concerning this type of calculation. De Marré's formula is a case in point:

$$\frac{WV_c^2}{D^3} = K \left(\frac{b}{D} \right)^{1.4} D^{-0.1} \tag{9.31}$$

where W is the weight and D the diameter of the projectile, V_c is the critical impact velocity for penetration, b is the target thickness and K is a constant involving material properties such as strength. The factor $D^{-0.1}$ this a correction factor for the scale effect. Because of the small power of 0.1 this is clearly assumed to be a secondary effect. Therefore, the formula is essentially an energy criterion for perforation:

$$\frac{WV_c^2}{D^3} = K\left(\frac{b}{D}\right)^m.$$

(9.32)

When $m = 1$, $WV_c^2 \propto bD^2$ implies that the kinetic energy of the projectile is mainly dissipated in the volume of the target, that is a cylinder of diameter D and height b. Compression of material ahead of the projectile is a result of this mode. If $m = 2$, then $WV_c^2 \propto b^2D$ implies the volume is much more dependent on the thickness. If the energy is consumed in an annular band of width w, then the energy balance should be $WV_c^2 \propto dDw$, $w = w(b, D, \ldots)$. Provided that $w \approx \sqrt{bD}$, then $m \approx 1.5$ and the energy criterion (9.32) is similar to de Marré's empirical relationship (9.31). This may be taken as an indication that shear bands are significant in perforation.

Backman and Goldsmith (1978) presented a comprehensive survey of the state of the art for penetration. Later, Wilkins (1980), Seaman (1983) and others carried out numerical simulations of penetration, consisting of indentation, compression, pushing material ahead of the projectile, plugged and so on, to examine the whole process of penetration and the relative importance of the various failure modes.

An estimation of perforation based upon adiabatic shear banding was proposed by Bai and Johnson (1982). The energy balance for this case is:

$$e = \frac{MV_c^2}{2} = \int_0^b \pi Db\tau dp$$

(9.33)

where τ is the shear stress at the periphery of the would-be plug and p is the current depth of penetration. If the following constitutive equation is assumed:

$$\tau = \tau_0(1 + \alpha\theta)\gamma^n$$

(9.34)

and the adiabatic approximation is made, the dependence of shear stress on shear strain becomes

$$\tau = \tau_m \left(\frac{\gamma}{\gamma_i}\right)^n \exp\left[\frac{n}{1+n}\left\{1 - \left(\frac{\gamma}{\gamma_c}\right)^{n+1}\right\}\right]$$

(9.35)

where γ_i and τ_m denote the critical shear strain and the maximum shear stress in adiabatic shear banding. Substitution of (9.35) into (9.33) gives the following expression for perforation

$$MV_c^2/\pi D^2 b\tau_m = f(b/(D/2))$$

$$= \int_0^{b/(D/2)} \left(\frac{1-n}{n}\frac{\xi}{\gamma_i}\right)^n$$

$$\times \exp\left[\frac{n}{1+n}\left\{1-\left(\frac{1-n}{n}\frac{\xi}{\gamma_i}\right)^{n+1}\right\}\right] d\xi \quad (9.36)$$

where $\xi = p/(D/2)$. Comparing this to (9.32), $f(b(D/2))$ corresponds to $(b/D)^{m-1}$. Some illustrative calculations are shown in Fig. 9.11. Clearly, titanium illustrates the worst dependence of target thickness because $f \approx$ constant, i.e. $m \approx 1$ when $b/(D/2)$ exceeds unity. Therefore, when designing target plates against perforations, adiabatic shear banding should be taken into account.

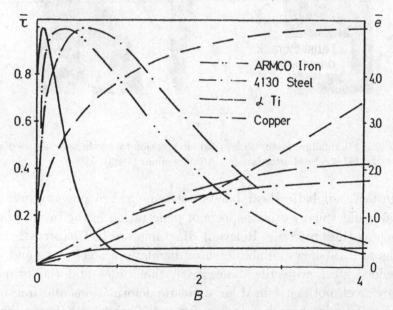

Fig. 9.11. Shear resistance and energy consumption for plugging versus the thickness of plug, according to a simplified adiabatic shear model. After Bai and Johnson (1981).

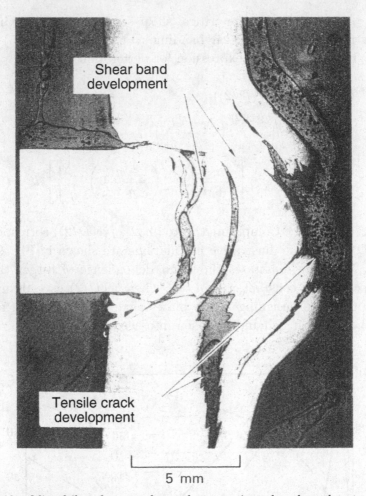

Fig. 9.12.   Microfailure features observed on a section taken through a steel plate and embedded steel rod after impact. After Seaman (1983).

In fact, adiabatic shear bands not only govern plugging failure but also affect many other aspects of penetration in the target plate as well as the projectile. Relevant diagrams are in Chapters 1 and 3. Figure 9.12 shows adiabatic shear bands in a steel plate and an embedded steel projectile. Since both the target and the projectile are steel, both sustained large plastic deformations and fracture where adiabatic shear bands had formed. Clearly, adiabatic shear bands are responsible for some microcracks and fragments formed in penetration.

Backman and co-workers (1986) investigated the scaling laws for adiabatic shear bands in penetration, by directly counting the shear bands observed in cross-sections of craters when the impact velocity was below the perforation velocity and counting fragments recovered after perforation above the perforation velocity. In their tests a scaling factor of 3 for the dimensions of the steel spheres and aluminium plates was assumed. The motivation for the scaling study as emphasized by Backman *et al.* (1986) is that it can provide insight into the mechanism for the initiation of fragmentation in penetration. For example, if the population of shear bands remains the same for any size of tests, initiation should be independent of any distribution of microstructures and it should be a scalar. Otherwise, initiation should be dependent on either the volumetric or surface distribution (cubic or square laws). However, these researchers found that the width of the adiabatic shear bands remains constant in their series of tests (see Section 7.5) and the ratio of the penetration depth to the projectile diameter conforms to this scaling. However, the population of the shear bands follows an approximate first-order scaling factor (Fig. 9.13) and fragmentation follows a square rule. The linear scaling

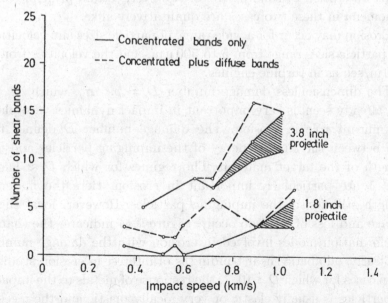

Fig. 9.13. Shear band counts as a function of impact speed for 1/8 and 3/8 inch scale thick-target systems. After Backman *et al.* (1986).

of the population of shear bands together with a roughly constant separation of slip surfaces may indicate that adiabatic shear bands nucleate in a one-dimensional way on the same surface, since the ratio of band spacing and band width can be expressed as an effective Prandtl number $Pr = V/\kappa$, a material parameter independent of size. If this is so, the square law in fragmentation should involve an additional mechanism.

9.4. Erosion

Erosion from surfaces of material by liquid or solid particles implies the granular removal of surface material by the flux of particles, continuously or intermittently. Although there is a monograph on erosion by Engel (1978), the subject of impact erosion, in particular the role of adiabatic shear bands in erosion, had not then been explored. Although there have been some observations of adiabatic shear failure occurring in polymers like PMMA and composites along the trajectories of maximum shear stress below the contact surface, the following discussions (Field and Hutchings, 1984) concentrate mainly on erosion by solid particles. However, certain principles and phenomena in these two cases are qualitatively alike.

Erosion may occur for a wide range of particle sizes and velocities. The particle sizes range from 5 to 500 μm, and the velocities from 1 to 10 m/sec as in turbine engines.

The dimensionless damage number $D = \rho v^2/\sigma_y$, which, as we have already seen, is very important in impact dynamics, also plays a significant role in erosion. The damage number D defines the ratio between the kinetic energy of the impinging particles and the strength of the target material. The regimes for which $D \approx 1$ and $D < 1$ are particularly important in erosion. Here the material strength still resists the impinging particles. However, local slight damage and loss of mass do occur. Figure 9.14 indicates the change of deformation modes involved in erosion with the damage number D. Clearly, adiabatic shear banding is of interest in erosion studies.

In cases for which $D < 10^{-5}$, the response of metals to the impacting particles is usually elastic or very locally plastic and the mechanism of erosion is of the fatigue type. However, for damage numbers in

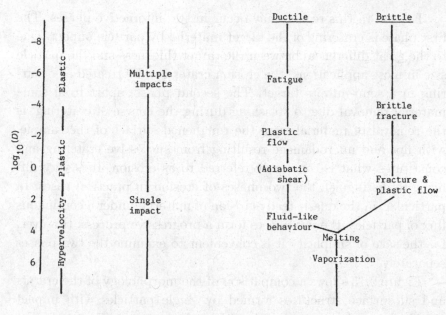

Fig. 9.14. Diagram indicating how as the damage number ($D = \rho v^2/\sigma_y$) is changed, the principal modes of deformation and mechanisms of erosion change.

the range $10^{-5} < D < 1$, corresponding to impact velocities of several hundred metres per second in metals, considerable plastic deformation can appear in the target materials. As far as the strain rate is concerned, one can expect a high value of the small sizes of the particles. Presumably the extent of plastically deformed material and velocity could be represented by the diameter and impact velocity of the impinging particle. A rough estimate of the plastic strain rate is $\bar{\dot{\epsilon}} \approx v/d$, provided the above-mentioned values of particle size and velocity are taken into account. Hutchings (1983) provides a formula for the strain rate

$$\bar{\dot{\epsilon}} \approx 0.18 \frac{v^{1/2}}{r} \left(\frac{3P}{2\rho}\right)^{1/4}$$

where r and ρ are the radius and density of the particle, respectively. P is the contact pressure. The calculated strain rate versus impact velocity for various sized particles ranges from 10^4 to 10^7/sec. Clearly, the high strain rates and local large plastic strain may favour the occurrence of adiabatic shear bands.

Erosion in this regime may occur in two interactive phases. The first phase is cratering of the target material by particle impact. Due to the great differences between the target thickness and the particle size in most applications, the erosion crater can be treated as occurring in a semi-infinite target. The second phase, apart from some material removal due to splashing during the course of cratering, is the removal of material from the roughened surface of the sample, with lips and microdamage resulting from successive cratering and constitutes what is commonly referred to as erosion. It is very difficult to distinguish the two phases of erosion in practical cases. In particular, in the quasi-steady erosion of material under a continuous flux of particles, the two phases form a progressive process. However, for the sake of simplicity it is convenient to examine the two phases separately.

Figure 9.15 shows a comparison of the morphology of the craters and subsurface structures formed by single particles with impact velocities ranging from quasi-static indentation to 7 km/sec. For impacts with velocities of several hundreds of metres per second against a material such as a titanium alloy, Fig. 9.15(c), a clear subsurface pattern of adiabatic shear bands can be seen. Adiabatic shear bands are also observed in some metals subjected to oblique impact (see Fig. 9.16).

Timothy and Hutchings (1984a,b, 1989) carried out a detailed study of adiabatic shear bands in a titanium alloy subjected to impact by a sphere, a cone and a wedge. When the impact was by a sphere, most of the shear bands lie on a conical surface. On the surface of the crater, there may be approximately circular steps indicating the intersection of shear bands with the surface. It was found that adiabatic shear bands were formed at nearly the same value of the ratio $a/r \approx 0.6$ for different sphere materials (see Fig. 9.17), where r is the radius of the sphere and a is the radius of the circular surface intersection of the crater. For wedge impact, Timothy and Hutchings (1989) obtained a similar threshold in terms of wedge angle for the occurrence of adiabatic shear bands (Fig. 9.18).

There are several important aspects to the pattern of adiabatic shear bands formed in erosion. Figures 9.15 and 9.16 clearly show that there are severe shear bands in the regions near the rim of the

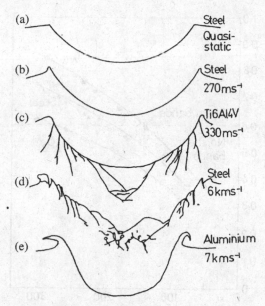

Fig. 9.15. Sections through indentations formed by the normal impact of 5 spheres single: (a) and (b) 9.35-mm-diameter hard-steel spheres, (c) 3.175 mm WC sphere, (d) water-filled polycarbonate sphere, (e) 0.25 g polyethylene projectile, diagrams not to scale. After Field and Hutchings (1984) by permission of the Institute of Physics.

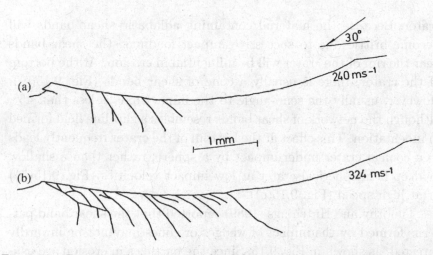

Fig. 9.16. Patterns of shear bands formed in Ti–6Al–4V by oblique impact (at 30°) of 3.175-mm hard-steel spheres: (a) at 240 m/sec and (b) at 324 m/sec. After Hutchings (1983).

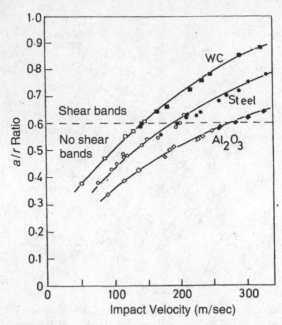

Fig. 9.17.   Variation of the ratio of the chordal radius of the indentation to the radius of the sphere with impact velocity for 3.175-mm spheres of different densities. Open symbols: no shear bands detected by optical metallography. Solid symbols: shear bands present. After Timothy and Hutchings (1981).

crater. Because the material containing adiabatic shear bands will become brittle due to successive impact loadings, the shear bands near the rim of the crater will be influential in erosion. At the bottom of the crater, there is usually a cone of shear bands (Fig. 9.15(c)). However, usually the semi-angle to the impact axis is less than 45°, although the network of shear bands resembles a slip-line field formed in indentation. This effect at the bottom of the crater frequently leads to a conical crater under impact by a sphere, rather than a shallow or deep hemispherical crater at low impact velocities (Fig. 9.15(b)) or at high speed (Fig. 9.15(e)).

Timothy and Hutchings (1989) reported that the shear band patterns formed by the impact of wedges or cones may not be inwardly directed, as shown in Fig. 9.15. Since the particles in erosion are usually angular, their observation may be important in practical erosion problems. After impacts by sharp wedges or cones, the adiabatic shear bands become outwardly directed to the free surface. If we

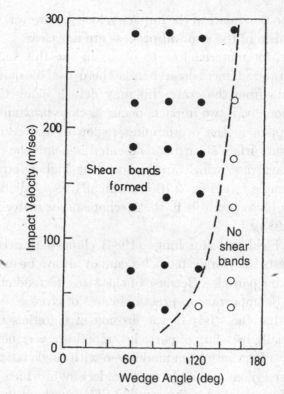

Fig. 9.18.   The threshold of shear banding induced by a wedge impact. After Timothy and Hutchings (1989).

visualize the relative positions of the surface of the wedge (or cone) and the shear bands, then the above phenomena become obvious.

At higher impact velocities, with $D > 1$, similar patterns of adiabatic shear bands and band-induced fracture are also observed (Field and Hutchings, 1984) (see Fig. 9.15(d)). However, it was reported by Shockey *et al.* (1975) that the pattern of shear bands seems to be dependent on the relative shock impedances of the impinging particle and target. In addition, the crater is usually rounded, whereas those closely related to shear bands are roughly conical, as discussed for cases of $D \approx 1$.

Removal of material by erosion is strongly dependent on the impact velocity. Generally, a good approximation for material loss is the so-called energy rule, i.e. the weight loss of material is proportional to the square of the impact velocity. However, the details of the mechanism of the removal process are not clear.

It is usual for material loss to occur in the thin extruded lip around the crater. Hence shear bands there can be quite influential. Fragments from the crater lip may detach along these shear bands. Additionally, if two impacts occur in close proximity, detachment can happen in the overlapping region of the craters. More severely, the subsurface shear bands located beneath the crater may form large fragments, which are removed at higher impact velocities (for example, see Fig. 9.16(b)). In all cases adiabatic shear bands play an important role in the mechanism for material removal (Hutchings, 1983).

However, Field and Hutchings (1984) claimed that conclusions drawn from tests with larger particles cannot always be extrapolated to erosion by fine particles. Because of this size effect, adiabatic shear is unlikely to be important in practical cases of erosion.

It seems that the study of the erosion of materials by particle impact, especially by a flux of particles, remains a very new field. So far, few observations have been made even with single large particles. However, there is some valuable work underway in which the effects of the flux of particles is being studied. The mechanism of erosion requires careful examination before it can be modeled accurately. Also, the role of adiabatic shear bands should be clarified as well, because the high strain rate and large local plastic strain favour their occurrence.

9.5. Ignition of Explosives

It is generally thought that the ignition of explosives is thermal in origin. Also, there is experimental evidence that the reaction occurs at local sites within the region undergoing bulk deformation. These local sites are called hot spots. These hot spots were presumed to be formed due to mechanical energy being converted into heat in these localized regions. Various mechanisms governing the conversion of mechanical work into heat, which produces localized high temperature, have been proposed.

The proposed mechanisms are: adiabatic compression of trapped gases, viscous dissipation, friction and adiabatic shear banding.

General introductions to this subject have been presented by Field *et al.* (1982) and Sandusky *et al.* (1984).

It has long been recognized that the ignition of explosives subjected to mechanical impact must occur at localized hot spots (Bowden and Yoffe, 1958). Bowden and Gurton (1949) even showed that the radii of the hot spots are typically between 0.1 and 10 μm. However, systematic experimental studies of the formation of hot spots together with relevant mechanisms were not made until some years later.

Heavens and Field (1974) reported on a series of drop-weight experiments on a sensitive pure explosive PETN (pentaerythritol tetranitrate) in which they studied the impact threshold of the explosive. In the tests high-speed photographs were taken and strain gauges were mounted on a hard steel roller to monitor the stress on the sample placed between other rollers. The recordings revealed a sudden drop in the gauge record, indicating mechanical failure of the sample had occurred at the moment of ignition of the explosive. Just before the failure, the radial velocity was only tens of metres per second. However, the expansion of the sample increased the radial velocity up to 300 m/sec before ignition. Their photographs showed that ignition occurs at localized spots. In addition, it was observed that ignition occurred during failure of the sample. Figure 9.19 shows a recording for RDX (cyclotetramethylene tetranitramine).

Based upon these observations and practical experience, it was proposed by Winter and Field (1975) and in the 1981 7th Symposium on Detonation by Swallowe and Field (1981), Coffey *et al.* (1981) and Frey (1981) that there are local zones that are subjected to higher strain rates than the bulk plastic zone. The shear bands become the ignition sites in the rapidly deforming explosives. Coffey (1984) even made further propositions on the ignition of explosives by adiabatic shear banding. He claimed that it was possible for hot spots produced in shear bands to serve as ignition sites for chemical reactions and it appears likely that at very high shear rates direct excitation of internal molecular vibrational modes is possible.

Figure 9.20 shows a schematic diagram after Elban and Armstrong (1981), which explains the connection between hot spots

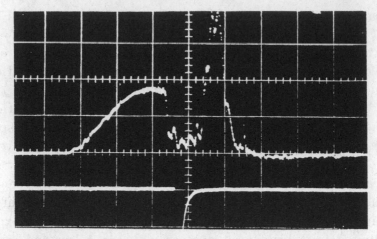

Fig. 9.19.　Stress–time relation during impact on a layer of RDX (upper trace), and the instant of explosion (lower trace), one square of trace = $100\,\mu$sec. After Field *et al.* (1982).

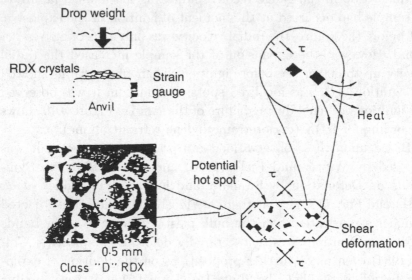

Fig. 9.20.　Microstructurally determined hot spots within impacted RDX explosive crystals. After Elban and Armstrong (1981).

and high strain rate shear bands in RDX crystals subjected to drop-weight loading. Although the combination of photographs and sketches gives a clear picture of the hot spots and shear bands, the evidence is still not conclusive.

Apart from the stress profile shown in Fig. 9.19, Field *et al.* (1982) provided more pictures that indicate the existence and role of adiabatic shear bands in impacted explosives. Figure 9.21 shows evidence of localized shear bands in a sample layer of the explosive PETN in a recovered unexploded sample (left) and there is a similarity between the pattern of shear bands formed and those formed in a similar sized disc of PMMA, which failed by shear under impact loading. Figure 9.22 provides further evidence for the adiabatic shear mechanism in the ignition of explosives with a series of photographs ranging from localized deformation to ignition. This impact test was also performed on a disc of PP surrounded by an annulus of PETN. There are some faint horizontal lines with a spacing of 0.25 mm illustrating shear banding from Fig. 9.22(b) onwards. Clearly the propagation of ignition is along bands. Therefore, the heat constrained in a shear band is assumed to be the trigger for the chemical reaction, which initially propagates rapidly along the shear band. Similar patterns of shear bands have also been observed in other impact tests with explosives in different configurations (Field *et al.*, 1982).

A further feature concerning the role of adiabatic shear bands in the ignition of explosives is that due to the effect of polymer additives to the explosive on the ignition itself (see Field *et al.*, 1984). It has been found that when explosives are mixed with certain polymers, they can become more sensitive to mechanical impact. This cannot be interpreted as a chemical effect of the polymer additive on the explosives. Rather, it is suggested that the thermal properties, deformation mode and its susceptibility to adiabatic shear banding affect hot-spot formation in explosives with polymer additives. As discussed in Section 2.5, there are two groups of polymers depending on their thermal properties and mechanical behaviour. Polymers in one group, like polycarbonate (PC), have low conductivities and low latent heats of fusion and often fail by shear banding or cracking. Polymers in the second group, like polypropylene (PP), have comparatively high conductivities and latent heats of fusion and usually deform in bulk. Obviously, the first group of polymers should be susceptible to adiabatic shear banding. Additionally, the thermal properties allow the 'hot spot' temperatures to reach much greater values than those of the softening points. A comparison of

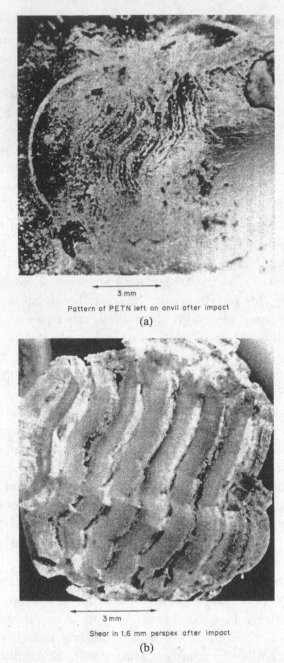

3 mm

Pattern of PETN left on anvil after impact

(a)

3 mm

Shear in 1.6 mm perspex after impact

(b)

Fig. 9.21.   A comparison of shear banding in impacted specimens: (a) a sample of PETN that failed to ignite; (b) a similar sized disc of PMMA viewed at the same magnification. After Field *et al.* (1982).

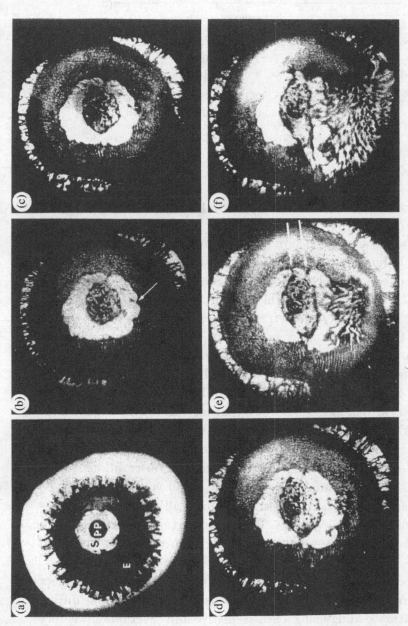

Fig. 9.22.   Impact on a PP disc surrounded by an annulus of PETN (E); the gas space is labelled S. Ignition is associated with instabilities on the inner boundary of the PETN, which cause jet formation. Examples are arrowed. Note the many shear bands, spaced about 0.25 mm throughout the PETN from frame (b) onwards. In the lower portion of frame (e) the propagation starts preferentially along a band. Mass of samples is 25 mg. Frame time is 5 μsec: (a) 0, (b) 75, (c) 82, (d) 89, (e) 96, (f) 103. After Field *et al.* (1982).

Table 9.3.   Comparison of the stresses and strains to
failure for HMX with two different binders.

Binder	% binder by weight	ϵ_f (%)	σ_f (MPa)
Polyurethane (PU)	2	0.50	1.22
	5	1.24	0.99
Polyethylene	2	0.09	0.67
	5	0.18	0.56

HMX; A: size range 50–1000 μm, median diameter *ca.*
400 μm. M: micronized (to pass 53 μm sieve). A/M:
ratio 63/35.

the stress–strain relation for the two groups of polymers was shown
in Fig. 2.17.

As modern explosives are bonded in polymer matrices (and
known as polymer-bonded explosives or PBX), it is important to
know the effects of polymer additives on the sensitivity of PBX.
Field *et al.* (1984) compared the stresses and strains to failure for
HMX (cyclotetramethylene tetranitramine) with different bonding
polymers. Here, two examples are cited in Table 9.3. Under the same
conditions, PBX with polyethylene as an additive is much more sensitive to mechanical impact than with polyurethane. After referring
to Section 2.5, it is clear that the susceptibility of some polymers to
adiabatic shear banding can have significant effects on the explosive-
ness of PBXs. Although the stress traces and photographic evidence
show that ignition of explosives occurs at local hot spots (Field *et al.*,
1982), no single mechanism can explain all the observations. Adiabatic shear banding, adiabatic heating of small volumes of trapped
gas, viscous flow, friction and so forth have all been supported as
mechanisms by some evidence. Therefore, further experiments and
theoretical analysis are required to study each possible mechanism
further.

The technique suggested by Coffey and Jacobs (1981) to detect
the temperature field of an inhomogeneously deformed sample during
short-duration loading using a heat-sensitive film (Section 4.5) is very
powerful. This technique has been used in the study of adiabatic
shear banding in polymers (see Section 2.5) and can be expected

to provide more information about the formation of hot spots in explosives.

From the theoretical point of view, Armstrong *et al.* (1982) and Coffey (1984) examined localized shear bands, assuming them to be produced by moving dislocations. According to the Orowan equation:

$$\dot{\gamma} = Nvb = N_{sp}\overline{N}vb \tag{9.37}$$

where b is the Burger's vector of a dislocation, v is the average dislocation velocity and N is the number of dislocations per unit area of the shear plane. After assuming that the shear band consists of N_{sp} slip planes, the thickness of the band is

$$2\delta_c \approx N_{sp}\,d \approx \dot{\gamma}_p/\bar{N}v \tag{9.38}$$

where d is the lattice spacing, $d \approx b$ and N is the number of dislocations per unit length of the slip plane. Elasticity theory of dislocations gives

$$\bar{N} \approx \frac{2(1-v)}{Gb}\tau\left(\frac{L}{l}\right)^{1/2} \tag{9.39}$$

where L is the size of the crystal and l is of the order of the distance of the lead dislocation from the edge. Then Coffey (1984) derived

$$2\delta_c \approx \frac{Gb}{2(1-v)\tau v}\left(\frac{l}{L}\right)^{1/2}\dot{\gamma}_p. \tag{9.40}$$

However, this estimation has not been verified by experiment so far.

Frey (1981) proposed a quasi-steady model of shear band development, which indicated there was a substantial increase in temperature within the shear band. Frey assumed the following one-dimensional quasi-steady, linear viscoplastic model:

$$r = \mu\gamma + \tau_0 \tag{9.41}$$

but the viscosity μ is presumed to be pressure and temperature dependent:

$$\mu = \mu_0 \exp(p/p_0) \exp(E/\theta - E/\theta_0) \tag{9.42}$$

where μ_0, p_0, θ_0 and E are material constants. The quasi-steady assumption leads to the following time-dependent heat conductive equation:

$$\frac{\partial \theta}{\partial t} = \frac{\lambda}{\rho c}\frac{\partial^2 \theta}{\partial y^2} + \frac{\dot{Q}}{c} + \frac{\tau \dot{\gamma}}{\rho c}. \tag{9.43}$$

$\dot{Q}$ is the heat generated per unit mass by chemical decomposition and is given by the Arrhenius equation:

$$\dot{Q} = QA\exp(-E_a/\theta) \tag{9.44}$$

where Q is the heat of reaction per unit mass and A is a frequency factor. The calculation was made for TNT for which the material parameters are:

$$\tau_0 = 0.07\,\text{GPa}, \quad \mu_0 = 1.39 \times 10^{-2}\,\text{kg/msec},$$
$$c = 1.29 \times 10^5\,\text{J/kg\,K}, \quad \lambda = 0.262\,\text{J/m\,K}, \quad \rho = 1.64 \times 10^3\,\text{kg/m}^3,$$
$$E = 3880\,\text{K}, \quad p_0 = 0.165\,\text{GPa}, \quad Q = 300\,\text{cal/kg},$$
$$A = 2.51 \times 10^{10}/\text{sec}, \quad E_a = 17,200°\text{C}.$$

Figure 9.23 shows one of the calculated results of the temperature rise due to the mechanical work done on the shear band. Clearly, the

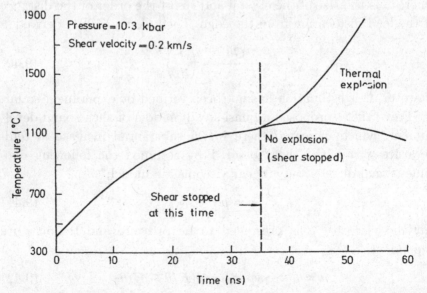

Fig. 9.23. Temperature versus time in a shear band, which proceeds to thermal explosion. If the shear is stopped before explosion occurs (at the time indicated by the dotted line), the reaction quenches. After Frey (1981).

local high temperature can initiate detonation at a driving velocity of 200 m/sec. The time required for ignition is less than a microsecond and the width of the band is typically less than one micron. Because of the exponential pressure dependence of the viscosity (Eq. (9.42)), an increase in pressure will greatly decrease the shear rate required to achieve a thermal explosion.

9.6. Explosive Welding

Since the discovery that two metal plates could be welded together by explosive loading in World War II, explosive welding has been recognized as a useful joining process in industry. In explosive welding, two plates are exploded together. A layer of explosive is placed on the back of one plate and the explosive is then detonated. The detonation causes a tremendous acceleration of the plates together. The plates impact and the bonded zone is usually thought to be hydrodynamic in nature. The bond often has a characteristic wavy pattern, which appears to be uniformly spaced along the length of the bond. The bond zone itself is fairly narrow. Although there are a number of theories for the formation of interfacial waves in the bonded region, generally the idea that explosive welding is essentially a solid-phase welding process, perhaps with a thin solidified layer at the interface, is now widely accepted.

The practical requirement in explosive welding is to obtain good welds with optimum quality. It is necessary to understand the structure of the bonding zones and the mechanisms of their formation.

There are still a number of unsolved problems concerning the formation and the structures of the bonding zone, e.g. does the wavy bond form just behind the point of collision of the flyer and base plates and is the formation of a jet a prerequisite for the bond? Due to the difficulties in making transient observations of the welding process and the complexities of the phenomenon, it is usual to obtain and use indirect evidence. This could be one reason for the existence of so many different theories of explosive welding.

In this section only facts relevant to the effects of adiabatic shear bands on the quality of the bonding zone will be discussed. Readers intending to obtain a comprehensive description of explosive welding,

Fig. 9.24.   Two types of adiabatic shear bands produced by explosive welding. Courtesy Zhang (1986).

should refer to the following sources: Crossland and Williams (1970), Johnson (1972), Carpenter (1981), and Cheng (1986).

Two types of adiabatic shear band have been observed in explosive welding, similar to those that occur in metal-cutting (see Fig. 9.24). One type of band is located in the bonding zone and secondary bands are inclined to the bond surface.

Detailed observations were made by Hammerschmidt and Kreye (1981) using a transmission electron microscope (TEM) and diffraction of the bonding zone in Al–3%Cu–Al–3%Cu claddings. Figure 9.25 is a schematic illustration of the structure of the bond together with the corresponding TEM photographs and diffraction patterns.

The centre of the bond zone consists of ultrafine equiaxed grains with diameter less than $0.3\,\mu m$. Similar fine equiaxed grains, 0.5 to about $1\,\mu m$ in diameter, were observed in the central primary shear band in the bonding zone of low-carbon steel (Zhang, 1986). The widths of the bands of fine grains are between 0.5 to $5\,\mu m$ and about $5\,\mu m$ in Al–3%Cu (Hammerschmidt and Kreye, 1981)

Fig. 9.25. Representation of the bonding zone microstructure in Al–3Cu–Al–3Cu claddings: (a) schematic illustration; (b) and (c) ultrafine grained centre zone U; (d) and (e) severely deformed zone containing extremely elongated particles E; (f) and (g) low deformation region D. 50 μm from collision plane. After Hammerschmidt and Kreye (1981).

and in low-carbon steel (Zhang, 1986). Outside the fine-grained zone in Al–3%Cu there is a severely plastically deformed zone containing extremely elongated grains oriented along the collision direction, reaching up to ten times the elongation of the average grain. Outside this region the material undergoes very little, if any, change in microstructure.

Similar severely plastically deformed narrow zones were also observed by Sek (1988) in tube welding of stainless steel and aluminium bronze. The thickness of the zone was between about 15 and 20 μm. Taking all these observations together leads us to the general conclusions that the centre of the bond consists of ultrafine and equiaxed grains, several microns in width. Outside this central region, there is a severely distorted zone where grains have undergone large plastic distortions.

Surprisingly, Hammerschmidt and Kreye (1981) found that in the outer distortion zone the grains are distorted by a factor of more than 3 in a direction transverse to the collision direction. This occurs without any measurable macroscopic contraction of either the flyer or base plate (Fig. 9.26). The severe plastic distortion can only be understood as complicated and severe shear deformation. Either the primary shear band is parallel to the collision direction, i.e. the bonding plane, or secondary inclined shear bands are also part of the distortion process. Therefore, plastic shear deformation and shear strength cannot be ignored in a discussion of bond formation. In fact, the existence of adiabatic shear bands usually make the material more brittle (see Chapter 3). Zhang (1986) found that when the inclined

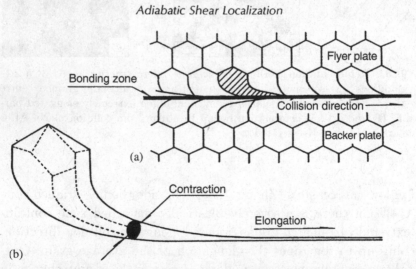

Fig. 9.26. Schematic representation of the deformation of grains forming the collision interface. After Hammerschmidt and Kreye (1981).

secondary shear bands were suppressed, the solid-state bond due to the fine equiaxed grain zone was more likely in welding.

Hammerschmidt and Kreye (1981) thought that the central zone or core consisting of ultrafine and equiaxed grains was formed by fast melting due to the deformation work and the subsequent extremely rapid cooling. The reasons are that the grains are equiaxed and have a random distribution of orientations. In addition, there is a distinct border between the equiaxed grain core and the outer elongated and aligned particles.

Hence, the equiaxed grain core plays a key role in the bonding zone. Because of the extremely fine grain size, the strength of the bonding zone can exceed that of the parent plates appreciably. This is certainly advantageous in practice.

Comparison of the fine grain bonding zone with transformed adiabatic shear bands is instructive. Both consist of fine grains less than $1\,\mu m$ in diameter. The width of transformed shear bands is about $10\,\mu m$ and the grains are nearly equiaxed (see Section 2.3). These similarities lead to the supposition that the bonding core is a special type of transformed shear band. However, this supposition does need more evidence.

Further Reading

Rosenberg, Z. and Dekel, E. (2012) *Terminal Ballistics*, Springer Verlag, Berlin.

Walley, S. M. (2012) 'Strain localization in energetic and granular materials'. In *Adiabatic Shear Localization: Frontiers and Advances* (eds Dodd, B. and Bai, Y.) pp. 267–310, Elsevier, London.

Walley, S. M., Field, J. E. and Greenaway, M. W. (2006) 'Crystal sensitivities of Energetic Materials', *Mater. Sci. Tech. Ser.*, **22**, 402–413.

Walley, S. M., Balzer, J. E., Proud, W. G. and Field, J. E. (2000) 'Response of thermites to dynamic high pressure', *Proc. Royal Soc. London, Ser. A*, **456**, 1483–1503.

Walley, S. M., Siviour, C. R., Drodge, D. R. and Williams, D. M. (2010) 'High-rate mechanical properties of energetic materials', *JOM — J Met*, **62**(1), 31–34.

Chapter 10

Selected Topics in Metalworking

We have seen from the early work of Tresca (1878) and Massey (1921) on hot forging, that high temperatures can be produced by large plastic strains. Very high temperatures can be produced for short times in certain processes. In most metalworking processes complex inhomogeneous deformation fields may be present. Much of the work of plastic deformation in bulk forming operations is caused by localized shear along fairly narrow bands. These bands are normally quite stable during plastic deformation and it is only under relatively extreme circumstances that they will embrittle and crack.

In sheet metal-forming operations, limitations to plastic deformation are normally set by tensile necking, either of the diffuse or localized type. These are well-recognized defects and they can be explained using isothermal plasticity theory. In bulk forming operations, such as forging, extrusion and rolling, necking is suppressed because the strains are predominantly compressive and very large plastic strains can be attained by a workpiece. It is when the plastic strains are large that shear bands are normally possible.

There are so many metalworking operations that it is almost impossible to classify them completely. However, attempts have been made to classify the major types of processes, for example, by Lange (1985).

10.1. Classification of Processes

Although there are a multiplicity of processes, they can be divided according to the differences in effective stresses. Processes are then grouped into: (a) compressive forming, (b) combined compressive and tensile forming, (c) tensile forming, (d) forming by bending, and (e) forming by shearing. Examples of compressive forming processes are open and closed die forging as well as rolling. Forming techniques utilizing compressive and tensile stresses are flange forming, deep drawing, spinning and bulging. Tensile forming methods include stretch forming and expanding. Bending is self-explanatory and shearing processes include blanking.

From this classification of forming processes it is clear that if adiabatic shear bands occur in any processes then they should appear in either compressive or combined compressive and tensile metalworking processes. That is, stress states which allow large accumulations of strain before unstable plastic flow occurs.

An alternative classification system for metalworking processes has been proposed by Kudo (1980). Kudo uses a number of categories including stress state as shown in Table 10.1. This method of classification is obviously more general than the one used by Lange, because it includes sequence of deformation, type of tooling, state of stress, deformation mode, working temperature, speed and starting shape of the workpiece. Kudo stresses that the state of strain undergone by a workpiece can be readily seen and understood, unlike the state of stress. In Table 10.1 the state of strain can, be designated according to directions 1, 2 and 3, which represent the three main directions in the workpiece in order of decreasing length. For example, for a plate the designations are 1 (length), 2 (width) and 3 (thickness).

A very generalized systematic approach has been made for metalforming processes by Backofen (1972), Lange (1985) and others. It is possible to divide any metal-forming system into a number of elements as shown in Fig. 10.1. These elements or areas are: (1) the work zone in which the workpiece undergoes plastic deformation, (2) the properties of the material being shaped, (3) the properties of the formed material, (4) the contact zone between the tool and the workpiece, (5) the tools themselves, (6) the workpiece and the surrounding atmosphere and (7) the forming machine.

Table 10.1.   Proposed classification system for forming operations.

Characteristic notation	Premises of operation — Purpose	Premises of operation — Workpiece material	Premises of operation — Shape of starting workp.	Mode of deformation — State of strain (1 2 3)	State of stress — Extent of deformation zone (1 2 3)	State of stress — Direct stress (1 2 3)	State of stress — Hydrostatic stress	Mode of deformation sequence — Direction of workg. of neighbg. part	Mode of deformation sequence — Transfer of workg. to ng. part	Mode of deformation sequence — Successive workg. of a part	Type of tool — Workg. if separate parts	Type of tool — Shape & size determining tool	Type of power transmission to workp.
(0)	General	General	General	General	General	General	General >0	General 1	General Intermt.	General Intermt. w. same tool	General Simultan. w. same tool	General Shapeless	General Gas & fluid pressure
(1)	Deformation		Bar & Wire	tension	Whole	Tension							
(2)	Separation		Tube & hollow bar	Compression	Partial	Compression	ca 0	2	Contin.	Intermt. w. dif. tool	Simultan. w. dif. tool	Shaped punch & die	Mechanism
(3)	Joining		Strip	Bending	Minute	Bending	0 to $-k_f$	3	Contin. & intermt.	Contin. w. same tool	Progres. w. same tool	Plain punch & die	Gravity
(4)	Reforming		Solid block	Shearing		Shearing	$< -k_f$	Simultan. combined 1 & 2		Contin w. dif. tool	Progres. w. dif. tool	Non-circular roll	Man-power
(5)	Surface finishing		Hollow block						One after another			Circular roll	Fuel & explosive
(6)	Property modification		Plate & thin flat piece						At random			Chuck & drum	Heat conduction
(7)			Ring & plate w. hole								Support & anvil		Electro-magnetic force

Abbreviations: contin. — continuous, dif. — different, intermt. — intermittent, k_f — flow stress of material, neighbg. — neighbouring, ng. — neighbouring, progres. — progressive, simultan. — simultaneous, w. — with, workg. — working, workp. — workpiece.

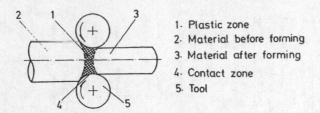

1. Plastic zone
2. Material before forming
3. Material after forming
4. Contact zone
5. Tool

Fig. 10.1.   A metal-forming system. After Pöhlandt (1989).

The first area is governed by plasticity theory and the behaviour of the workpiece material subjected to plastic strains. Area 2 deals with the properties of the workpiece before it has undergone deformation. Clearly, the mechanical properties of the material are important in this area. Area 3 is concerned with the properties of the workpiece after deformation. Once again, mechanical properties are important, but so is surface finish and dimensional accuracy. Area 4 is concerned with the contact zone between the tool and the workpiece and includes tribology. Area 5 is the tool itself.

From the point of view of fracture and shear localization in area 1 plastic flow is important. Likewise in area 4 high interfacial friction, such as may occur in upsetting, can lead to shear localization and fracture. One important point should be made here before we examine the details of some individual metalworking processes. This is, that although the overall cross-head speed or strain rate may be low, or even quasi-static, the strain rates along the narrow zones of shear localization can be much higher. This is the underlying reason why it is very important to pay detailed attention to adiabatic shear banding in what at first appears to be quite slow processes.

It is not surprising that adiabatic shear banding predominates in the high-speed machining of metals, in particular, cutting and blanking. The reason for this is that the plastic zone undergoing fast deformation is very narrow, thus providing a high localized strain rate.

Amongst the vast number of metalworking processes we choose to discuss only three in this chapter. These are forging, cutting and blanking. This does not mean that adiabatic shear banding does not occur in other metalworking processes.

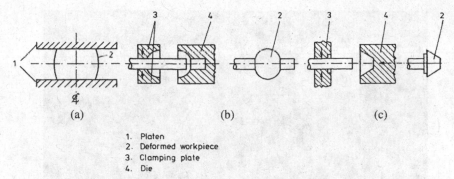

1. Platen
2. Deformed workpiece
3. Clamping plate
4. Die

Fig. 10.2. (a) Upsetting of the whole of the length of a workpiece; (b) upsetting of part of the length of a workpiece; (c) heading.

10.2. Upsetting

Upsetting or upset forging is the axial compression of the whole or part of the whole length of a workpiece to enlarge the cross-sectional area of the part. A further configuration of upsetting, normally referred to as heading, is when only one end of the workpiece is enlarged. Figure 10.2 illustrates the three operations. Upsetting is one of the commonest processes industrially. Also, in the form of compression of a cylinder with a lubricant, it is used as a standard laboratory test to obtain the stress–strain behaviour of the material in compression.

The workability of a material in upsetting is usually called the upsettability. Upsettability is dependent on the free surface ductility of the material. Moreover, the reduction in height of a workpiece is governed by the onset of cracking. Tests in industry and laboratories have shown clearly that the frictional constraint between the platens and the workpiece can lead to severe barrelling, which results in induced tensile stresses on the surface of the sample, especially at the equator. This is the general concept of free surface ductility.

It is clear that well-lubricated compression will produce insignificant barrelling; the sample will gradually decrease in height and increase in cross-section uniformly, suppressing surface cracking.

Since Latham's pioneering work of 1963, a number of further significant observations have been made. These observations have

Fig. 10.3.   General view of fracture of a magnesium alloy compression specimen, aspect ratio 1:1/2. The material failed in an intergranular manner. After Dodd *et al.* (1985).

greatly increased our understanding of upsettability or free surface ductility. Some of these important observations are as follows.

(1) There are two classes of failure in compression testing in aluminium alloys and brass (Latham, 1963). For ductile metals, fracture is due to barrelling whilst for less ductile metals fracture occurs through the whole body of the specimen (see Fig. 10.3).

(2) There are three types of surface cracking (Fig. 10.4): longitudinal, mixed and oblique cracks. Kudo and Aoi (1967), in tests on 0.45% carbon steel, controlled the degree of barrelling by using different die sets, lubricants and sample aspect ratios (i.e. height–diameter ratios). They observed that the three types of surface cracking occurred at different height reductions for the same material (Fig. 10.5).

(3) It was observed by Kudo and Aoi (1967) and Kudo and co-workers (1968) that there is a linear fracture line given by:

$$\epsilon_{\theta f} = a - \epsilon_{zf}/2$$

where $\epsilon_{\theta f}$ is the hoop fracture strain, ϵ_{zf} is the axial fracture strain and a is the plane strain fracture strain. This fracture

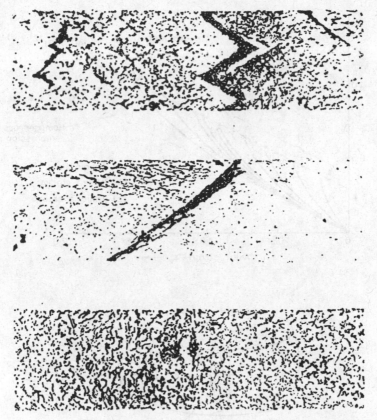

Fig. 10.4. Three modes of fracture observed in ductile materials. After Korhonen (1981). Originals by Kivivuori and Sulonen (1978).

line is parallel to the line for homogeneous compression (see Fig. 10.5).

(4) The three types of surface cracking fall on the same fracture locus in the $(\epsilon_\theta, \epsilon_z)$ plane. Longitudinal cracks occur when there is a high frictional constraint and appear at relatively small strains. On the other hand, oblique cracks occur at larger strains on the right-hand side of the diagram.

(5) A strain perturbation is observed just prior to fracture (Lee, 1972). This perturbation is equivalent to $d\epsilon_z = 0$, as shown in Fig. 10.6. The perturbation corresponds to a transient plane strain state at the surface of the upset sample, just prior to fracture.

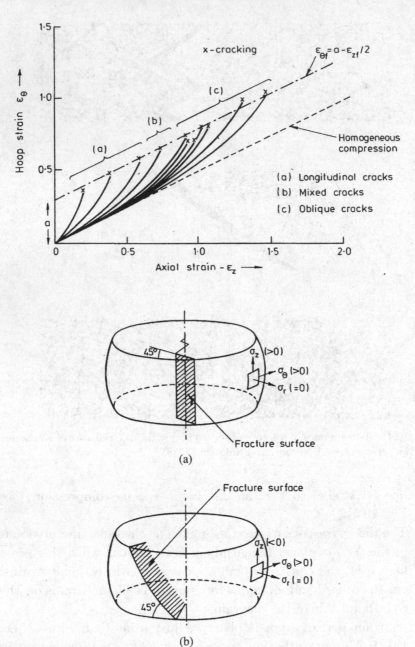

Fig. 10.5.   (a) Straight-line fracture condition for a medium-carbon steel. By Kudo and Aoi (1967). (b) Corresponding observed fracture planes.

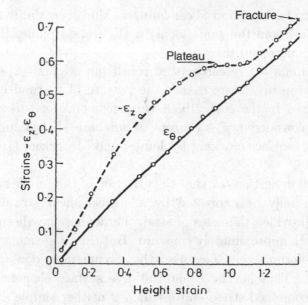

Fig. 10.6. Strain perturbation corresponding to $d\epsilon_z = 0$. Observed by Lee (1972). After this perturbation there is a sudden void coalescence at the cylinder equator.

The above observations, particularly those for the types of cracks and the linear fracture locus, have been confirmed by several authors, e.g. Thomason (1969) and Kobayashi (1970). Therefore, for a successful upset component to be processed, the surface strains should always be below this fracture line. To achieve this, the following approaches can be taken: (i) improve the lubrication between the die and the workpiece to make the strain path less severe so that it does not deviate very much from homogeneous compression, (ii) improve the material to give a higher plane strain fracture strain and (iii) avoid excessive barrelling of the material. The second approach raises important questions: what mechanical properties underlie the positioning of the fracture line and the three modes of cracking?

Experimental observations cannot be interpreted using the theory of maximum stress because all of the three cracking modes occur in shear. Also, the observations cannot be explained in terms of a maximum shear stress or strain. Essentially, the three types of crack correspond to two different shear failures. Moreover, the two modes of shear failure join the same locus in the $(\epsilon_\theta, \epsilon_z)$ plane. How is it possible to understand this?

The dilemma led researchers to recall the original experiments and intuition in upsetting; that is the pattern of thermal lines and the heat cross. In the early 1980s the occurrence of surface shear failures was considered. The concept of adiabatic shear banding was introduced to surface cracking by Jenner and co-workers (1981) and Staker (1981).

We noted in Section 6.1 that the criterion of thermoplastic instability can usually be expressed by a critical shear strain γ_i (see Table 6.1), provided the rates of strain hardening and thermal softening are both approximately constant. But this assessment is made for shear deformation. Of course, the circumstances in upsetting are different. That is, the equatorial free surface element is subjected to a combined stress state, which is neither simple shear nor plane strain. The necessary conditions for thermoplastic instability to occur on the surface of an upset specimen are twofold. Firstly, there must exist the possibility of pure shear deformation. In the combined stress state under consideration, this requires a state of plane strain as a prerequisite. For example, for pure shear on the (θ, r) plane (see Fig. 10.6) the requirement is $d\epsilon_z = 0$. Secondly, a certain shear instability condition, such as Eq. (6.3) or those in Table 6.1, should be satisfied within the surface element (see Jenner *et al.*, 1981).

Based on these requirements, it is possible to represent the conditions for three modes of shear instability on the $(\epsilon_\theta, \epsilon_z)$ plane as follows:

$$\epsilon_\theta = \frac{\gamma_i}{2} - \frac{\epsilon_z}{2}, \quad d\epsilon_z = 0, \quad \text{mode I} \tag{10.1}$$

$$\epsilon_\theta = \gamma_i + \epsilon_z, \quad d\epsilon_r = 0, \quad \text{mode II} \tag{10.2}$$

$$\epsilon_\theta = \gamma_i \pm 2\epsilon_z, \quad d\epsilon_\theta = 0, \quad \text{mode III}. \tag{10.3}$$

Mode III instability is completely suppressed in upsetting, while mode II failure appears on the upper part of the linear fracture line, which corresponds to the formula for mode I shear instability.

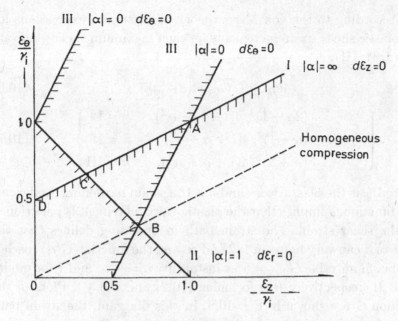

Fig. 10.7. Bai–Dodd theory applied to free surface cracking.

All the equatorial strain paths in upsetting can be described by

$$\begin{cases} \alpha = \dfrac{d\epsilon_\theta}{d|\epsilon_z|} \geq \dfrac{1}{2} & (10.4) \\[4mm] \dfrac{d^2\epsilon_\theta}{d|\epsilon_z|^2} \geq 0 & (10.5) \end{cases}$$

and α increases with a reduction in height.

For mode III instability to initiate requires $d\epsilon_\theta = 0$ and $\alpha = 0$; see Eq. (10.3). This is impossible for any strain path in upsetting as shown in Eq. (10.4). This is why mode III shear failure is suppressed. Also from Fig. 10.7, the segment CD can only be caused by mode I shear failure. This is in agreement with Kudo and Aoi's experimental observations.

It is still unclear what is required for mode II instability. Combining the shear instability analysis and plasticity theory, Dodd and Bai (1985) explored the mechanism underlying these shear instabilities. Here only the outline is cited.

According to the von Mises theory of plasticity, expressions for the plastic shear strain increments on each maximum shear plane are ($k = $ I, II, III):

$$d\gamma_k \sim \sqrt{G_k(\alpha)}(dW_p)^{\frac{n}{n+1}} \tag{10.6}$$

$$\left\{ \begin{array}{ll} (2\alpha+1)^2/4(1+\alpha+\alpha^2), & k = \text{I} \\ G_k(\alpha) = (\alpha-1)^2/4(1+\alpha+a^2), & k = \text{II} \\ (\alpha+2)^2/4(1+\alpha+\alpha^2), & k = \text{III} \end{array} \right\} \tag{10.7}$$

where W_p is the plastic work and n is the strain hardening coefficient. The function G implies that the plastic work increment is partitioned for the shear strain. The strain path in upsetting defines that the value of α can vary between $1/2$ and ∞ and the function $G(\alpha)$ reaches its maximum value $G_{\max} = 1$ at instability $\alpha = -\infty$ and 1 for modes I and II, respectively (see formulae (10.2) and (10.3)). Plots of the function G are shown in Fig. 10.8. In this diagram, the strain path starts from the right-hand side of the axis $1/\alpha = 2$ and tends to the origin $1/\alpha = 0$.

For mode I, function G_I increases with increasing α steadily until $G_\text{I} = 1$ and mode I shear instability must be initiated eventually. However, mode II instability is not the same. At $\alpha = 1$,

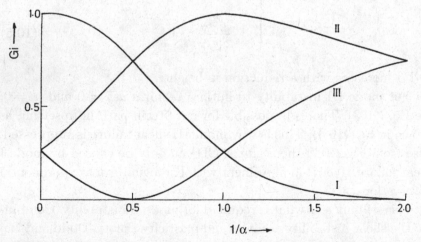

Fig. 10.8.   The functions $G_k(\alpha)$, showing the state of instabilities in the three possible shearing modes. After Dodd and Bai (1985).

$G_{\mathrm{II}}(1) = G_{\mathrm{max}} = 1$. But with increasing α, i.e. a further reduction in height, G passes through its maximum and then decreases. This means that this shear deformation may become metastable, provided the shear instability does not occur at $\alpha = 1$, due to either the absence of a trigger or the non-fulfilment of condition (10.2). With increasing α, as mentioned before, mode I shear instability must occur eventually. Of course, this does not exclude the reappearance of $\alpha = -1$ and the initiation of mode II instability. In fact, this does happen after mode I instability.

Now let us examine Lee's observation more closely (Fig. 10.6). There is a strain perturbation $d\epsilon_z = 0$ before fracture. This is an indication of the possible onset of mode I shear instability. After that, there is another perturbation $d\epsilon_r = 0$. This indicates the reappearance of the onset of mode II instability. This implies that the onset of mode I shear instability triggers the onset of mode II instability, which sustains the shear strain far beyond its threshold in the CB segment of the fracture line in Fig. 10.7. The mixed type of surface cracking may be the remnant of a mode I trigger for a mode II shear failure. The fracture mode has to be mode II, i.e. an oblique fracture, if fracture occurs far from point C on the fracture line, because mode II shear deformation accumulates a higher strain.

Thus far, the theory of adiabatic shear instability explains, in a unified way, the five significant observations as well as the thermal cross in upsetting.

In order to apply the idea of adiabatic shear banding, which governs the upsettability of metals in an industrial process, a considerable amount of work has been carried out.

Semiatin and Jonas (1984) suggested a workability map for the occurrence of shear bands in isothermal side-pressing for a titanium alloy, Ti-6242Si. As mentioned in Section 6.3, they proposed a variable $\alpha = \Delta \ln \dot{\bar{\epsilon}} / \Delta \bar{\epsilon}$ as an indication of shear localization. They reported that, with the exception of two points, the loci corresponding to $\alpha_{\mathrm{max}} \geq 5$ were separate regimes over which adiabatic shear bands are and are not observed for this alloy. Therefore, these maps give a guide to safe side-pressing with no shear bands (Fig. 10.9).

Turner (1988a,b) developed an alternative approach to adiabatic shear in the heading process. He noticed that the most

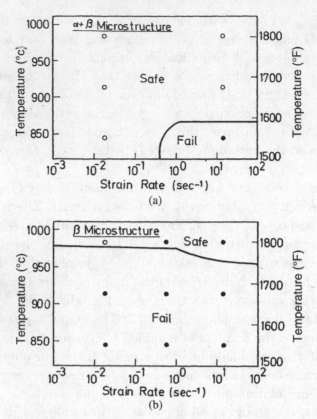

Fig. 10.9. Workability maps for the occurrence of shear bands in isothermal side-pressing of Ti-6242Si with (a) an $\alpha + \beta$ microstructure and (b) a β-microstructure. Workability prediction (—) and forging conditions in which shear bands were (•) and were not (○) observed are noted. After Semiatin and Lahoti (1982), published by permission of the American Soc. of Metals and the Minerals and Materials Soc.

easily controllable parameter in processing might be the temperature dependence of the flow stress of metals at elevated temperature.

Turner (1988a,b) explained the occurrence of ductile fracture in cold forging using thermally aided instability. Despite the fact that heat is conducted away into the tooling, Turner found that the concept of thermally aided instability is helpful. Because of conduction, the most important feature of a mechanical working process will be the speed of the operation, as this will govern the heat loss. Using an exponential cubic temperature dependence of the flow curve, the instability limit locus does not need to be exceeded if the heat loss produces a strain path similar to that shown in Fig. 10.10. Because of the potential heat loss variation, an infinite number of strain paths can occur.

Affouard (1984) carried out a large series of dynamic experiments on steels using a compression Hopkinson bar. Depending on the dimensions and hardness of the starting cylinder and the applied strain rate, Stelly and Dormeval (1986) obtained different patterns of transformed adiabatic shear bands. From their experimental observations, they managed to sketch the development of the adiabatic shear surfaces within them. They found at least two different patterns of surfaces, depending on the material hardness.

The concept of three-dimensional adiabatic shear surfaces as opposed to the sections of bands of shear is very important and this requires much more experimental work. One aspect that is unclear at the moment is, are the mechanisms for the progressive development of shear bands in upsetting and forging similar to that in torsion testing? Answers to questions such as this can only be obtained from large numbers of experiments.

10.3. Metal-cutting

Metal-cutting is used in fabrication to remove metal in the form of chips or swarf from a workpiece to achieve a desired shape. It is clear that metal-cutting must be carried out beyond the threshold defined by ductile fracture. Although it has long been clear from the geometry of metal-cutting that severe shear localization occurs in the narrow primary shear zone and therefore adiabatic shear bands are possible in high-speed cutting, there are still important problems, which require proper solutions. For example, when and how is a chip formed and what is the relation between the chip shape and the occurrence and geometry of adiabatic shear bands?

There are a great number of monographs on metal-cutting. Shaw (1984) gives a comprehensive view of the process, Childs and Rowe (1973) concentrate on the physics of cutting and Oxley (1989) on analytical approaches. Dodd and Bai (1987) and Semiatin and Jonas

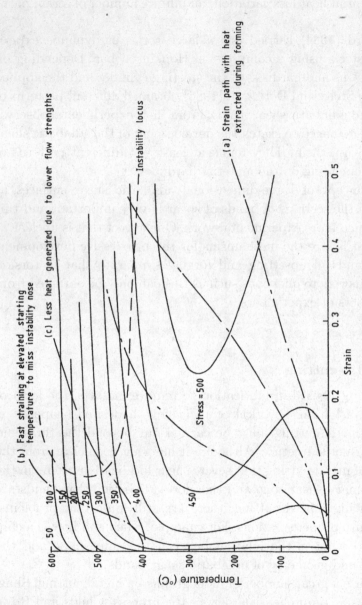

Fig. 10.10. Schematic illustration of variation in strain paths in cold forging. After Turner (1988a,b). Graph reproduced with permission from British Steel Technical.

(1984) examined the process more closely from the point of view of adiabatic shear banding.

Early in 1964 Recht carried out a number of very revealing experiments in which titanium was cut at various speeds. Recht predicted a shear zone temperature of 650°C at a cutting speed of 43 m/min. He also noted the importance of adiabatic shear banding in metal-cutting. Dao and Shockey (1979) showed clearly adiabatic shear bands and discontinuous chips in steel and aluminium. Various chip types are shown in Fig. 10.11. Dao and Shockey measured the temperature in orthogonal cutting of the above metals by using an infrared microscope. Peak temperatures of 180°C for the steel and 100°C for the aluminium were recorded. By taking into account that the measured area is greater than the band width, these temperatures were corrected to about 500°C for the steel and 120°C for the aluminium.

The cutting technique that has been studied the most from the theoretical and practical points of view is two-dimensional orthogonal cutting. Also, practically, if the cutting edge is large compared to the depth of cut, the cutting process may be treated as orthogonal. Because this process is in a state of plane strain then it is amenable to theoretical modelling and can provide much useful information.

A typical distribution of temperature in orthogonal machining, produced using infrared photography, clearly shows the situation in the primary and secondary shear zones (Fig. 10.12). The high temperature rise in these zones clearly indicates the significance of thermally assisted shear banding in metal-cutting. Also, it is reasonable to assume that there is some correlation between the formation and morphology of chips and localized shearing.

The morphology of chips can vary. Predominantly, there are two chip morphologies: the continuous chip and the periodically serrated chip, which can break into separate segments. For most steels, at moderate cutting speeds, continuous ribbon chips form, which are reasonably ductile, whereas the individual segments are produced by fracture, with a gross fracture surface coinciding with the primary shear zone. Serrated chips are formed, for example, when titanium alloys are cut and some steels are machined at high speeds. The formation of these chips normally produces a poor surface finish and

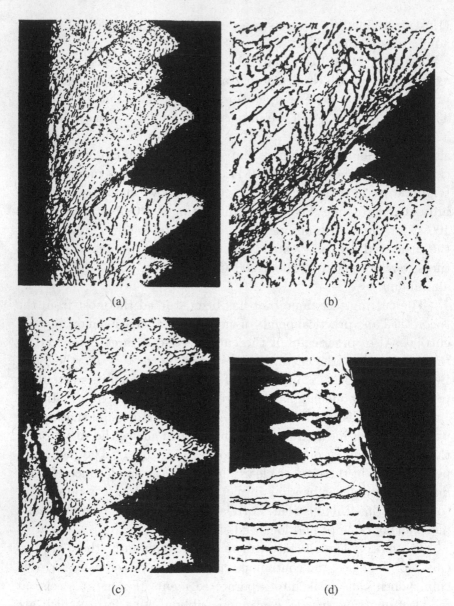

(a)   (b)

(c)   (d)

Fig. 10.11.   Continuous chips with inhomogeneous shear: (a) titanium cut at a high speed (53 m/min), adiabatic shear; (b) enlargement of (a); (c) titanium cut at a low speed (25 mm/min), periodic fracture, gross sliding and rewelding; (d) 60–40 cold-rolled brass (60% reduction in area) cut with a high-speed steel tool having minus 15° rake angle, cutting speed 0.075 m/min. Undeformed chip thickness 0.16 mm. Sketches based on a figure in Shaw (1984).

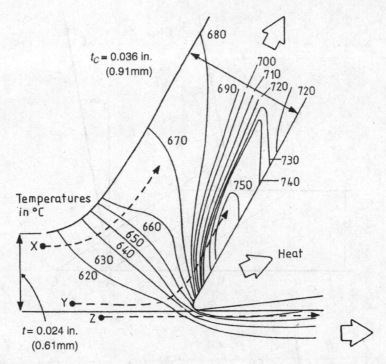

$t_c = 0.036$ in.
(0.91mm)

680
700
710
720 720
690
670
730
740
750

Temperatures
in °C

660
650
640
630
620

X

Heat

Y
Z

$t = 0.024$ in.
(0.61mm)

Fig. 10.12. Temperature distribution in workpiece and chip during orthogonal cutting for free cutting mild steel where the cutting speed is 75 ft/min
(0.38 m/min), the width of cut is 0.25 in (6.35 mm), the working rake is 30°
and the workpiece temperature is 611°C. After Boothroyd (1963). Reprinted by
permission of the Council of the Institution of Mechanical Engineers.

final workpiece shape. Therefore, a simple analysis, even though it
may be rather approximate, would be very helpful.

Let us consider Fig. 10.13, which is a simplified configuration
for orthogonal cutting. The newly created surface is parallel to the
cutting velocity v. The material that makes up the depth of cut d is
converted into a chip and moves upwards with a velocity v_s, roughly
parallel to the tool rake. The transition from the depth of cut to
the chip takes place in the primary shear zone, which is inclined
to the cutting velocity v by a shear angle φ. Hence the appearance
of the primary shear zone is a prerequisite for the process of cutting.
There are two significant parameters here, the shear angle φ and the
width of the primary shear zone δ, which corresponds to segment AF

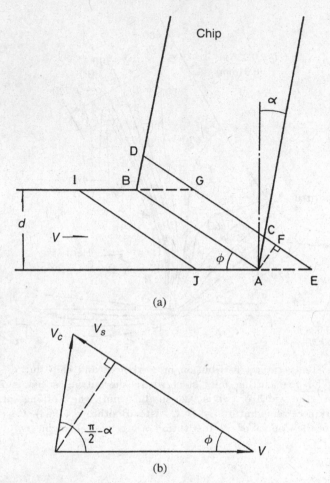

Fig. 10.13.   Determination of shear strain: (a) shear strain in orthogonal cutting; (b) hodograph.

in Fig. 10.13. We will now consider a method for the characterization of the primary shear zone.

For the condition of plane strain deformation in orthogonal cutting, pure shear, and therefore shear banding, can be initiated along directions of zero extension. This makes the angle φ dependent on both the cutting configuration and the force. A simple description of shear angle φ is

$$\varphi = \varphi(\alpha, d, v, \beta, k)$$
$$= \Phi(\alpha, \beta) \tag{10.8}$$

where α is the rake angle, d is the depth of cut, v is the cutting speed, β is the friction and k is the flow stress in shear. Formula (10.8) is valid approximately for perfectly plastic materials and the final expression of (10.8) is derived using Buckingham's π theorem. A number of explicit expressions based on Eq. (10.8) have been deduced, each one derived from different assumptions such as a maximum in the shear stress. The details of these expressions can be found in the book by Dodd and Bai (1987).

There are other parameters characterizing the primary shear zone as well as the shear angle φ. These parameters are the shear strain and the shear strain rate within the region.

Measurements of the primary shear zone made by Stevenson and Oxley (1970) showed that the distribution of the maximum shear strain rate is approximately symmetrical with respect to the shear line AB and the shear strain rate remains nearly constant along it. This observation suggests that the one-dimensional shear model discussed in Section 6.2 may be applicable to orthogonal cutting.

The shear strain and shear strain rate within the shear zone support shear banding. The shear strain can be estimated as follows. Deformation near the shear line AB in Fig. 10.13 can be expressed by a transformation of ABIJ or ABGE to ABCD. Then the shear strain is expressed by

$$\gamma \approx \frac{\text{CF} + \text{FE}}{\text{AF}} = \tan(\varphi - \alpha) + \cot\varphi$$

$$= \frac{\cos\alpha}{\sin\varphi\cos(\varphi - \alpha)}. \tag{10.9}$$

Since the rake angle α is usually between -10 and $+10°$, the shear strain may be of the order of 1 for large shear angles ($\varphi \approx 30$ or $40°$) and of the order of 10 for small angles ($\varphi \approx 10°$).

In a similar way, the shear strain rate can be estimated by:

$$\dot{\gamma} = \frac{V_s}{\text{AF}} = \frac{V_s}{\delta} = \frac{\cos\alpha}{\cos(\varphi - \alpha)}\frac{v}{\delta}$$

$$= \gamma\sin\varphi v/\delta \tag{10.10}$$

where the band width $\delta = \text{AF}$. As an example, if we take $\alpha \approx 0°$, $\varphi \approx 30°$, $v \approx 30\,\text{m/min}$, $\delta \approx t/20$ and $t \approx 0.1\,\text{mm}$ we obtain $\dot{\gamma} \approx$

10^5/sec. This is a very high strain rate and adiabatic shear banding is certainly possible.

Let us now examine the temperature rise. As in the experimental observation of shear strain rate, the observed temperature distribution in metal-cutting (Fig. 10.12) is similarly symmetrical with respect to the shear line AB. Once again this supports the one-dimensional approximation of the primary shear zone. However, amongst the special aspects of metal-cutting there is the heat converted from plastic work, which is conducted away in two ways. One portion is transferred to the tool and workpiece and the other portion is convected away by the chips. So, as an approximation, the average temperature rise $\Delta\theta$ can be written as:

$$\Delta\theta \approx \int \frac{\tau d\gamma}{\rho c}(1 - \eta) \tag{10.11}$$

where η is a special factor for metal-cutting representing the fraction of heat transferred to the tool, workpiece and chip. In fact, η should be a function characterizing the heat transfer related to the primary shear zone. Hence:

$$\eta = \eta\left(\frac{\delta^2}{\kappa t_k}\right) \tag{10.12}$$

where t_k is the characteristic timescale in metal-cutting. The argument in (10.12) can be expressed approximately by

$$\frac{\delta^2}{\kappa t_k} \approx \frac{vd}{\kappa}\frac{\delta}{t_k v_s} = \frac{vd}{\kappa}\frac{1}{t_k\dot\gamma} = \frac{vd}{\kappa}\frac{1}{\gamma} \sim R\tan\varphi \tag{10.13}$$

where $R = vd/\kappa$ is called the thermal number, which is an indication of the thermal conductivity in metal-cutting. In the derivation of Eq. (10.13), the conservation of mass was applied, i.e. $vd = v_s\delta$. Several methods have been used to evaluate the function, e.g. see Loewen and Shaw (1954), Weiner (1955) and Boothroyd (1963). A comparison of these approaches is shown in Fig. 10.14. The data shows that η is about a half for $R\tan\varphi \approx 1$. Therefore, although there is metal flow in metal-cutting, still half of the heat produced is retained within the primary shear zone, which increases the temperature.

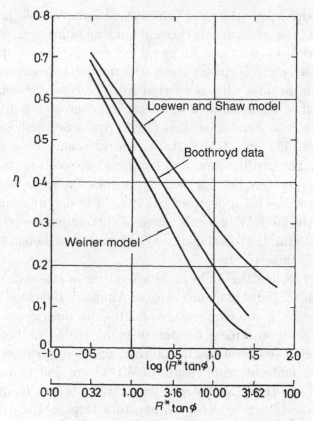

Fig. 10.14. Comparison of model predictions of η to experimental measurements. After Semiatin and Rao (1983).

Until now, pictures, measurements and estimations of the primary shear zone have all supported the idea that segmented chips are formed by adiabatic shear banding. In fact, shear localization occurs in continuous chips in the primary shear zone. We need to identify the transition from continuous to segmented chips, according to thermoplastic shear instability and localization.

Semiatin and Jonas (1984) applied the idea of shear localization to this problem (see Section 6.3). According to data for AISI-4340 steel, they found that when the localization parameter $\alpha \sim 5$ serrated chips form. This may well be another preliminary quantitative prediction for the formation of serrated chips by making use of adiabatic shear localization, since Recht's pioneering work.

As for upsetting, the most sensible and controllable parameter amongst all the mechanical, thermal and metallurgical properties of the workpiece material in metal-cutting is the temperature-dependent strength. For many steels, the temperature dependence of the strength includes what is referred to as blue brittleness, namely the strength at some elevated temperature, over a certain range of temperature. Wang and co-workers (1990) performed orthogonal cutting tests on $45^{\#}$ and 37Mn5 steels, which usually show this blue brittleness. The cutting conditions for these two steels are as follows: at room temperature ($25°$C), the cutting velocity ranges between 50 and $300\,\mathrm{m/min}$, with a depth of cut $d \approx (0.1-0.4)\,\mathrm{mm}$ and a rake angle $\alpha \approx (0-20°)$. Within this range of cutting parameters no serrated chips form due to adiabatic shear banding, although the shear strain rate attains 10^5/sec.

However, when the ambient temperature is raised from 25 to 210 and $420°$C, the chip mode changes. Although the chips remain a continuous ribbon at a temperature of $210°$C for both steels, serrated chips appear at workpiece temperatures of $420°$C for both steels. The calculated temperatures in the primary shear zone for the two steels at an ambient temperature of $420°$C are $589°$C and $597°$C in the $45^{\#}$ and the 37Mn5 steels, respectively. These temperatures are above the blue brittleness temperature threshold and therefore thermal softening occurs $\partial\tau/\partial\theta < 0$. Thus, adiabatic shear banding becomes possible.

Although adiabatic shear banding can explain some of the special features in metal-cutting and provide some guidance for the improvement of the process, they are generally qualitative or descriptive evaluations. It is clear that much more research is required in this area.

10.4. Blanking

Blanking and related processes such as cropping, punching, etc., belong to a group of processes by which a specified part of the metal is removed. Unlike metal-cutting, the shear plane in blanking and related processes is usually specified beforehand and is approximately parallel to the direction of the punch velocity. In this sense blanking is simpler than cutting where the shear plane angle φ is not specified apparently by the processing configuration. However, there are other aspects of blanking, which can make the process complicated. For example, the clearance and sharpness of the punch and the die add new factors to the process. In some cases, visible cracks may appear ahead of the die. Therefore, there may be many possible mechanisms governing the process.

The schematic representation of the punch force–punch displacement autographic diagram provided by Johnson and Slater (1967) shows some features of the process (Fig. 10.15). There are several distinctive steep drops in the load–displacement curve. They may originate from different mechanisms: shear instability, initiation of micro- and macrocracks and frictional resistance due to the blank and hole or the punch and hole. There are two important features in the process which require measuring. These are the energy consumption and the major load drop. These two factors are closely related to each other and to the post-instability shearing and cracking.

Atkins (1981) suggested that geometrical softening, i.e. the reduction in cross-section due to the indentation of the punch in blanking, should dominate plastic instability and the load drop (Fig. 10.14).

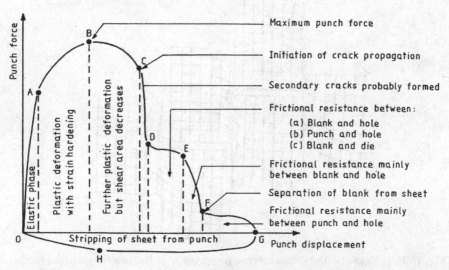

Fig. 10.15. Schematic representation of punch force versus punch displacement autographic diagram. After Johnson and Slater (1967).

As discussed in Chapter 9, Bai and Johnson (1982) supposed that a non-uniform shearing in the workpiece adjacent to the punch edge may eventually evolve into adiabatic shear bands. Later, Dodd and Atkins (1983) combined the two softening mechanisms and considered softening due to the presence of microvoids. All these approaches have been helpful in beginning to understand what is clearly a complex process. However, a quantitative model is required that can combine all the relevant mechanisms in blanking. In particular, a theory is required which can combine shear banding with micro- and macrocracking simultaneously. Figure 10.16 illustrates the progression of

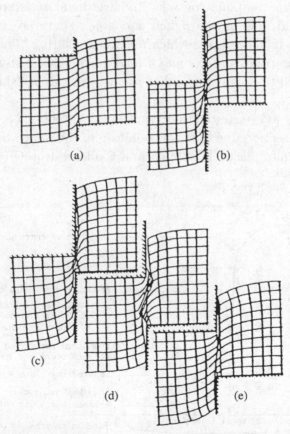

Fig. 10.16.   Network diagrams showing progression of fracture in shearing metal bars (clearance nil): (a) lead 22% penetration; (b) lead 74% penetration; (c) aluminium 86% penetration; (d) copper 55% penetration; (e) brass 53% penetration. After Chang and Swift (1950).

fracture in different materials in blanking. Separation is often a complicated mixture of shear localization and macrocracking.

Further Reading

Childs, T. H. C. (2012) *Adiabatic Shearing in Metal Machining*, CIRP Encyclopedia of Production Engineering, Springer, New York.

Childs, T. H. C., Maekawa, K., Obikawa, T. and Yamane, Y. (2004) *Metal Machining: Theory and Applications*, Arnold Publishers, London.

Trent, E. M. and Wright, P. K. (2000) *Metal Cutting* 4th edition, Butterworth–Heinemann, Woburn.

Appendix A

Quick Reference

This quick reference appendix is meant to provide a fast acquaintance
with adiabatic shear banding and it is not a handbook of formulae.
Before this section is used it is recommended that the context is
thoroughly read and understood.

I. Parameters describing adiabatic shear bands

1. Critical condition for the onset of adiabatic shearing:

$$\frac{\beta\left(-\frac{\partial \tau}{\partial \theta}\right)\tau}{\rho c\left(\frac{\partial \tau}{\partial \gamma}\right)} = 1 \tag{6.5}$$

$$\gamma_i = \frac{n\rho c}{\beta\left(-\frac{\partial \tau}{\partial \theta}\right)}. \tag{6.6}$$

(For details see Table 6.1)

2. Conditions for localization:

$$\gamma_l = \frac{\rho c(n+m)}{v}\frac{\theta}{\tau} \tag{6.46}$$

$$n + m - v < 0. \tag{6.54}$$

3. Width of shear band

$$\delta \approx \sqrt{\kappa\frac{\gamma}{\dot{\gamma}}} \tag{7.42}$$

$$\delta \approx \sqrt{\frac{\lambda\theta_*}{\beta\tau_*\dot{\gamma}_*}} \tag{7.28}$$

363

$$\delta \approx \frac{1}{\dot{\gamma}_*} \sqrt{\frac{\lambda \theta_*}{\beta \eta}} \tag{7.30}$$

$$\delta = \frac{1}{\sqrt{2\beta\tau}} \int_{\theta_\delta}^{\theta_m} \frac{\lambda(\xi)d\xi}{\sqrt{\int_\xi^{\theta_m} \lambda(\eta)g(\tau,\eta)d\eta}}. \tag{7.27, 7.35}$$

4. Spacing of shear bands (Section 9.1)

$$l \approx y_{k3} \approx \frac{1}{\dot{\gamma}_*} \sqrt{\frac{c\theta_*}{\beta}} \tag{9.1}$$

$$l \approx \frac{1}{\dot{\gamma}_0} \sqrt{2\epsilon_0}. \tag{9.22}$$

II. Parameters dominating adiabatic shear banding
Thermal softening $P = (-\partial\tau/\partial\theta)$

- decisively determines the occurrence of adiabatic shear banding (Sections 6.1, 6.2, 8.2, 8.4)
- accelerates the growth rate (Section 6.2)

Strain hardening $Q = \left(\frac{\partial\tau}{\partial\gamma}\right)$, $n = \frac{\gamma}{\tau}\left(\frac{\partial\tau}{\partial\gamma}\right)$

- retards the occurrence and formation of a shear bands (Section 8.2)

Strain-rate hardening $R = \left(\frac{\partial\tau}{\partial\dot{\gamma}}\right)$, $m = \frac{\dot{\gamma}}{\tau}\left(\frac{\partial\tau}{\partial\dot{\gamma}}\right)$

- retards the formation of shear bands (Section 8.2)
- retards the early growth of disturbances (Section 6.2)

Strain rate $\dot{\gamma}$

- a high enough strain rate is a prerequisite for adiabatic shear banding (Sections 6.1, 8.4)
- a high strain rate leads to a narrow shear band (Sections 7.5, 8.4)

Heat conduction λ

- a low heat conduction is a prerequisite for adiabatic shear banding (Section 2.4)
- a material with a low heat conduction (e.g. titanium) is susceptible to adiabatic shear bands (Section 2.4)

- expands the width of adiabatic shear bands (Section 7.5)
- retards the early growth of the disturbance (Section 6.2)
- a high heat conduction leads to a high cooling rate inertia, characterized by the density ρ (Section 1.2)
- retards early shear band development (Sections 6.2, 8.4)

Volumetric specific heat ρc

- retards adiabatic shear banding
- high-strength metals are more prone to adiabatic shear than low strength ones (Section 6.4)

Disturbance amplitude

- decisively shortens the time between occurrence and the full formation of a shear band (Section 8.2)

Disturbance width

- may influence the location of shear band (Section 8.2)

Disturbance distribution

- quicker coalescence shortens the time from occurrence to full formation of a shear band (Section 8.2)

Imperfections

- not necessary for a shear instability (Sections 6.2, 8.2, 8.4)
- may initiate the disturbance leading to the occurrence of a shear band (Section 8.2)

Multiaxial hardening characterization

- secondary effects of adiabatic shear banding (Section 8.4)

III. Major non-dimensional parameters (Sections 2.4, 7.2, 9,1, 9.3)

- effective Prandtl number $Pr = t_h/t_v$
  $Pr \approx$ ratio between timescales of heat diffusion and rate-dependent diffusion

$$\approx \left(\frac{l}{\delta}\right)^2 \approx (\text{band spacing/band width})^2$$

- number $B = \frac{\beta(-\partial\tau/\partial\theta)\tau}{\rho c(\partial\tau/\partial\gamma)}$

  $B = 1$ critical condition for occurrence of adiabatic shear bands

- strain hardening number $n = \frac{\gamma}{\tau}\left(\frac{\partial \tau}{\partial \gamma}\right)$
- strain-rate hardening number $m = \frac{\dot{\gamma}}{\tau}\left(\frac{\partial \tau}{\partial \dot{\gamma}}\right)$
- Prandtl number $Pd = \left(\frac{\partial \tau}{\partial \dot{\gamma}}\right)\Big/ \rho\kappa = Pr\, m$
- damage number $D = \frac{\rho v^2}{\sigma_y}$

Appendix B

Specific Heat and Thermal Conductivity

The Dulong–Petit law states that the molar heat capacity at constant pressure is a constant, approximately 6.4 cal/K for all solid elements. Using the Debye theory for the thermal behaviour of solids gives the value of 6 cal/(K mole) for the molar heat capacity at constant volume for solids at temperatures higher than the Debye temperature. Provided the difference between the two capacities under the different conditions is taken into account, the two theories are consistent. At low temperatures the heat capacity decreases with decreasing temperature.

Electrons are the carriers of thermal energy in metals. Therefore, it is natural for there to be a close correlation between electrical and thermal conductivities for metals. This is the Wiedemann–Franz relation:

$$\frac{\text{thermal conductivity}}{\text{electrical conductivity}} = \text{constant } L$$

where the Lorenz number $L = 1.6$ to 2.5 $(\text{volt/K})^2 \times 10^{-8}$ for most metals at 293 K.

Like electrical conductivity, the thermal conductivity decreases rapidly when alloying elements are added to a metal, especially in small quantities. This tendency can be seen in the data listed below, such as the thermal conductivities from ingot iron to 8630 steel.

As the environmental temperature increases, the thermal conductivity usually decreases.

Table B.1.   Solid elements — mechanical and thermal properties.

Element	Density (10³ kg m⁻³)	Young's modulus (10¹⁰ Pa)	Shear modulus (10¹⁰ Pa)	Thermal expansion (10⁻⁶ K⁻¹)	Heat capacity (J kg⁻¹ K⁻¹)	Heat conductivity (W m⁻¹ K⁻¹)	Melting point (K)	Heat of fusion (10³ J kg⁻¹)
Aluminium	2.70	6.9	2.6	23.2	903	238	933	397
Antimony	6.69	7.7	2.1	11.0	207	18	904	163
Arsenic	5.73			6	330		883	433
Barium	3.5	1.3	0.5	16.4	190		998	57
Beryllium	1.85	30	14	11.5	1825	230	1551	1384
Bismuth	9.75	3.1	1.2	13.5	122	8.5	544	52
Cadmium	8.65	5.0	2.0	31.5	232	92	594	57
Calcium	1.55	2.0	0.8	22.3	658	98	1115	228
Carbon, diamond	3			1.1	509	1.56	>3820	17000
graphite	2.25			8.8	711	0.49	>3820	17000
Chromium	7.2	2.5		8.5	448	87	2160	280
Cobalt	8.9	20	8.0	13.7	425		1768	280
Copper	8.96	12	4.6	16.8	385	400	1356	205
Dysprosium	8.54	0.64		9	173		1680	
Europium	5.26	0.15			176		1099	
Gadolinium	7.90			4	230		1585	
Gallium	5.91			19.2	375		303	80.1
Germanium	5.32	8.1	3.1	5.7	322		1211	480
Gold	19.32	7.9	2.7	14.1	129	311	1336	56
Hafnium	13.29			6.0	144	22	2420	140

(Continued)

Table B.1. (*Continued*)

Element	Density (10^3 kg m^{-3})	Young's modulus (10^{10} Pa)	Shear modulus (10^{10} Pa)	Thermal expansion (10^{-6} K^{-1})	Heat capacity (J kg^{-1} K^{-1})	Heat conductivity (W m^{-1} K^{-1})	Melting point (K)	Heat of fusion (10^3 J kg^{-1})
Indium	7.31	1.1	0.4	31.9	234	25	430	28
Iodine	4.93			87	215		387	62
Iridium	22.42	52	20	6.5	133	148	2680	144
Iron	7.87	21	8.4	11.7	449	82	1808	276
Lead	11.35	1.6	0.54	28.9	130	35	601	24.7
Lithium	0.534	0.49		47.0	3570	71	452	420
Magnesium	1.74	4.4	1.7	25.6	1024	150	424	368
Manganese	7.3	20	8	22.8	479		1517	270
Molybdenum	9.01	33	13	5.0	248	140	2880	253
Nickel	8.90	20	8	12.7	444	90	1726	310
Niobium	8.57	10	3.7	7.1	267	52	2688	261
Osmium	22.57	56	22	4.7	130		3300	140
Palladium	12.02	11	4.4	11.6	244	70	1825	162
Phosphorus	1.82			127	750		317	21
Platinum	21.45	16	6.1	8.9	138	69	2042	113
Plutonium	19.84			57		8	913	39
Potassium	0.862			83	757	99	337	59.7
Rhenium	21.02	47	17.4		138	71	3450	180
Rhodium	12.41	37	15	8.3	242	150	2239	210
Rubidium	1.53			90	361		312	26
Ruthenium	12.41			6.7	240		2520	252

(*Continued*)

Table B.1.   (*Continued*)

Element	Density (10^3 kg m^{-3})	Young's modulus (10^{10} Pa)	Shear modulus (10^{10} Pa)	Thermal expansion (10^{-6} K^{-1})	Heat capacity (J kg^{-1} K^{-1})	Heat conductivity (W m^{-1} K^{-1})	Melting point (K)	Heat of fusion (10^3 J kg^{-1})
Selenium	4.79	10		26	322		490	66
Silicon	2.33	10	3.3	2.5	707		1680	165
Silver	10.50	7.8	2.8	19.2	236	418	1234	105
Sodium	0.971			69.6	1230	135	371	113
Strontium	2.54				301		1042	105
Sulphur	2.07			61	736	0.20	392.2	38
Tantalum	16.6	18	7	6.5	141	54	3269	170
Tellurium	6.24	4.1	1.6	18.2	202	1.7	723	140
Thallium	11.85	0.8	0.3	29.2	129	41	577	21
Thorium	11.7	8.0	3.1	11.1	118	41	$\approx$2000	83
Tin, grey	5.75	5.4	2.0	21.2	222	63	505	59
Titanium	4.54	11	4	8.5	522	19	1948	400
Tungsten	19.3	38	15	4.5	133	170	3653	192
Uranium	18.9	18	7.2	13.5	116	25	1405	53
Vanadium	5.87	13	4.7	7.8	486	32	2160	330
Yttrium	4.45	0.66			280	15	1768	190
Zinc	7.13	9.8	4	29.7	389	120	693	117
Zirconium	6.53	7.0	2.5	5.4	275	21	2125	220

Table B.2. Other solids — mechanical and thermal properties.

Name	% compositon by weight	Density (10³ kg m⁻³)	Expansion coefficient (10⁻⁶ K⁻¹)	Young's modulus (10¹⁰ Pa)	Shear modulus (10¹⁰ Pa)	Heat capacity (10³ J kg⁻¹ K⁻¹)	Heat conductivity (W m⁻¹ K⁻¹)	Melting point (K)
1 Aluminium bronze 5%	Cu 94.6, Al 5, Mn 0.4	8.1	18	12		0.42	84	1333
2 Argentan 18%	Cu 60, Zn 22, Ni 18	8.7	17	12–15		0.40	23	1375
3 Brass	Cu 62.7, Zn 37.3	8.4	21	10.5	3.5	0.38	79	1188
4 Cast iron	<4%C	7.3	11	10		0.50	30–45	1475
5 Constantan	Cu 58, Ni 41, Mn 1	8.9	15	11		0.41	22	1545
6 Duraluminium	Cu 3–4, Mg 0.5, Mn 0.25–1, rest Al	2.8	24	7.2		0.93	160	925
7 Electrolytic ion	Mg 92, Al 5, Zn 3	1.8	25	4.4	1.7	1.00	115	900
8 Invar	Fe 64, Ni 36	8.1	2.0	14.5		0.50	16	1723
9 Kanthal Al	Fe 67.5, Cr 25, Al 5.5, Co 2	7.1	17				17	
10 Manganin	Cu 84.5, Mn 12.5, Fe 1, Ni 2	8.5	16	12.6		0.41	22	1275
11 Silumin	Al 87, Si 13	2.6	19	8.5	3	0.88	161	845
12 Steel	C 0.85	7.8	11.5	20	8.1	0.46	45	1625
13 Tin bronze 10%	Cu 90.75, Sn 9, P 0.25	8.9	19	10–12		0.38	46	1285
14 Wood's metal	Bi 44.5, Pb 35.5, Sn 10, Cd 10	9.7				0.15	13	344
15 Wrought iron	C 0.04–0.4	7.6		22			60	
1 Acrylic		1.2	70–100			1.4–2.1	0.2	
2 Araldite (epoxy)		1.2	60			1.7	0.2	

(*Continued*)

Table B.2. (*Continued*)

Name	% composition by weight	Density (10^3 kg m^{-3})	Expansion coefficient (10^{-6} K^{-1})	Young's modulus (10^{10} Pa)	Shear modulus (10^{10} Pa)	Heat capacity (10^3 J kg^{-1} K^{-1})	Heat conductivity (W m^{-1} K^{-1})	Melting point (K)
3 Asbestos		0.6				0.84	0.2	
4 Brick		1.4–1.8	8–10			0.8	0.6–0.8	
5 Concrete (dry)		1.5–2.4	12			0.92	0.4–1.7	
6 Cork		0.20–0.35				1.7–2.1	0.045–0.06	
7 Ebonite		1.2	85			1.67	0.2	
8 Fibre board (porous)		0.3					0.06	
9 Glass (common)		2.5	8			0.84	0.9	
10 Granite		2.7	8			0.80	3.5	
11 Gypsum		1.0	25			1.1	1.3	
12 Ice (−4°C)		0.917	50			2.2	2.1	
13 Marble		2.5–2.8	5–16			0.9	3	
14 Mica		2.8	3			0.88	0.5	
15 Paper		0.7–1.2					0.2	
16 Paraffin		0.85	100–200			2.1–2.9	0.21–0.26	
17 Polyamide (nylon)		1.1	100–140			1.8	0.2	
18 Polyethene		0.92	100–200			2.1	0.23–0.29	
19 Polystyrene		1.05	60–80			1.3	0.07–0.08	
20 Polyvinyl chloride (PVC)		1.2–1.5	150–200			1.3–2.1	0.16	
21 Porcelain		2.3–2.5	2–5			0.8	1.0–1.7	
22 Quartz (fused)		2.2	0.4			0.8	0.2	
23 Rubber		0.92–0.96	150–200			2	0.13–0.16	
24 Teflon		2.1–2.3	60–100			1.0	0.2	
25 Wood (pine)		0.52	5–30			0.4	0.14	

Table B.3.

Specific heat low-carbon steels.

Temperature range (°F)	Specific heat (cal/gm/C)
122–212	0.115
302–392	0.125
392–482	0.130
482–572	0.133
572–662	0.136
662–752	0.142
842–932	0.158
1022–1112	0.177
1202–1292	0.205
1292–1382	0.272
1382–1472	0.229
1562–1652	0.195

Material — 0.082C, 0.31Mn,
0.07Ni, 0.045Cr, 0.02Mo, TrCu.

Specific heat medium-carbon steels.

Alloy temperature range (°F)	Specific heat (cal/gm/C)	
	1023	1040
122–212	0.116	0.116
302–392	0.124	0.123
392–482	0.127	0.126
482–572	0.133	0.131
572–662	0.137	0.136
662–752	0.143	0.140
842–932	0.158	0.155
1022–1112	0.179	0.169
1202–1292	0.202	0.184
1292–1382	0.342	0.378
1382–1472	0.227	0.149
1562–1652	—	0.131

Table B.4. Thermal conductivity of steels and aluminium alloys.

	(W/m.K)
Thermal conductivity of irons and steels	
Ingot iron	65.8
Wrought iron	59.7
Wrought steels:	
1020	46.7
1030	50.2
1040	46.7
1060	46.7
1117	46.7
4130	42.6
4340	37.6
4620	44.1
8630	37.6
9Ni4Co25	26.0
9Ni4Co45	23.0
18Ni Marage 200	20.9
18Ni Marage 250	25.3
Thermal conductivity of wrought stainless steels	
201	16.3
304	14.7
310	14.3
400	24.9
410	24.9
440	24.2
501	36.6
Thermal conductivity of wrought aluminium alloys	
1100	222.1
2024	121.1
3004	163.0
4032	138.5
5050	193.3
6061	167.3
7075	129.8

(*Continued*)

Table B.4.   (*Continued*)

Alloy	20° λ, (cal/cm.sec.K)	(W/m.K)
Thermal conductivity of aluminium alloys		
Al + 4%Cu	0.33–0.34	137.9–142.1
Al + 8%Cu	0.26–0.29	108.7–121.2
Al + 10%Mg	0.20	83.6
Al + 4%Mg	0.32	133.8
Al + 5%Si	0.33–0.37	137.9–154.7
Al + 13%Si	0.34–0.38	142.1–158.8
Al + 10%Cu; 1%Fe; 0.25%Mg	0.31–0.34	129.6–142.1
Al + 4%Cu; 1%Mg; 2%Ni	0.31–0.32	129.6–133.8
Al + 5%Si; 1.25%Cu; 0.5%Mg (aged)	0.33–0.35	137.9–146.3
	0.40 (annealed)	167.2
Al+ 12–14%Si; 1%Mg; 1%Cu; 1–2% Ni	0.28–0.34	117.0–142.1
Al + 1%Si; 0.6%Mg	{ 0.50 annealed	209.0
	0.40 quenched	167.2

	At 0°C (cal/cm.sec.K)
Thermal conductivity of irons at 0°C	
Electrolytic iron	0.2254
Armco iron 0.23%C; 0.007%Si; 0.025%Mn; 0.007%P; 0.020%S	0.1770
Steel 0.02%C; 0.03%Mn; 0.042%P; 0.005%S	0.174
Steel 0.06%C	0.153
Iron	0.1340

Table B.5.   Effect of temperature on thermal conductivity of iron and steel.

C	Mn	λ (cal/cm.sec.K)				
		100°C	200°C	300°C	400°C	500°C
		0.208	0.184	0.157	0.134	0.120
0.065	0.40	0.193	0.165	0.140	0.123	0.109
0.29	0.84	0.180	0.154	0.125	0.105	0.0907
0.52	0.63	0.162	0.132	0.109	0.0832	0.0750
0.85	0.65	0.160	0.122	0.102	0.0875	0.0695
1.10	0.55	0.156	0.119	0.100	0.0800	0.0650
1.40	0.53	0.152	0.119	0.0961	0.0760	0.0601

(*Continued*)

Table B.5.   (*Continued*)

Steel

No.	%				λ (cal/cm.sec.K)				
	C	Mn	Ni	Cr	100°	200°	300°	400°	500°
1	0.13	0.40	4.5	1.1	0.137	0.134	0.130	0.128	0.125
2	0.30	0.60	1.5	0.5	—	0.073	0.063	0.060	0.058
3	0.35	0.60	4.5	1.3	0.109	0.101	0.090	0.080	0.071
4	0.12	—	18.0	27.0	0.049	0.050	0.052	0.053	0.054
5	1.12	13.50	—	—	0.0442	0.0448	0.0455	0.0478	—

Thermal conductivity low-carbon steel

Temp. (°F)	Conductivity (Btu/sec/ft^2/°F/in)
32	0.114
212	0.111
392	0.102
572	0.095
752	0.088
932	0.079
1112	0.071
1292	0.064
1472	0.055
1832	0.053
2192	0.057

Material — 0.08C, 0.31Mn,
0.045Cr, 0.07Ni, 0.02Mo.

Thermal conductivity medium-carbon steel

Alloy Temperature (°F)	Thermal conductivity (Btu/sec/ft^2/°F/in)	
	1023	1040
32	0.100	0.100
212	0.098	0.098
392	0.094	0.093
572	0.089	0.088
752	0.082	0.081
932	0.076	0.073
1112	0.068	0.065
1292	0.061	0.058
1472	0.050	0.048
1832	0.052	0.052
2192	0.057	0.057

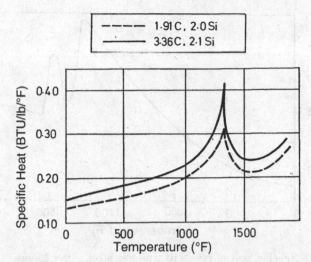

Fig. B.1. Effect of temperature on the specific heat of grey iron. After Moore (1973).

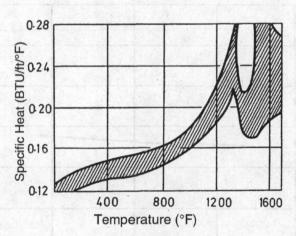

Fig. B.2. Effect of temperature on the specific heat of several white irons. After Moore (1973).

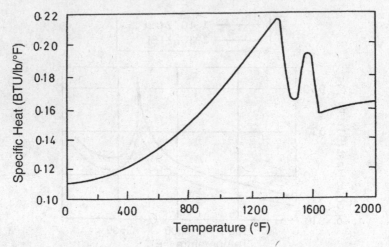

Fig. B.3. Specific heat of type 410 stainless steel. After Moore (1973).

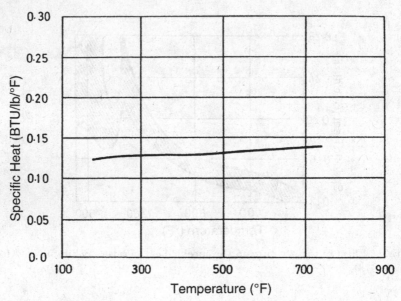

Fig. B.4. High-temperature specific heat of type 304 stainless steel. After Moore (1973).

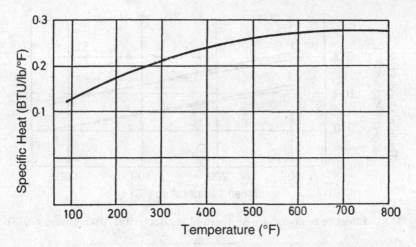

Fig. B.5. Specific heat for 6061 aluminium.

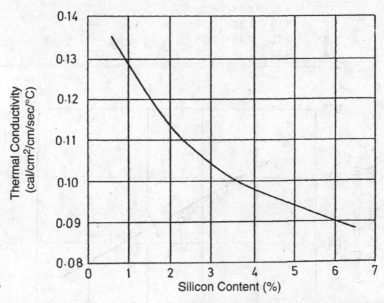

Fig. B.6. Effect of silicon on the thermal conductivity of iron. After Moore (1973).

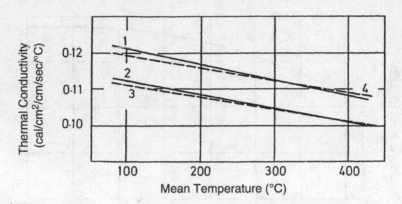

Fig. B.7.   Effect of temperature on thermal conductivity. After Moore (1973).

Iron no.	Chemical analysis %					
	C	Si	Mn	Cr	Cu	Mo
1	3.19	1.57	0.70	—	—	—
2	3.19	1.57	0.70	—	1.58	—
3	3.12	2.28	0.38	—	—	—
4	3.12	2.28	0.38	0.54	—	0.77

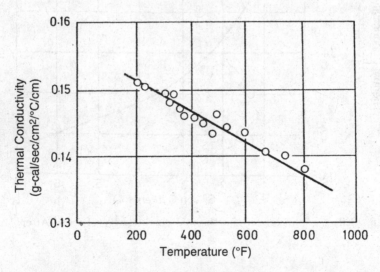

Fig. B.8.   Thermal conductivity of malleable cast iron. After Moore (1973).

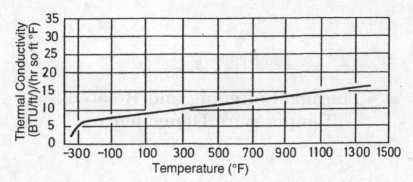

Fig. B.9. High-temperature thermal conductivity of type 304 stainless steel. After Moore (1973).

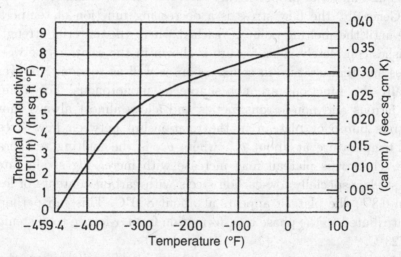

Fig. B.10. Low-temperature thermal conductivity of type 304 stainless steel. After Moore (1973).

Appendix C

Thermal Softening and Related Temperature Dependence

Data for the thermal softening of materials is not readily available in the literature. Also discussions about the dependence of thermal softening on various parameters such as strain, temperature, composition and material properties are even less readily available.

Generally, the flow stress is a decreasing function of temperature until the melting point is reached where the material strength vanishes completely. This feature is shown in the relations between stress and temperature in compression as well as torsion. However, usually the functions are not simple or monotonous. For example, ferrous and non-ferrous metals and alloys almost always show a stress bump or plateau in the relationship between flow stress and temperature at about θ_m, where θ_m is the melting temperature. The stress plateau may increase with increasing strain rate. For steels, especially low-carbon steels with carbon contents of less than 0.8%, the plateau appears at about 800°C. This can perhaps be attributed to the phase transformation from α-ferrite to austenite at 723°C.

At even higher strain rates, for example 10^3/sec, it appears that the plateaus become unclear or disappear completely for certain metals at about 0.5 θ_m. The temperature dependence on the flow stress can be simply described by an Arrhenius law:

$$\sigma \approx \exp(B/\theta).$$

The mechanisms which govern this dependence are recovery, recrystallization and grain growth if the strain rate is not high. However, the question of what mechanism dominates in determining the temperature dependence of the flow stress at high strain rates, remains open.

The strain-rate sensitivity parameter,

$$m = \frac{\partial \ln \tau}{\partial \ln \dot{\gamma}}$$

and the rate of strain hardening n are also temperature dependent. Limited data show that the former appears to increase with temperature whereas the latter decreases with temperature, as is expected physically.

However, it is interesting to see that large changes of these temperature dependences may occur in the range of half the homologous temperature (θ/θ_m).

It is worth noting that for iron the abrupt increase in the ratio of dynamic to static yield strength appears to be approximately in the same range of homologous temperatures as that between the phase transformation $\alpha \rightarrow \gamma$ from bcc to fcc and the Curie temperature.

Thermal softening curves are shown for the following materials:

(a) *Shear*

6060–T6 Al	(C.1)
6061–T6 Al	(C.2)
4340 steel	(C.3)
4140 steel	(C.4)
304 stainless steel	(C.5)
OFHC copper	(C.6)
IMI 130 Ti	(C.6, C.7)
Mild steel EN1B	(C.6, C.8)
4340 VAR steel	(C.9)
1100–0 Al	(C.10, C.11, C.12, C.13)

$$\frac{\Delta \ln \tau}{\Delta \ln \dot{\gamma}} \approx \theta$$

$$n \approx 0.$$

(b) *Compression*

Steel	(C.14)
Ti	
Ti–5A1	
TH–6A1	
Ti–Mo	(C.15)
Al	
Al–Mg	
Al–Cu	
Duralumin	
Non-ferrous metals	(C.16)–(C.24)
Steels and iron	(C.25)–(C.31)

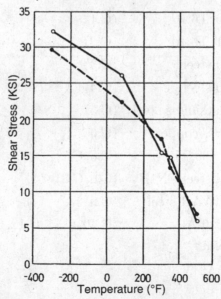

All specimens bonded at 54 psi
pressure at 1040°F for 60 minutes

Fig. C.1.   Shear strength of 6060–T6 aluminium simple lap joints. After Moore (1973).

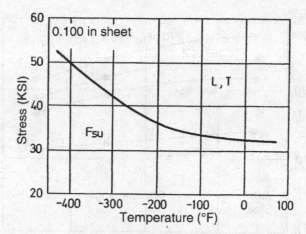

Fig. C.2. Shear strength of 6061–T6 aluminium sheet. After Moore (1973).

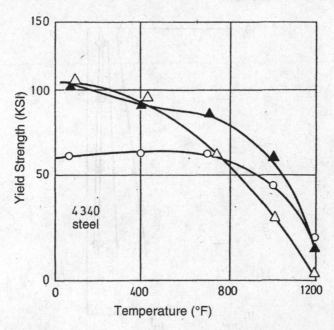

Fig. C.3. Yield strength of 4340 steel in torsion. After Moore (1973).

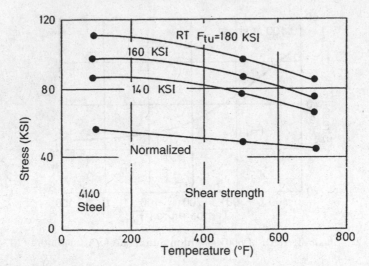

Fig. C.4. Effect of test temperature on shear strength of bar and forgings of 4140 steel. After Moore (1973).

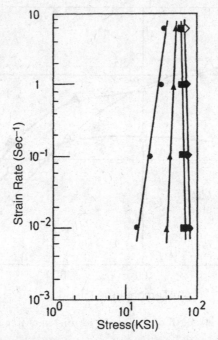

Fig. C.5.   Effect of temperature and strain rate of torsional equivalent flow stress for type 304 stainless steel. Torsional strain = 0.4. ◆ 400°C ■ 600°C ▲ A 800°C • 1000°C. After Moore (1973).

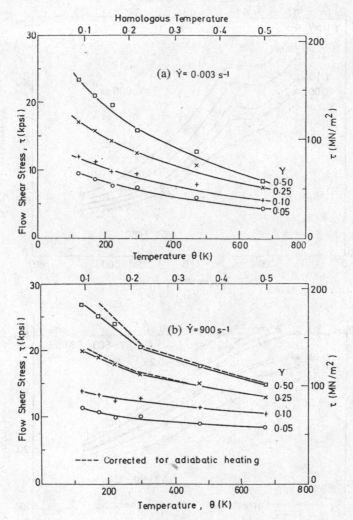

Fig. C.6. Temperature dependence of flow shear stress of copper at (a) low and (b) high strain rates. After Eleiche and Campbell (1976a,b).

Material	As supplied	Heat treatment of specimens	Grain density (mm^{-2})
Copper (OFHC)	Hot-rolled annealed rod	$400°C \times 2$ hrs @ $10^{-4} - 10^{-5}$ torr	360
Titanium (IMI 130)	Hot-rolled annealed rod	$700°C \times 1\frac{1}{2}$ hrs @ $10^{-4} - 10^{-5}$ torr	75
Mild Steel (EN1B)	Hot-rolled	None	690

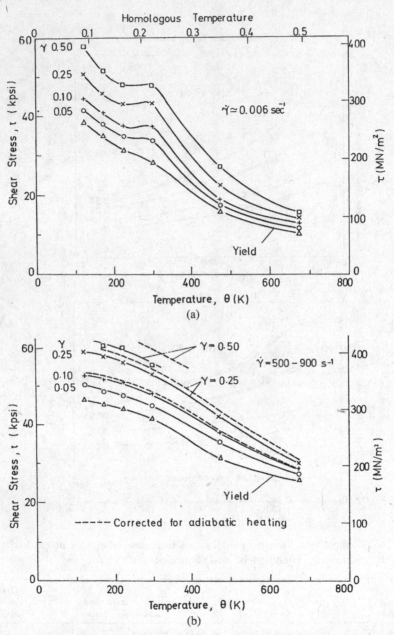

Fig. C.7.   Temperature dependence of flow shear stress of titanium at (a) low, and (b) high strain rates. After Eleiche and Campbell (1976).

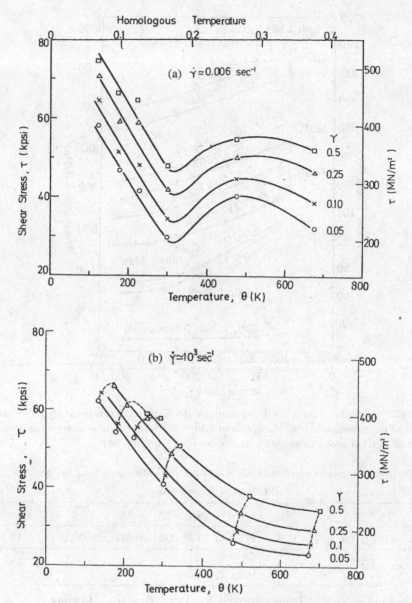

Fig. C.8.   Temperature dependence of shear stress of mild steel at (a) low and (b) high strain rates. After Eleiche and Campbell (1976).

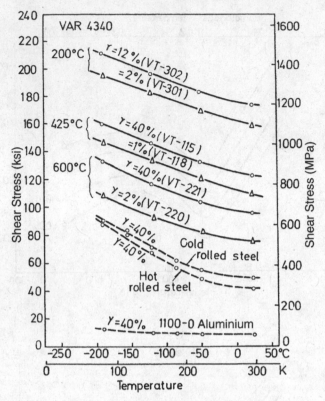

Fig. C.9.   Results of tests involving multiple decrements in temperature. All tests start at room temperature. The strain at which the temperature decrements occur is indicated on each curve. After Tanimura and Duffy (1984).

Chemical Composition of 4340 VAR Steel from Republic Steel: Heat No. 3841687

(Wt. % of 4340 Steel Alloy)

C	Mn	P	S	Si	Cu	Ni	Cr	Mo	Al	N	O	H ppm
0.42	0.46	0.009	0.001	0.28	0.19	1.74	0.89	0.21	0.031	0.005	0.001	1.0

Heat Treatment

	Temperature (°C)	Time (Hours)	Cooling
Normalize	900	1/2	Argon cool
Austenitize	845	1/2	Oil quench
Temper	600	1/2	Oil quench
	425	1/2	Oil quench
	200	1/2	Argon cool

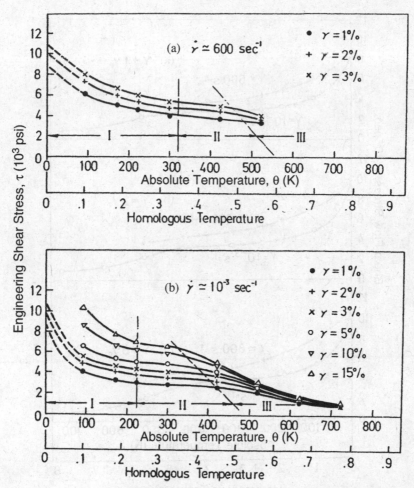

Fig. C.10.   Flow stress as a function of temperature, with strain as a parameter, at two different strain rates. Melting temperature $\theta_m = 933\,\text{K}$. CP aluminium. After Eleiche and Duffy (1975).

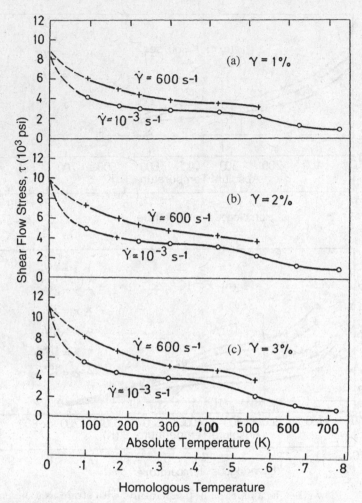

Fig. C.11.   Flow stress as a function of temperature, with strain as a parameter at three different strains. Melting temperature $\theta_m = 930\,\mathrm{K}$. CP aluminium. After Eleiche and Duffy (1975).

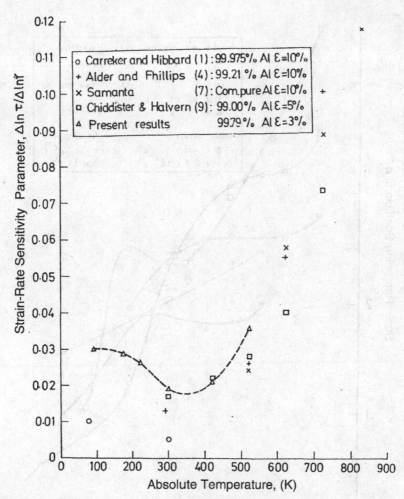

Fig. C.12.   Relation between strain-rate sensitivity parameter and temperature of 1100-0 aluminium in torsion. After Eleiche and Duffy (1975).

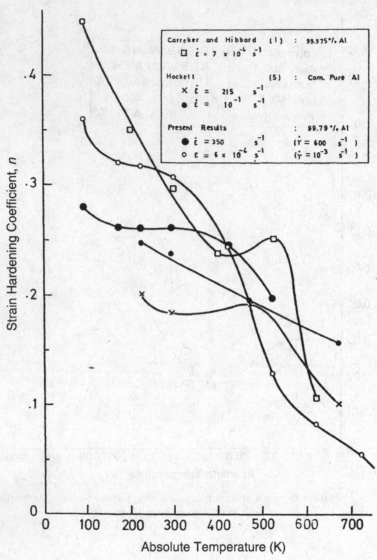

Fig. C.13. Strain hardening coefficient as a function of temperature of aluminium. After Eleiche and Duffy (1975).

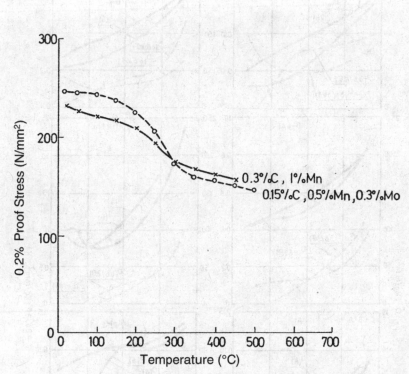

Fig. C.14. Variation of 0.2% proof stress with temperature of steel. After Turner *et al.* (1984).

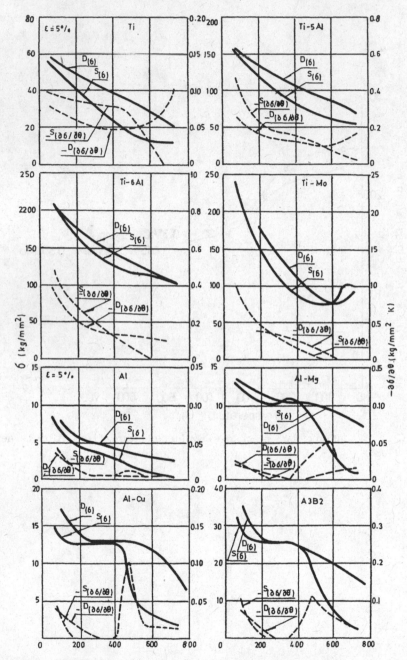

Fig. C.15.   Effect of temperature (x axis) on the flow stress (y axis) and the temperature sensitivity. After Tanaka *et al.* (1978).

Material	Chemical composition	Heat treatment	Vickers hardness	Specimen
Ti		Annealed at 650°C for 2 hours	150	$7^\varphi \times 7$
Ti–5Al	5.07%Al 1.29%Sn 0.32% Fe	Annealed at 800°C for 2 hours	330	$5^\varphi \times 7$
Ti–6Al	5.99%Al 4.14%V 0.22%Fe	Annealed at 700°C for 2 hours	435	$5^\varphi \times 7$
Ti–Mo	14.8%Mo 5.09%Zr 0.04%Fe	Annealed at 800°C for 2 hours	330	$5^\varphi \times 7$
Al	99.997%Al	Annealed at 500°C for 1 hour	20	$8^\varphi \times 8$
Al–Mg	1.08%Mg 0.01%Cu 0.06%Si 0.01%Fe	Annealed at 500°C for 1 hour	35	$8^\varphi \times 8$
Al–Cu	4.01%Cu	Solution treated at 530°C for 1 hour Quenched in ice water Aged at 350°C for 36 hours	34	$5^\varphi \times 7$ $6^\varphi \times 7$
A3B2 (2017)	(Duralumin)	Solution treated at 500°C for 1 hour Quenched in ice water Aged at 350°C for 2 hours	67	$6^\varphi \times 6$

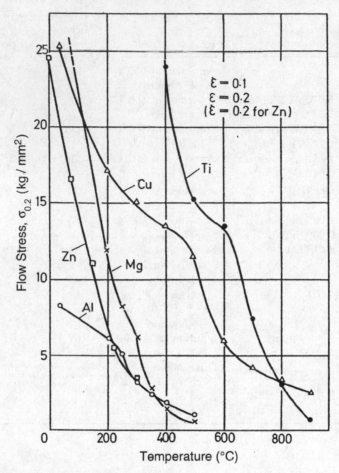

Fig. C.16. Temperature dependence of the flow stress of commercial purity metals. After Suzuki *et al.* (1968).

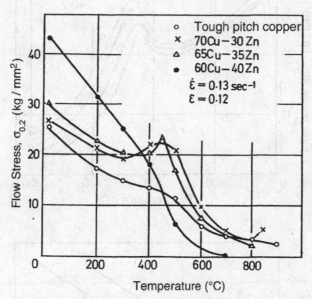

Fig. C.17.   Temperature dependence of the compression stress of copper and copper–zinc alloys at $\epsilon = 0.2$, strain rate $= 0.13/\text{sec}$. After Suzuki *et al.* (1968).

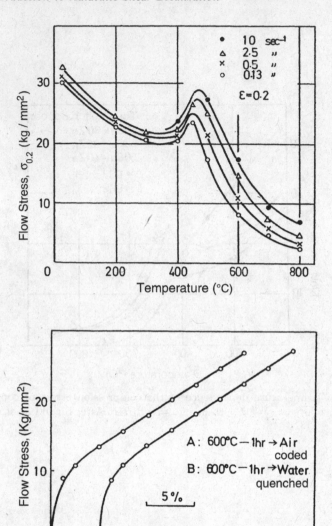

Fig. C.18.   Temperature dependence of the compression stress of 65%Cu–35%Zn alloys at $\epsilon = 0.2$. After Suzuki *et al.* (1968).

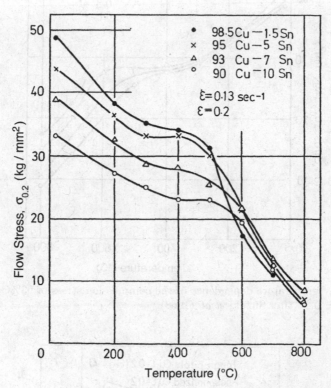

Fig. C.19.   Temperature dependence of the compression stress of Cu–Sn alloys at $\epsilon = 0.2$. After Suzuki *et al.* (1968).

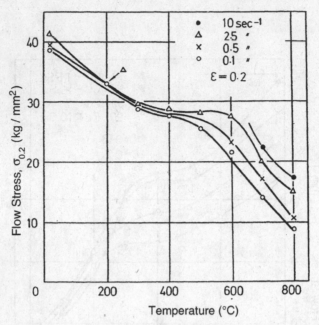

Fig. C.20.   Temperature dependence of the compression stress of 93%Cu–7%Sn alloys at $\epsilon = 0.2$. After Suzuki *et al.* (1968).

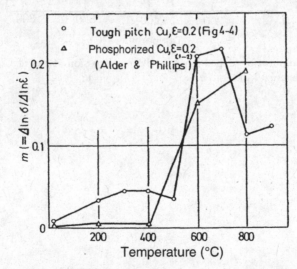

Fig. C.21.   *m* value versus temperature relation for copper. After Suzuki *et al.* (1968).

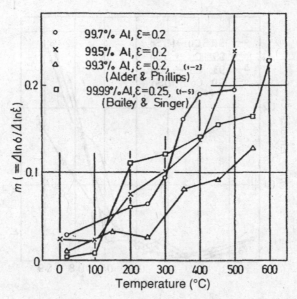

Fig. C.22.   *m* value versus temperature relation for aluminium. After Suzuki *et al.* (1968).

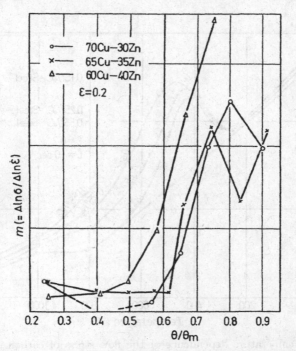

Fig. C.23.   *m* value versus reduced temperature relation for Cu–Zn alloys. After Suzuki *et al.* (1968).

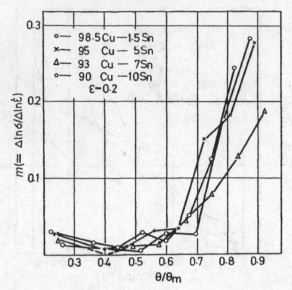

Fig. C.24. *m* value versus reduced temperature relation for Cu–Sn alloys. After Suzuki *et al.* (1968).

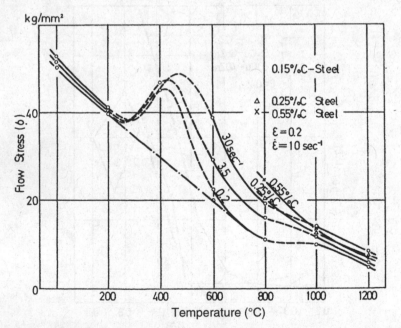

Fig. C.25. Temperature dependence of the flow stress of carbon steel. Strain rate range: 0.2–30/sec. After Suzuki *et al.* (1968).

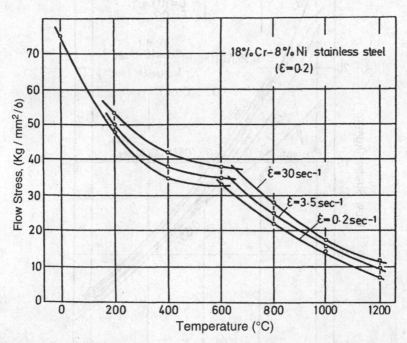

Fig. C.26. Temperature dependence of the flow stress of 18%Cr–8%Ni stainless steel. After Suzuki *et al.* (1968).

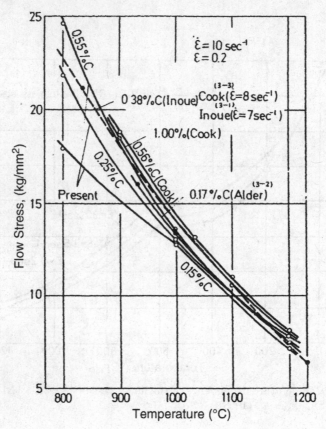

Fig. C.27.   Comparison of the flow stress of carbon steels measured by various investigators. After Suzuki *et al.* (1968).

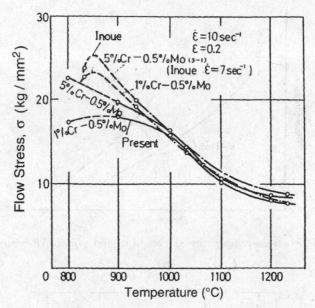

Fig. C.28. Comparison of the flow stress of low alloy steels. After Suzuki *et al.* (1968).

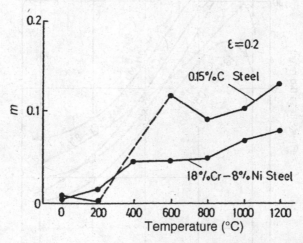

Fig. C.29. *m* value versus temperature curves for 0.15%C steel and 18%Cr–8%Ni stainless steel. After Suzuki *et al.* (1968).

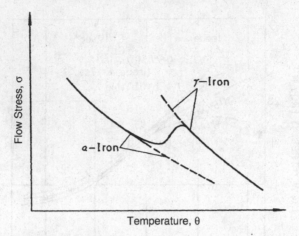

Fig. C.30. Flow stress–temperature curves of α- and γ-iron. After Suzuki *et al.* (1968).

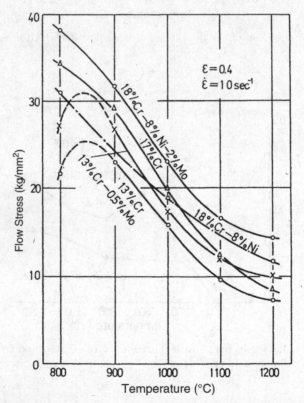

Fig. C.31. Composition dependence of the flow stress of stainless steels. After Suzuki *et al.* (1968).

Appendix D

Materials Showing Adiabatic Shear Bands

Materials		Method	Loading rate	Remarks	References
Al alloys					
2014-T6 (145 HV)	P	Impact	$v = 200$–900 m/sec	Shear band including cracks. Knobbly fracture surface	Stock and Thompson (1970)
2014-T6	O	Machining	$\dot{\gamma} = 160$/sec	Deformed bands, $w = 5$–$200\,\mu$m	Dao and Shockey (1979)
2014 (75 HV) (overaged)	P	Impact	$v = 200$–900 m/sec	Shear bands including cracks	Stock and Thompson (1970)
2024-T4	P	Impact	—	Similarity of trajectories of shear bands	Backman and Finnegan (1973)
6061-T6	D	Torsion	$\dot{\gamma} = 74$/sec	Deformed bands, $w = 500\,\mu$m	Culver (1973)
6061-T6	S	Compression	$\dot{\gamma} = 10^8$/sec	$w = 0.1 \sim 0.3\,\mu$m, spacing $= 2$–$10\,\mu$m	Grady and Kipp (1987)
7039	D	Compression	$\epsilon = 8000$/sec	Diagonal shear banding with cracks	Wulf (1979)
7039	D	Torsion	$\dot{\gamma} = 167$/sec	Multiple shear banding	Johnson *et al.* (1983b)
7039 (80, 120, 150 HV)	P	Impact	$v = 320$ m/sec	Knobbly fracture surface $w = 20$–$90\,\mu$m	Leech (1985)
7039-T6	P	Impact	$v = 200$–350 m/sec	Shear banding with cracks	Woodward *et al.* (1984)
7039-T6	P	Impact	$v = 325$ m/sec	Separation of dead zone	Leech (1985)
7075-T651	P	Impact	$v = 80$–180 m/sec	Dimple pattern and knobbles on deformed surface	Chrisman and Shewmon (1979a,b)

(Continued)

(Continued)

Materials		Method	Loading rate	Remarks	References
Aluminium alloy		Impact			Wingrove (1973a,b)
Ti and Ti alloys					
Titanium	P	Impact	$v = 0.72\,\text{m/sec}$		Pak et al. (1986)
Commercial purity titanium	O	Machining		Intense shear bands, $w = 2.5\,\mu\text{m}$	Recht (1964)
	D	Torsion	$\dot{\gamma} = 320/\text{sec}$	Intense shear bands	Culver (1973)
	P	Impact	$v = 330\,\text{m/sec}$	Shear bands including cracks and voids	Winter (1975)
	P	Impact	$v = 120\text{–}175\,\text{m/sec}$	Intense shear bands. Dimple patterns	Winter and Hutchings (1974)
	D	Compression	$\dot{\epsilon} = 8000/\text{sec}$	Intense shear bands $w = 10\,\mu\text{m}$	Wulf (1979)
	O	Machining	Slow speed	Intense shear bands	Turley et al. (1982)
	P	Impact	$v = 600\text{–}800\,\text{m/sec}$	Shear bands including cracks and voids	Grebe et al. (1985)
	D	Torsion	$\dot{\gamma} = 0.011\text{–}224/\text{sec}$	$w = 0.12\text{–}0.22\,\text{mm}$	Kobayashi (1987)
	D	Compression	$\dot{\epsilon} = 8000/\text{sec}$	Transformed bands ('T-bands')	Wulf (1979)
Ti–6Al–4V	O	Machining	$v = 2 \times 10^5$, $5.1\,\text{m/sec}$	T-bands, $w = 6\,\mu\text{m}$	Komanduri (1982)
	P	Impact	$v = 150\text{–}300\,\text{m/sec}$	T-bands, $w = 5\text{–}10\,\mu\text{m}$	Timothy and Hutchings (1981)
	P	Impact	$\dot{\gamma} = 3 \times 10^5/\text{sec}$	T-bands, $w = 3\text{–}40\,\mu\text{m}$	Me-Bar and Shechtman (1983)

(Continued)

(Continued)

Materials		Method	Loading rate	Remarks	References
	P	Impact	$v = 330$ m/sec	Tested at -100, $500°$C. T-bands, $w = 5$–$15\,\mu$m	Timothy and Hutchings (1984a,b)
	P	Impact	$v = 247$ m/sec	T-bands including voids and cracks	Woodward et al. (1984)
	P	Impact	$v = 500$–850 m/sec	T-bands including voids and cracks	Grebe et al. (1985)
	P	Impact	$v = 315$ m/sec	T-bands including voids and cracks	Timothy and Hutchings (1985)
Ti-8Al-7Mo-1V	D	Compression	$\dot{\epsilon} \sim 4000$/sec	$w = 10$–$50\,\mu$m	Xu et al. (1989)
Ti-13V-10C-3Al	D	Compression	$\dot{\epsilon} = 8000$/sec	T-bands	Wulf (1979)
	C	Explosion	$\dot{\gamma} = 10^5$–5×10^5/sec	Intense shear bands, $w = 10$–$30\,\mu$m	Stelly et al. (1981)
β-Ti alloy TB-2	D	Compression	$\dot{\gamma} = 2000$–3000/sec	$w = 10$–$100\,\mu$m	Lu et al. (1986)
TA6V titanium alloy, XC48 steel, maraging steel TA6V titanium alloy		Torsion			Chiem et al. (1986)
Steels					
CRS-1018 (0.18C, 0.77Mn)	D	Torsion	$\dot{\gamma} = 500$/sec	Deformed bands (D-bands), $w = 200$–$250\,\mu$m	Costin et al. (1979)
CRS-1018	D	Torsion	$\dot{\gamma} = 1000$/sec	D-bands, $w = 250\,\mu$m	Hartley et al. (1987)
Low-carbon steel	D	Torsion	$\dot{\gamma} = 0.007$–166/sec	Measuring temp. profile, $w = 260$–$340\,\mu$m	Kobayashi (1987)

(Continued)

(*Continued*)

Materials	Method	Loading rate	Remarks	References
HRS-20#	D Torsion	$\dot{\gamma} \sim 1500$/sec	D-bands, $w \sim 400\,\mu m$	Huang (1987)
HRS-1020(0.26C, 0.5Mn)	D Torsion	$\dot{\gamma} =$	D-bands, $w = 200\,\mu m$, Temp, profile in bands	Hartley *et al.* (1987)
SEA-1020	C Explosion	2000–5000/sec	Intense shear bands with cracks	Backman and Finnegan (1973)
AISI-1040	P Impact	—	Transformed bands (T-bands) with HAZ	Rogers (1983)
AISI-1215 (0.07C, 1.04Mn)	D Torsion	$\dot{\gamma} =$ 2000–3000/sec	Multiple shear banding D-bands	Hartley and Duffy (1984)
SEA-4130	P Impact	$v = 1240$ m/sec	T-bands	Backman and Finnegan (1973)
Low alloyed steel, 0.34% C–0.5% Mn 1045 steel, 1015–1040–4142–4340 steels	Expansion of shells (spheres, cylinders)			Lamborn *et al.* (1974); Thornton and Heiser (1971); Woodward and Aghan (1978); Pearson and Finnegan (1981)
4340 steel	P Impact			Pak *et al.* (1986)
AISI-4340	P Impact	$v = 700$ m/sec	— T-bands with cracks	Mescall and Papirno (1974)
AISI-4340 (400, 450 HV)	D Compression	$\dot{\epsilon} = 10^4$/sec	Shear failure	Wulf (1979)
AISI-4340 (40 HRC)	O Machining	$\dot{\gamma} = 240$/sec	T-bands, $w = 20\,\mu m$	Dao and Shockey (1979)

(*Continued*)

(Continued)

Materials		Method	Loading rate	Remarks	References
AISI-4340	C	Explosion	—	T-bands, thermal instability strain	Staker (1981)
AISI-4340 (21, 40, 52 HRC)	C	Explosion	—	T-bands with cracks	Shockey and Erlich (1981)
4340	D	Torsion	—	$w = 20\,\mu m$	Giovanola (1987)
AMS-6418	D	Torsion	$\dot{\gamma} = 98/\sec$	T-bands, voids on fracture surface	Lindholm and Hargreaves (1976)
HF-1	C	Explosion	—	T-bands with cracks, $w = 5-15\,\mu m$	Erlich *et al.* (1977)
Armco iron	C	Explosion	—	Deformed bands (D-bands) with cracks	Seaman and Shockey (1975)
Mild steel (annealed)	D	Torsion	$\dot{\gamma} = 105/\sec$	D-bands	Culver (1973)
Mild steel	P	Impact	$v = 160\,m/\sec$	D-bands	Winter and Hutchings (1974)
Carbon steel (0.5C)	C	Explosion	—	Transformed bands (T-bands), $w = 10-16\,\mu m$	Woodward and Aghan (1978)
Carbon steel (1C)	D	Shearing	$v = 4-14\,m/\sec$	T-bands	Stock and Thompson (1970)
Carbon steel (1C)	P	Impact	$v = 300-1000\,m/\sec$	T-bands including cracks	Wingrove and Wulf (1973)
Steel (363 HB)	D	Blanking	$\dot{\gamma} = 2000/\sec$	T-bands	Zener and Hollomon (1944)

(Continued)

(*Continued*)

Materials		Method	Loading rate	Remarks	References
Steel (315, 425 HV) (0.9C, 1.2Mn, 0.5Cr)	D	Blanking	$v = 10\,\mathrm{m/sec}$	T-bands, $w = 4-20\,\mu\mathrm{m}$	Balendra and Travis (1970)
Steel (0.52C, 18.5Ni)	O	Machining	$v = 0.003-0.008\,\mathrm{m/sec}$	T-bands, 3 types of chip formation	Lemaire and Backofen (1972)
Steel (1C, 1Cr)	P	Impact	$v = 300-1000\,\mathrm{m/sec}$	T-bands including cracks	Wingrove and Wulf (1973)
Steel (350, 420 HV) (0.22C, 0.65Mn)	P	Impact	$v = 300-1000\,\mathrm{m/sec}$	T-bands including cracks	Wingrove and Wulf (1973)
Ni–Cr steel (0.24C, 3.2Ni, 1.12Cr)	P	Impact	$v = 4000-7400\,\mathrm{m/sec}$	T-bands, $w = 1-5\,\mu\mathrm{m}$	Shockey et al. (1975)
Ni–Cr steel (0.22C, 3.15Ni, 1.06Cr)	D	Blanking	$\dot{\gamma} = 9.4 \times 10^7/\mathrm{sec}$	T-bands, $w = 35\,\mu\mathrm{m}$	Moss (1981)
S-7 tool steel	D	Torsion	$\dot{\gamma} = 153/\mathrm{sec}$	D-bands including voids	Johnson and Slater (1967)
Hoffman steel ball-bearing (140 HV)	P	Impact	$v = 600, 1200\,\mathrm{m/sec}$	T-bands	Carrington and Gayler (1948)
Martensitic steel, Martensitic steel, TA6V titanium alloy		Compression			Affouard (1984)
28CND8, Z50CDV5, 35NCDV16, 4340 steels					Meyer and Manwaring (1986)
Armoured steels					Affouard et al. (1984)

(*Continued*)

(Continued)

Materials	Method		Loading rate	Remarks	References
18-4-1 alloy steel CrMo V steel TA6V alloy, 35NCD16 steel					Doraivelu et al. (1981) Hartmann et al. (1981) Dormeval and Ansart (1985)
Non-ferrous materials					
Copper	P	Impact	$v = 330\,\text{m/sec}$	Deformed bands, $w = 500\,\mu\text{m}$	Winter (1975)
OFHC copper	D	Torsion	$\dot{\gamma} = 330/\text{sec}$	Deformed bands, $w = 340\,\mu\text{m}$	Lindholm et al. (1980)
Magnesium alloy (Mg–8Al–0.5Zn–0.3Mn)	P	Impact	$v = 200{-}320\,\text{m/sec}$	Deformed bands with cracks	
Tungsten alloy (W–7Ni–3Fe)	P	Impact	$v = 850\,\text{m/sec}$	Deformed bands with cracks	Erlich et al. (1977)
Uranium alloy (U–2Mo)	P	Impact	—	Deformed bands with voids	Irwin (1972)
Uranium alloy (U–1.5Mo)	C	Explosion	$\dot{\gamma} = 10^5{-}5 \times 10^5/\text{sec}$	Deformed bands with cracks, $w = 10{-}30\,\mu\text{m}$	Stelly et al. (1981)
Al–Li alloy as received	D	Torsion	$\dot{\gamma} = 0.008{-}237/\text{sec}$		Kobayashi (1987)
as received (8090)	D	Torsion	$\dot{\gamma} = 1300{-}2600/\text{sec}$	$w = 146{-}445\,\mu\text{m}$	Tian and Bai (1985)

P Impact
O Machining
D Torsion
S Compression shock wave compression

D Compression
D Blanking
D Shearing
C Explosion

Specification of Selected Materials Showing Adiabatic Shear Bands

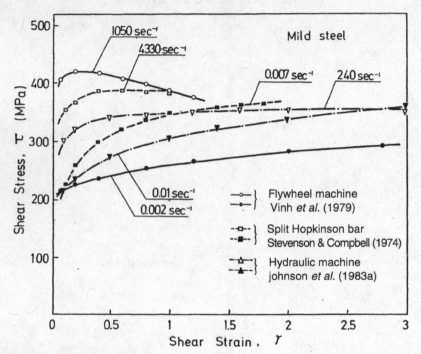

Fig. E.1. Comparison of shear stress–strain curves for mild steel. After Kobayashi (1987).

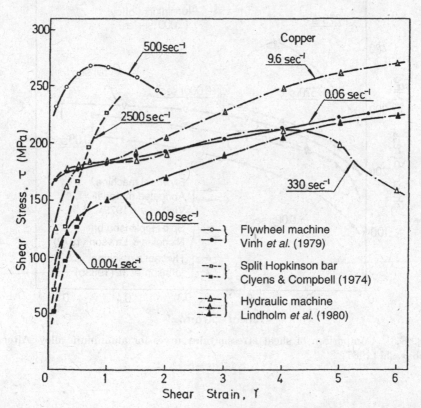

Fig. E.2. Comparison of shear stress–strain curves for OFHC copper. After Kobayashi (1987).

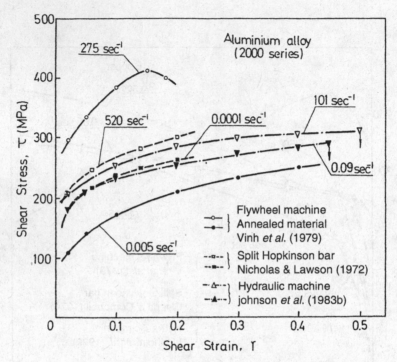

Fig. E.3.   Comparison of shear stress-strain curves for aluminium alloy. After Kobayashi (1987).

Table E.1.

Material	Thermal properties			Constitutive constants				Temperature @ γ_c (°C)	Critical strain	
	Density ρ (kg/m3)	Specific Heat C_p (J/kg K)	Melting Temp. θ_M (K)	A (MPa)	B (MPa)	n	C		Theoretical	Experimental
OFHC copper (CDA 101)	8950	383	1355	69	106	0.32	0.027	252	5.3	5.8
Cartridge brass (CDA 260)	8520	385	1189	62	186	0.34	0.007	241	3.7	3.0
Aluminium 2024-T 351	2770	875	775	152	202	0.34	0.015	89	0.66	0.50
Aluminium 7039	2770	875	877	193	157	0.41	0.010	100	0.77	0.55
Nickel 200	8900	446	1726	138	234	0.32	0.008	148	0.03	0.18
Armco IF iron	7890	452	1811	76	196	0.25	0.028	296	4.3	4.1
Carpenter electrical iron	7890	452	1811	193	109	0.43	0.028	327	4.4	5.8
1006 steel	7890	452	1811	200	129	0.36	0.022	269	3.3	3.5
RHA steel	7840	477	1793	455	237	0.37	0.006	192	1.2	1.1
AMS 6418 steel	7750	477	1763	896	200	0.18	0.010	65	0.16	0.20
S-7 tool steel	7750	477	1763	883	248	0.18	0.012	67	0.16	0.50
Tungsten alloy (7%Ni, 3%Fe)	17000	134	1723	862	94	0.12	0.016	38	0.03	0.18
Depleted uranium (0.75 Ti)	18600	117	1473	621	561	0.25	0.007	65	0.23	0.25

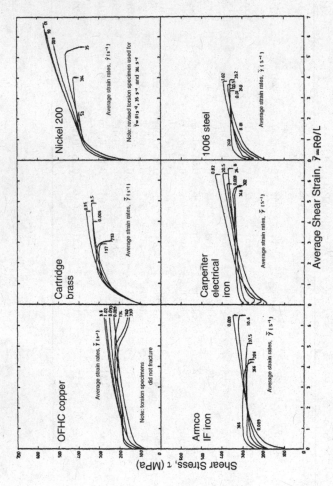

Fig. E.4.   Torsional stress–strain test data at various strain rates. After Lindholm and Johnson (1983).

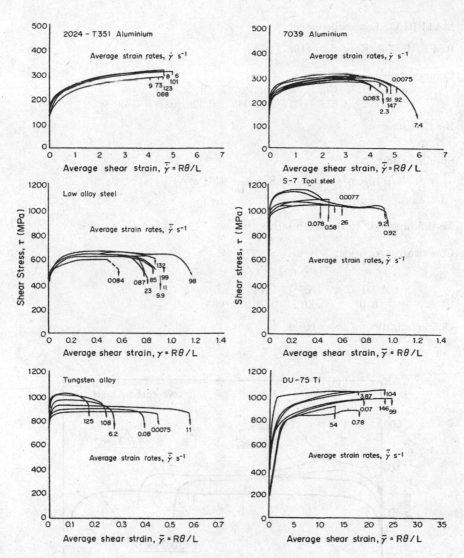

Fig. E.4. (*Continued*)

MATERIAL: Low-carbon steel

REFERENCE: Kobayashi (1987)

Composition:

	C	Mn	P	S	Si	Pb
wt %	0.073	1.11	0.0085	0.32	0.03	0.19

Heat treatment:

Specification:

ρ (g/cm^3)	c (J/Kg.K)	$\frac{\partial \tau}{\partial \theta}$ (MPa/K)
7.89	437	−0.423

Testing method: Torsion with lathe

Constitutive equation:
$$\tau = A\dot{\gamma}^m \gamma^n (1 - a\theta)$$

A(MPa)	a (K^{-1})	m	n
640	0.0012	0.012	0.05

Characterization of shear band:
$\gamma_i \sim 0.28$
$w \sim 260 - 340\,\mu m$

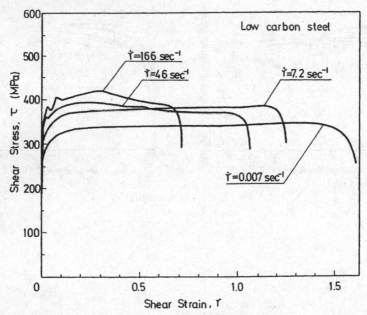

Fig. E.5.   Torsional stress–strain curves for low-carbon steel. After Kobayashi (1987).

MATERIAL: AISI 1018 cold-rolled steel (CRS)

REFERENCE: Costin *et al.* (1979)

Shawki *et al.* (1983)

Hartley *et al.* (1987)

Composition:

	C	Mn	P	S
wt %	0.18	0.71	0.020	0.022

Heat treatment: Annealed in an argon atmosphere at 900°C for 1 hr, then cooled

at 25°C/hr to 700°C, finally furnace cooled.

Specification:

ρ (g/cm^3)	c (J/Kg.K)	λ (W/m.K)
7.9	500	54

Testing method: Split Hopkinson torsional bar technique

$$\dot{\gamma} \sim (700-1200)/\text{sec}$$

Constitutive equation:

$$\tau = c(1 - a\theta)(1 + \dot{\gamma})^m \gamma^n$$

c (MPa)	a (K^{-1})	b (sec)	m	n
614	0.0015	10^4	0.025	0.05

$$\tau = A\dot{\gamma}^m \gamma^n \theta^{-\gamma}$$

A (MPa)	m	n	γ
436	0.019	0.015	0.38

Characterization of shear band:

$$\gamma_i \sim 0.1-0.2$$
$$w \sim 250\,\mu\text{m}$$

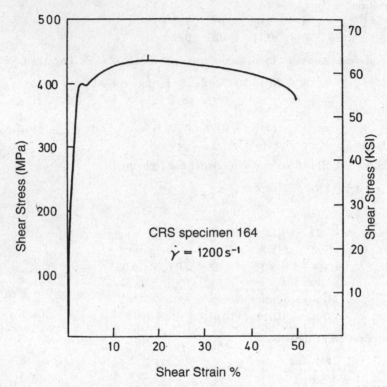

Fig. E.6. Torsional stress–strain curve for AISI 1018 cold-rolled steel (CRS). After Hartley *et al.* (1987).

MATERIAL: AISI 1020 hot-rolled steel (HRS)

REFERENCE: Hartley *et al.* (1987)

    Costin *et al.* (1979)

Composition:

	C	Mn	P	S
wt %	0.26	0.50	0.018	0.030

Heat treatment:

Specification:

ρ (g/cm^3)	c (J/Kg.K)	λ (W/m.K)
7.9	500	54

Testing method: Split Hopkinson torsional bar technique

  $\dot{\gamma} \sim (2000-5000)/\mathrm{sec}$

Constitutive equation:

  $\tau = c(1 - a\theta)(1 + b\dot{\gamma})^m \gamma^n$

c (MPa)	a (K^{-1})	b (sec)	m	n
531	0.0012	10^4	0.034	0.20

  $\tau = A\dot{\gamma}^m \gamma^n \theta^{-\gamma}$

A (MPa)	m	n	γ
261	0.0133	0.12	0.51

  (Shawki *et al.*, 1983)

Characterization of shear band:

  $\gamma_i \sim (0.5-1.4)$

  $w \sim 150\,\mu\mathrm{m}$

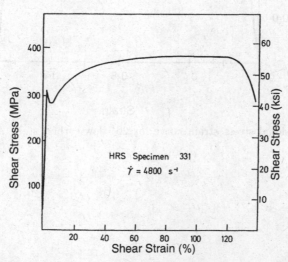

HRS Specimen 331
$\dot{\gamma} = 4800\ \mathrm{s}^{-1}$

Fig. E.7. Torsional stress–strain curve for AISI 1020 hot-rolled steel (HRS). After Hartley *et al.* (1987).

MATERIAL: 20$^{\#}$ low-carbon steel

REFERENCE: Huang (1987)

Composition:

	C	Mn	P	S	Si
wt %	0.21	0.52	0.016	0.011	0.34

Heat treatment: Normalization at 900°C, hot-rolled.

Specification:

ρ (g/cm^3)	c (cal/g.K)	(J/Kg.K)	λ (cal/cm.sec.K.)	(W/m.K)
7.82	0.112	468.2	0.121	50.6
	(20–100°C)		(100°C)	
	0.115	480.7	0.116	48.5
	(20–100°C)		(200°C)	

Testing method: Split Hopkinson torsional bar technique

$\dot{\gamma} \sim 1500/\text{sec}$

Characterization of shear band:

$\gamma_1 = 0.82 \pm (2\%)$

$w \sim 400\,\mu\text{m}$

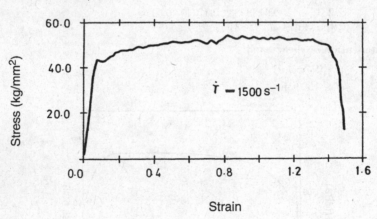

Fig. E.8.   Torsional stress–strain curve for 20$^{\#}$ low-carbon steel. After Huang (1987).

MATERIAL: HY 100 steel

REFERENCE: Marchand and Duffy (1988)

Composition:

	C	Si	S	Mn	P	Ni	Cr	Mo	Cu	Ti	Va
wt %	0.18	0.20	0.005	0.25	0.003	2.51	1.63	0.43	0.037	0.001	0.005

Heat treatment: Tempered martensite

Specification:

Testing method: Split Hopkinson torsional bar technique

$$\dot{\gamma} \sim (300-5000) \text{ sec} \quad \theta \sim (-190 - +250)^\circ C$$

Characterization of shear band:

$$\gamma_i \sim 0.27$$
$$w \sim 20\text{--}40\,\mu m$$

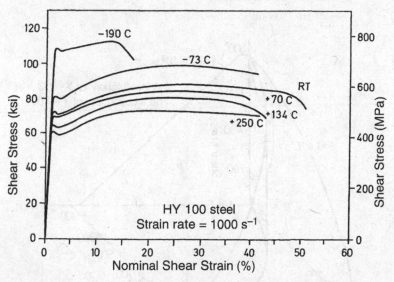

Fig. E.9.   Torsional stress–strain curve for HY 100 steel. After Marchand and Duffy (1988).

MATERIAL: 4340 steel

REFERENCE: Giovanola (1987)

Composition:

Heat treatment: Vacuum arc remelted, quenched and tempered to hardness of
HRC 40

Specification:

Testing method: Split Hopkinson torsional bar technique

Characterization of shear band:

$\gamma_i = 0.13$ (first localization)

$\quad\quad 1.0$ (second localization)

$w = 60\,\mu$m (first localization)

$\quad\quad 20\,\mu$m (second localization)

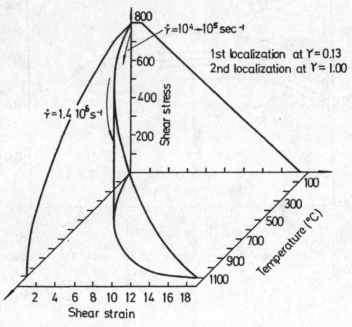

Fig. E.10. Torsional stress–strain–temperature curve for 4340 steel. After Giovanola (1987).

MATERIAL: CP titanium (commercial purity titanium)

REFERENCE: Kobayashi (1987)

Composition:

	O	N	H	C	Fe	Al	Pb	S	V
wt %	0.195	0.004	0.0015	0.037	0.038	0.01	0.01	0.01	0.01

Heat treatment:

Specification:

ρ (g/cm^3)	c (J/Kg.K)	$\frac{\partial \tau}{\partial \theta}$ (MPa)
4.51	528	-0.595

Testing method: Torsion with lathe

Constitutive equation:
$$\tau = A\dot{\gamma}^m \gamma^n (1 - a\theta)$$

A (MPa)	a (K^{-1})	m	n
800	0.0015	0.033	0.15

Characterization of shear band:

$\gamma_i = 0.5$

$w = 120-200\,\mu\text{m}$

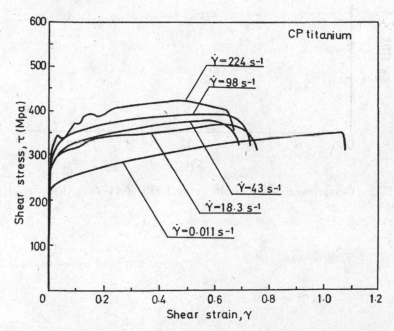

Fig. E.11. Torsional stress–strain curve for CP titanium. After Kobayashi (1987).

MATERIAL: Ti–6Al–4V

REFERENCE: Xu *et al.* (1989),
Heat treatment: 750°C for 1 hr, quenched in air, $\alpha + \beta$ phase

Specification:

Testing method: Split Hopkinson compression bar

$\dot{\epsilon} \sim 4000/\text{sec}$

Constitutive equation:

Characterization of shear band:

$w = 10{-}50\,\mu\text{m}$

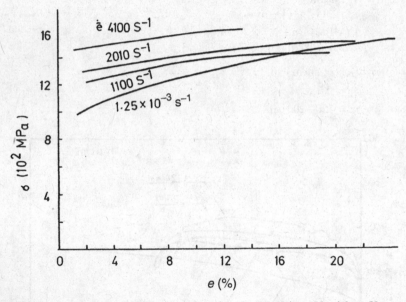

Fig. E.12.   Compression stress–strain curve for Ti–6A1–4V. After Xu *et al.* (1989).

MATERIAL: TB-2 β-Ti alloy

REFERENCE: Lu *et al.* (1986)

Composition:

	Mo	V	Cr	Al
wt %	4.8–5.4	5.25–5.28	8.02–8.55	2.76–2.99

Heat treatment: Solid solution, β-pbase

Specification:

ρ (g/cm^3)	c (J/Kg.K)
4.5	530

Testing method: Impulsive compression

Constitutive equation:

Characterization of shear band:

$$w = 10-10^2 \, \mu m$$

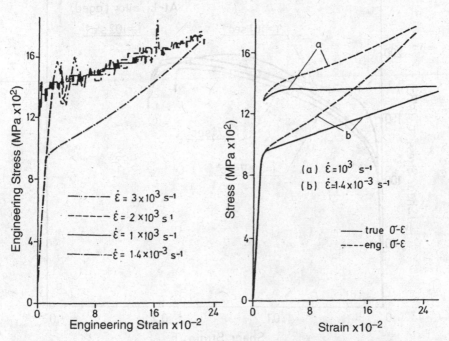

Fig. E.13. Compression stress–strain curves for TB-2 β-Ti alloy. Left: Engineering curves, —·— $\dot{\epsilon} = 3 \times 10^3$/sec, - - - - - $\dot{\epsilon} = 2 \times 10^3$/sec, ———— $\dot{\epsilon} = 1 \times 10^3$/sec, —·—·— $\dot{\epsilon} = 1.4 \times 10^3$/sec. Right: True σ versus ϵ curves (————) and engineering curves (- - - - -): (a) $\dot{\epsilon} = 10^3$/sec and (b) $\dot{\epsilon} = 1.4 \times 10^{-3}$/sec. After Lu *et al.* (1986).

MATERIAL: Al–Li Alloy (aged, Alcan Inter. Ltd)

REFERENCE: Kobayashi (1987)

Composition:

	Li	Fe	Si	Cu	Mg
wt %	2.76	0.08	0.03	0.01	0.01

Heat treatment: Aging for one or two hours at 190°C.

Specification:

Testing method: Torsion with lathe

Constitutive equation:

Characterization of shear band:

$$\gamma_i = 140 - 200 \, \mu\text{m}$$

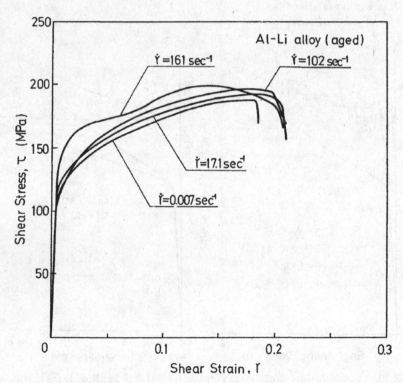

Fig. E.14.   Torsional stress–strain curves for Al-Li alloy (aged). After Kobayashi (1987).

MATERIAL: Al–Li alloy (as received, Alcan Inter. Ltd)

REFERENCE: Kobayashi (1987)

Composition:

	Li	Fe	Si	Cu	Mg
wt %	2.76	0.08	0.03	0.01	0.01

Heat treatment: Solution treated and stretched

Specification:

Testing method: Torsion with lathe

Constitutive equation:

Characterization of shear band:

$$\gamma_i = 420{-}460\,\mu\text{m}$$

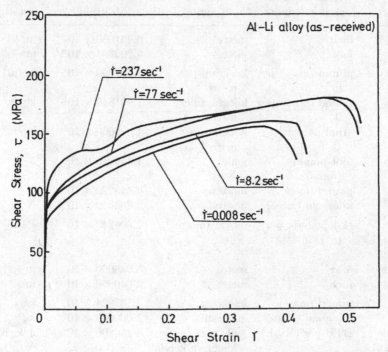

Fig. E.15.   Torsional stress–strain curves for Al–Li alloy (as received, Alcan Inter. Ltd.). After Kobayashi (1987).

Appendix F

Conversion Factors

Property	To convert from imperial or metric units	To SI units	Multiply by	SI notation
Area	inch^2	metre^2	6.451600×10^{-4}	m^2
	foot^2	metre^2	9.290304×10^{-2}	m^2
Density	pound (mass) per in^3	kilogram per metre^3	2.767990×10^4	kg/m^3
	pound (mass) per ft^3	kilogram per metre^3	1.601846×10^1	kg/m^3
	gram per cm^3	kilogram per metre^3	1.000000×10^3	kg/m^3
Energy	foot-pound (force)	joule	1.355818×10^0	J
Force	pound (force)	newton	4.448222×10^0	N
	kilogram (force)	newton	9.806650×10^0	N
Stress intensity factor	(kilopound$_f$, per in^2) (in$^{1/2}$)	pascal (metre)$^{1/2}$	1.098843×10^6	Pa(m)$^{1/2}$
Length	foot	metre	3.048000×10^{-1}	m
	inch	metre	2.540000×10^{-2}	m
Mass	pound (mass)	kilogram	4.535924×10^{-1}	kg
	kilogram (mass)	kilogram	1.000000×10^0	kg
Specific heat	BTU per (1b$_m$F)	joule per kilogram-kelvin	4.186800×10^3(b)	J/kg K
		joule per kilogram-kelvin	4.184000×10^3(a)	J/kg K
Stress or pressure	pound (force) per inch^2	pascal	6.894757×10^3	Pa

(*Continued*)

(*Continued*)

Property	To convert from imperial or metric units	To SI units	Multiply by	SI notation
	kilopound (force) per inch^2	pascal	6.894757×10^6	Pa
	kilogram (force) per metre^2	pascal	9.806650×10^0	Pa
Thermal conductivity	BTU ft per (hr ft^2 F)	watt per (metre-kelvin)	1.730735×10^0 (b)	W/mK
Thermal diffusivity	foot^2 per hour	metre^2 per second	2.580640×10^{-5}	m^2/sec
Thermal expansion	in per in per F	metre per metre per kelvin	5.555556×10^{-1}	1/K
Torque or bending	inch-pound (force)	newton-metre	1.129848×10^{-1}	Nm
moment	foot-pound (force)	newton-metre	1.355818×10^0	Nm
Velocity	inches per minute	metre per second	4.233333×10^{-4}	m/sec

(a) Thermochemical BTU values.
(b) International steam table BTU values.

References

Abbott, K. H. (1960) Report WAL-TR-161.85.1 PB 161800, Watertown Arsenal Laboratories.

Affouard, J. L. (1984) Doctor's thesis, ISMCM.

Affouard, J. L., Dormeval, R., Stelly, M. *et al.* (1984) In *3rd Conf. Mechanical Properties of Materials at High Rates of Strain* (ed. Harding, J.) pp. 533–554, Inst. of Physics Conf. Series No. 70, Oxford.

Anand, L., Kim, K. H. and Shawki, T. G. (1987) *J. Mech. Phys. Solids*, **35**, 407–429.

Andrew, J. H., Lee, H. and Bourne, L. (1950) *J. Iron Steel Inst.*, **165**, 374–376.

Ansart, J. R. (1986) In *Proc. Int. Conf. on Fragmentation, Form and Flow in Fractured Media*, Nevellan, Israel.

Armstrong, R. W., Coffey, C. S. and Elban, W. L. (1982) *Acta Metall.*, **30**, 2111–2116.

Asay, J. R. and Chhabildas, L. C. (1981) *Shock Waves and High-Strain-Rate Phenomena in Metals* (eds. Meyers, M. A. and Murr, L. E.) pp. 417–431, Plenum Press, New York.

Ashby, M. F. (1981) *Prog. Mat. Sci.*, Bruce Chalmers Anniv. Vol., pp. 1–25.

Atkins, A. G. (1981) *Phil. Mag.*, **43**, 627–641.

Backman, M. E. (1969) US Naval Weapons Center, NWC TP4853, pp. 111–128.

Backman, M. E. and Finnegan, S. A. (1973) *Metallurgical Effects at High Strain Rates* (eds Rohde, R. W., Butcher, B. M., Holland, J. R. and Karnes, C. H.) pp. 531–543, Plenum Press, New York.

Backman, M. E., Finnegan, S. A., Schulz, J. C. *et al.* (1986) *Metallurgical Applications of Shock Wave and High-Strain-Rate Phenomena* (eds Murr, L. E., Staudhammer, K. P. and Meyers, M. A.) pp. 675–688, Marcel Dekker Inc., New York.

Backman, M. E. and Goldsmith, W. (1978) *Int. J. Engng. Sci.*, **16**, 1–100.

Backofen, W. A. (1972) *Deformation Processing*, Addison-Wesley Publ. Co., Menlo Park.

Bai, Y. (1981) *Shock Waves and High-Strain-Rate Phenomena in Metals* (eds Meyers, M. A. and Murr, L. E.) pp. 277–284, Plenum Press, New York.

Bai, Y. (1982) *J. Mech. Phys. Solids*, **30**, 195–207.

Bai, Y. (1989) In *4th Oxford Int. Conf. Mechanical Properties of Materials at High Rates of Strain* (ed. Harding, J.) pp. 99–110, Inst. of Physics Conf. Series No. 102.

Bai, Y. (1990) *Res. Mechanica*, **31**, 133–203.

Bai, Y., Cheng, C. and Ding, Y. (1987) *Res. Mechanica*, **22**, 313–324.

Bai, Y., Cheng, C. and Yu, S. (1986) *Acta Mechanica Sinica*, **2**, 1–7, also presented at *16th ICTAM*, Denmark, 1984.

Bai, Y. and Johnson, W. (1981) In *2nd Meet. Explosive Working of Materials*, pp. 245–250, Novosibirsk, USSR.

Bai, Y. and Johnson, W. (1982) *Metals Technology*, **9**, 182–190.

Bailey, J. A., Hass, S. L. and Shah, M. K. (1972) *Int. J. Mech. Sci.*, **14**, 735–754.

Baker, W. E. and Yew, C. H. (1966) *J. Appl. Mech.*, **33**, 917–923.

Balakrishnan, V. and Bottani, C. E. (1986) *Mechanical Properties and Behaviour of Solids: Plastic Instabilities*, World Scientific, Singapore.

Balendra, R. and Travis, F. W. (1970) *Int. J. Mach. Tool Design Res.*, **10**, 249–271.

Barker, L. M. (1968) *Behavior of Dense Media under High Dynamic Pressures*, p. 483, Gordon and Breech, New York.

Basinski, Z. S. (1957) *Proc. Royal Soc. London*, **A240**, 229–242.

Batra, R. C. (1987a) *Int. J. Plasticity*, **3**, 75–89.

Batra, R. C. (1987b) *Int. J. Solids and Structures*, **23**, 1435–1446.

Batra, R. C. (2012) 'Analysis of adiabatic shear bands by numerical methods'. In *Adiabatic Shear Localization: Frontiers and Advances* (eds. Dodd, B. and Bai, Y.) pp. 173–214, Elsevier, London.

Bedford, A. J., Wingrove, A. L. and Thompson, K. R. L. (1974) *J. Aust. Inst. Metals*, **19**, 61–73.

Beetle, J. C., Rinnovatore, J. V. and Corrie, J. D. (1971) *Scanning Electron Microscopy*, pp. 137–144, Chicago.

Bell, J. F. (1968) *The Physics of Large Deformation of Crystalline Solids*, Springer-Verlag, Berlin.

Bell, J. F. (1973) *Mechanics of Solids I, The Experimental Foundations of Solid Mechanics* (ed. Truesdell, C.), Springer-Verlag, Berlin.

Bhambi, A. K. (1979) *Microstructural Science*, **7**, 255–264.

Bitans, K. and Whitton, P. W. (1971) *Proc. Inst. Mech. Engrs.*, **185**, 1149–1158.

Blazynski, T. Z. (1987) *Materials at High Strain Rates* (ed. Blazynski, T. Z.), pp. 71–132, Elsevier Applied Science, Barking.

Bodner, S. R. and Merzer, A. (1978) *J. Eng. Mat. Tech.*, **100**, 388–394.

Bodner, S. R. and Partom, Y. (1975) *J. Appl. Mech.*, **42**, 385–389.

Boothroyd, G. (1963) *Proc. Inst. Mech. Engrs.*, **177**, 789–802.

Boothroyd, G. (1981) *Fundamentals of Metal Machining and Machine Tools*, McGraw-Hill, Kogakusha Ltd., New York.

Bowden, P. B. (1970) *Phil. Mag.*, **22**, 455–462.

Bowden, P. B. and Gurton, O. A. (1949) *Proc. Royal Soc. London*, **A198**, 337.

Bowden, P. B. and Yoffe, A. D. (1958) *Fast Reactions in Solids*, Butterworths, London.

Bridgman, P. W. (1952) *Studies in Large Plastic Flow and Fracture*, McGraw-Hill, New York.

Burns, T. J. (1986) *Metallurgical Applications of Shock-Wave and High-Strain-Rate Phenomena* (eds Murr, L. E., Staudhammer, K. P., Meyers, M. A.) pp. 741–747, Marcel Dekker Inc., New York.

Burns, T. J. and Trucano, T. G. (1982) *Mech. of Matls*, 1, 313–324.

Calvert, N. G. (1955) *Proc. Inst. Mech. Engrs*, 169, 897–911.

Campbell, J. D. (1970) *Dynamic Plasticity of Metals*, course held at Udine, Course No. 46, Springer-Verlag, New York.

Campbell, J. D. (1973) *J. Mat. Sci. Engng*, 12, 3–21.

Campbell, J. D. and Dowling, A. R. (1970) *J. Mech. Phys. Solids*, 18, 43–63.

Campbell, J. D., Eleiche, A. M. and Tsao, M. C. C. (1977) *Fundamental Aspects of Structural Alloy Design* (eds Jaffee, R. I. and Wilcox, B. A.) pp. 545–563, Plenum Press, New York.

Campbell, J. D. and Ferguson, W. G. (1970) *Phil. Mag.*, 21, 63–83.

Campbell, J. D. and Lewis, J. L. (1969) Univ. of Oxford Report 1080/69.

Carpenter, S. H. (1981) *Shock Waves and High-Strain-Rate Phenomena* (eds Meyers, M. A. and Murr, L. E.) pp. 941–959, Plenum Press, New York.

Carrington, W. E. and Gayler, M. L. V. (1948) *Proc. Royal Soc. London, Ser. A*, 194, 323–331.

Chang, T. M. and Swift, H. W. (1950) *J. Inst. Metals*, 78, 119–148.

Chatani, A. and Hosei, A. (1978) *Trans. Japan Soc. Mech. Engrs*, 44-I, 1445–1453.

Cheng, C. M. (1986) *Proc. Int. Symp. on Intense Dynamic Loading and its Effects*, Science Press, Beijing.

Chiem, C. Y., Rouxel, A. and Dormeval, R. (1986) *Proc. Journées de Détonique*, 37, Establishment Technique de Bourges, DGA.

Childs, T. H. C. (2012) *Adiabatic Shearing in Metal Machining*, CIRP Encyclopedia of Production Engineering, Springer, New York.

Childs, T. H. C. and Rowe, G. W. (1973) *Reports on Progress in Physics*, 36, 223–288.

Childs, T. H. C., Maekawa, K., Obikawa, T. and Yamane, Y. (2004) *Metal Machining: Theory and Applications*, Arnold Publishers, London.

Chin, G. Y., Hosford, W. F. and Backofen, W. A. (1964a) *Trans. Met. Soc. AIME*, 230, 437–449.

Chin, G. Y., Hosford, W. F. and Backofen, W. A. (1964b) *Trans. Met. Soc. AIME*, 230, 1043–1048.

Chrisman, T. and Shewmon, P. G. (1979a) *Wear*, 52, 57–70.

Chrisman, T. and Shewmon, P. G. (1979b) *Wear*, 54, 145–155.

Clifton, R. J. (1980) *Material Response to Ultra High Loading Rates*, NRC Report No. 356, US National Material Advisory Board.

Clifton, R. J., Duffy, J., Hartley, K. A. and Shawki, T. G. (1984) *Scripta Metall.*, 18, 443–448.

Clyens, S. and Campbell, J. D. (1974) In *Mechanical Properties of Materials at High Rates of Strain* (ed. Harding, J.) pp. 62–79, Inst. of Physics Conf. Series No. 21, Bristol.

Coddington, E. A. and Levinson, N. (1955) *Theory of Ordinary Differential Equations*, McGraw-Hill, New York.

Coffey, C. S. (1984) In *Mechanical Properties of Materials at High Rates of Strain* (ed. Harding, J.) pp. 519–524, Inst, of Physics Conf. Series 42, Bristol.

Coffey, C. S. and Armstrong, R. W. (1981) *Shock Waves and High-Strain-Rate Phenomena in Metals* (eds Meyers, A. M. and Murr, L. E.) pp. 313–324, Plenum Press, New York.

Coffey, C. S. and Jacobs, S. J. (1981) *J. Appl. Phys.*, **52**, 6991–6993.

Coffey, C. S., Frankel, M. J., Liddiard, T. P. *et al.* (1981) In *Proc. 7th Int. Symp. Deton. ONR*, pp. 970–975.

Considere, A. G. (1885) *Annals des Ponts et Chaussées*, **9**, 574–775.

Costin, L. S., Crisman, E. E., Hawley, R. H. *et al.* (1979) In *Inst. Phys. Conf. Ser. No. 47* (ed. Harding, J.) pp. 90–100.

Crossland, B. and Williams, J. D. (1970) *Met. Rev.*, **15**, 79–100.

Culver, R. S. (1972) *Exp. Mech.*, **12**, 398–405.

Culver, R. S. (1973) *Metallurgical Effects at High Strain Rates* (eds Rohde, R. W., Butcher, B. M., Holland, J. R. *et al.*) pp. 519–530, Plenum Press, New York.

Curran, D. R. (1979) *Computational Model for Armor Penetration*, SRI Report, CA. USA.

Curran, D. R. and Seaman, L. (1985) *J. de Physique C5*, **46**, 395–401.

Curran, D. R., Seaman, L. and Shockey, D. A. (1987) *Physics Reports*, **147**, 253–388.

Dao, K. C. and Shockey, D. A. (1979) *J. Appl. Physics*, **50**, 8244–8246.

Davidge, R. W. (1979) *Mechanical Behaviour of Ceramics*, Cambridge University Press, Cambridge.

Dawidenkow, N. and Mirohibov, I. (1935) *Technical Physics of the USSR*, **2(1)**, 281–291.

Derep, J. L. (1987) *Acta Metall.*, **35**, 1245–1249.

Dieter, G. E. (1986) *Mechanical Metallurgy*, 3rd Edition, McGraw-Hill Co., New York.

Dodd, B. and Atkins, A. G. (1983) *Acta Metall.*, **31**, 9–15.

Dodd, B. and Bai, Y. (1985) *Mat. Sci. Tech.*, **1**, 38–40.

Dodd, B. and Bai, Y. (1987) *Ductile Fracture and Ductility*, Academic Press, Orlando and London.

Dodd, B. and Bai, Y. (1989) *Mat. Sci. Tech.*, **5**, 557–560.

Dodd, B., Stone, R. C. and Bai, Y. (1985) *Res. Mechanica*, **13**, 265–273.

Doraivelu, S. M., Gopinathan, V. and Venkatesh, V. C. (1981) *Shock Waves and High-Strain-Rate Phenomena in Metals* (eds Meyers, M. A. and Murr, L. E.) pp. 363–375, Plenum Press, New York.

Dormeval, R. (1987) *Materials at High Strain Rates* (ed. Blazynski, T. Z.) pp. 47–70, Elsevier Applied Science, Barking.

Dormeval, R. and Ansart, J. P. (1985) *Int. Conf. on Mech. and Phys. Behav. of Matls. under Dynamic Loading (Dymat 85)*, p. 299.

Dormeval, R. and Stelly, M. (1981) *Study of Adiabatic Shear Bands by means of Dynamic Compressive Tests* In *7th Int. Conf. on High Energy Rate Fabrication*, Leeds, 1981.

Douglas, A. S. and Chen, H. Tz. (1985) *Scripta Metall.*, **19**, 1277–1280.

Douglas, A. S., Chen, H. Tz. and Malek-Madini, R. (1989) In *Mechanical Properties of Materials at High Rates of Strain* (ed. Harding, J.) pp. 275–282, Inst. of Physics Conf. Series No. 102, Bristol.

Douglas, J. J. and Dupont, T. (1970) *SIAN J. Numer. Anal.*, **7**, 575–626.

Drew, D. A. and Flaherty, J. E. (1984) *Phase Transformations and Material Instabilities in Solids* (ed. Gurtin, M. E.) pp. 37–60, Academic Press, Orlando.

Drucker, D. C. (1951) In *Proc. 1st U.S. Natl. Congress Appl. Mech.*, pp. 487–491, ASME, New York.

Duffy, J. (1979) In *Mechanical Properties of Materials at High Rates of Strain* (ed. Harding, J.) pp. 1–15, Inst. of Physics Conf., London.

Duffy, J. (1981) In *Proc. Workshop on Shear Localization*, pp. 19–29, Brown Univ. Report MRL-E-127.

Duffy, J. (1984) *Mechanics of Material Behavior* (eds Dvorak, G. J. and Shield, R. T.) pp. 75–86, Elsevier Science Publishers, B.V., Amsterdam.

Duffy, J., Campbell, J. D. and Hawley, R. H. (1971) *J. Appl. Mech.*, **38**, 83–91.

Elban, W. L. and Armstrong, R. W. (1981) In *Proc. 7th Int. Symp. Deton. ONR*, pp. 976–985.

Eleiche, A. M. (1972) AAFML-TR-72-125.

Eleiche, A. M. and Campbell, J. D. (1974) Oxford Univ. Engng. Lab. Report No. 1106/74.

Eleiche, A. M. and Campbell, J. D. (1976a) *Expt. Mech.*, **16**, 281–290.

Eleiche, A. M. and Campbell, J. D. (1976b) AAFML-TR-76-90.

Eleiche, A. M. and Duffy, J. (1973) AAFML-TR-73-105.

Eleiche, A. M. and Duffy, J. (1975) *Int. J. Mech. Sci.*, **17**, 85–95.

Engel, P. A. (1978) *Impact Wear of Materials*, Elsevier Scientific Publishing Co., Amsterdam.

Erlich, D. C., Curran, D. R. and Seaman, L. (1977) Final Report for contract DAAD05-76-C-0762, US Army.

Erlich, D. C., Curran, D. R. and Seaman, L. (1980) SRI Report AMMRC-TR80-3.

Ferguson, W. G., Kumar, A. and Dorn, J. E. (1967) *J. Appl. Phys.*, **38**, 1863–1869.

Farren, W. S. and Taylor, G. I. (1925) *Proc. Royal Soc. London*, **A107**, 422–451.

Field, J. E. and Hutchings, I. M. (1984) In *Mechanical Properties of Materials at High Rates of Strain* (ed. Harding, J.) pp. 349–371, Inst. of Physics Conf. Series No. 70.

Field, J. E., Swallowe, G. M. and Heavens, S. N. (1982) *Proc. Royal Soc. London*, **A382**, 231–244.

Field, J. E., Swallowe, G. M., Pope, P. H. *et al.* (1984) In *Mechanical Properties of Materials at High Rates of Strain* (ed. Harding, J.) pp. 381–388, Inst. of Physics Conf. Series No. 42.

Fressengeas, C. and Molinari, A. (1987) *J. Mech. Phys. Solids*, **35**, 185–211.

Frey, R. B. (1981) In *Proc. 7th Int. Symp. Deton, ONR*, 36–42.

Frost, H. J. and Ashby, M. F. (1982) *Deformation Mechanism Maps*, Pergamon Press, Oxford.

Gandhi, C. and Ashby, M. F. (1979) *Acta Metall.*, **27**, 1565–1602.

Gilman, J. J. (1969) *Micromechanisms of Flow in Solids*, McGraw-Hill, New York.

Giovanola, J. H. (1987) In *Impact 87*, Bremen, Germany.

Giovanola, J. H. (1988a) *Mech. of Materials*, **7**, 59–71.

Giovanola, J. H. (1988b) *Mech. of Materials*, **7**, 73–87.

Goods, S. H. and Brown, L. M. (1979) *Acta Metall.*, **27**, 1–15.

Grady, D. E. (1977) *High Pressure Research: Applications in Geophysics* (eds Manghnani, M. and Akimoto, S.) pp. 389–438, Academic Press, New York.

Grady, D. E. (1980) *J. Geophys. Res.*, **85**, 913–924.

Grady, D. E. and Asay, J. R. (1982) *J. Appl. Phys.*, **53**, 7350–7354.

Grady, D. E., Asay, J. R., Rohde, R. W. *et al.* (1983). In *Materials Behaviour under High Stress and Ultrahigh Loading Rates* (eds Mescall, J. and Weiss, V.), pp. 81–100, 29th Sagamore Army Materials Conference, Plenum Press, New York.

Grady, D. E., Hollenbach, R. E. and Schuler, K. W. (1978) *J. Geophys. Res.*, **83**, 2839–2849.

Grady, D. E. and Kipp, M. E. (1985) In *Dymat 85. Int. Conf. on Mech. and Phys. Behaviour of Matls Under Dynamic Loading*, pp. 291–298, Les Editions de Physique.

Grady, D. E. and Kipp, M. E. (1987) *J. Mech. Phys. Solids*, **35**, 95–118.

Grady, D. E., Murri, W. J. and DeCarli, P. (1974) *EOS Trans. AGU*, **55**, 417.

Grebe, H. E., Pak, H.-R. and Meyers, M. A. (1985) *Met. Trans.*, **16A**, 761–775.

Hammerschmidt, M. and Kreye, H. (1981) *Shock Waves and High Strain-Rate Phenomena in Metals* (eds Meyers, M. A. and Murr, L. E.) pp. 941–959, Plenum Press, New York.

Harding, J. (1987) *Materials at High Strain Rates* (ed. Blazynski, T. Z.) pp. 133–186, Elsevier Applied Science, Barking, Essex.

Harding, J. and Huddart, J. (1979) *Mechanical Properties of Materials at High Rates of Strain* (ed. Harding, J.) pp. 49–61, Inst. Physics Conf. Series No. 47.

Hargreaves, C. R. and Hoegfeldt, J. M. (1976) In *Proc. Int. Conf. Mech. Behav. of Matls. (ICM2)*, pp. 1806–1810, ASM, Metals Park, OH.

Hargreaves, C. R. and Weiner, L. (1974) Report AD/A-006490.

Hartley, K. A. (1986) PhD thesis, Brown University.

Hartley, K. A. and Duffy, J. (1984) In *Mechanical Properties of Materials at High Rates of Strain* (ed. Harding, J.) pp. 21–30, Inst. of Physics Conf. Series, No. 70.

Hartley, K. A.; Duffy, J. and Hawley, R. H. (1987) *J. Mech. Phys. Solids*, **35**, 283–301.

Hartmann, K. H., Kunze, H. D. and Meyer, L. W. (1981) *Shock Waves and High-Strain-Rate Phenomena in Metals* (eds Meyers, M. A. and Murr, L. E.) pp. 325–337, Plenum Press, New York.

Heavens, S. N. and Field, J. E. (1974) *Proc. Royal Soc. London*, **A338**, 77–93.

Hill, R. (1962) *J. Mech. Phys. Solids*, **11**, 1–16.

Hill, R. and Hutchinson, J. W. (1975) *J. Mech. Phys. Solids*, **23**, 239–264.

Hockett, J. E. (1967a) *Appl. Polym. Symposia*, **5**, 205–225.

Hockett, J. E. (1967b) *Trans. AIME, Metall. Soc.*, **239**, 969–976.

Hodierne, F. A. (1963) *J. Inst. Met.*, **91**, 267–273.

Hosford, W. F. and Caddell, R. M. (1983) *Metal Forming-Mechanics and Metallurgy*, Prentice-Hall, Englewood Cliffs, NJ.

Huang, X. L. (1987) Thesis, Institute of Mechanics, Beijing.

Hutchings, I. M. (1983) In *Material Behaviour under High Stress and Ultrahigh Loading Rates* (eds Mescall, J. and Weiss, V.) pp. 161–196, 29th Sagamore Army Materials Conf., Plenum Press, New York.

Hutchinson, J. W. (1984) *Scripta Metall.*, **18**, 421–458.

Irwin, C. J. (1972) DREV, R-652/72, Canada.

Jenner, A., Bai, Y. and Dodd, B. (1981) *J. Strain Anal.*, **16**, 159–164.

Johnson, G. R. (1981) *J. Engng. Matl. and Tech.*, **103**, 201–206.

Johnson, G. R. and Cook, W. H. (1983) In *Proc. 7th Int. Symp. on Ballistics*, pp. 541–547, The Hague.

Johnson, G. R., Hoegfeldt, J. M., Lindholm, U. S. *et al.* (1983a) *J. Engng. Matl. Tech.*, **105**, 42–47.

Johnson, G. R., Hoegfeldt, J. M., Lindholm, U. S. *et al.* (1983b) *J. Engng. Matl. Tech.*, **105**, 48–53.

Johnson, G. R. and Holmquist, T. J. (1988) *J. Appl. Phys.*, **64**, 3901–3910.

Johnson, W. (1972) *Impact Strength of Materials*, Edward Arnold, London.

Johnson, W. (1987) *Int. J. Mech. Sci.*, **29**, 301–310.

Johnson, W., Baraya, G. L. and Slater, R. A. C. (1964) *Int. J. Mech. Sci.*, **6**, 409–414.

Johnson, W. and Mellor, P. B. (1983) *Engineering Plasticity*, Ellis Horwood, Chichester.

Johnson, W. and Slater, R. A. C. (1967) In *Proc. CIRP-ASTME*, pp. 825–851.

Johnson, W. and Travis, F. W. (1968) In *Engineering Plasticity*, (eds Heyman, J. and Leckie, F. A.), pp. 385–400, Cambridge University Press, Cambridge.

Joule, J. P. (1859) *Phil. Trans. Royal Soc.*, London, **149**, 91–131.
    bibitemKivivuori, S. and Sulonen, M. (1978) *Ann. CIRP*, **27**, 141–145.

Klepaczko, J. R. (1968) *Int. J. Mech. Sci.*, **10**, 297–313.

Klepaczko, J. R. (1987) *J. Mech. Working Technology*, **15**, 143–165.

Klepaczko, J. R. (1988) *J. de Physique*, Coll. C3, **49** (DYMAT 1988), 553-560.

Klepaczko, J. R. (1989) In *Mechanical Properties of Materials at High Rates of Strain* (ed. Harding, J.) pp. 283–298, Inst. of Physics Conf. Series No. 102.

Klepaczko, J. R. and Chiem, C. Y. (1986) *J. Mech. Phys. Solids*, **36**, 29–54.

Kobayashi, H. (1987) *Shear Localization and Fracture in Torsion of Metals*, PhD thesis, University of Reading.

Kobayashi, H. and Dodd, B. (1988) *J. Japan Soc. Tech. Plasticity*, **29**, 1152–1158.

Kobayashi, H. and Dodd, B. (1989) *Int. J. Impact Engng.*, **8**, 1–13.

Kobayashi, S. (1970) *Trans. ASME, J. Engng. Ind.*, **92**, 391–399.

Kocks, U. F., Argon, A. and Ashby, M. F. (1975) *Prog. Mat. Sci.*, **19**, 1–288.

Kolsky, H. (1949) *Proc. Phys. Soc. London*, **B62**, 676–700.

Komanduri, R. (1982) *Wear*, **76**, 15–35.

Kondo, K. (1987) *High Pressure Explosive Processing of Ceramics* (eds Graham, R. A. and Sawaoka, A. B.) pp. 277–282, Trans. Tech. Publications, Switzerland.

Korhonen, A. S. (1981) Doctor of Tech. thesis, Helsinki University of Technology.

Kou, S. Q. (1979) *Advances in Mechanics* (in Chinese) pp. 52–59.

Kudo, H. (1980) *Annals of CIRP*, **29**, 469–476.

Kudo, H. and Aoi, K. (1967) *J. Japan Soc. Tech. Plasticity*, **8**, 17–27.

Kudo, H., Sato, K. and Aoi, K. (1968) *Annals of CIRP*, **16**, 309–318.

Kuriyama, S. and Meyers, M. A. (1987) *Macro- and Micro-Mechanisms of High Velocity Deformation and Fracture* (eds Kawata, K. and Shioiri, J.) pp. 203–212, Springer-Verlag, Berlin.

Kwon, Y. W. and Batra, R. C. (1988) *Int. J. Engng. Sci.*, **26**, 1177–1187.

Laible, R. C. (1980) *Ballistic Materials and Penetration Mechanics*, Elsevier Scientific Publishing Co., Amsterdam.

Lamborn, I. R., Bedford, A. S. and Walsh, B. E. (1974) In *Mechanical Properties of Materials at High Rates of Strain* (ed. Harding, J.) pp. 251–261, Inst. of Physics Conf. Series No. 21.

Lange, K. (1985) *Handbook of Metal Forming*, McGraw-Hill, New York.

Lataillade, J. L. (1989) In *Mechanical Properties of Materials at High Rates of Strain* (ed. Harding, J.) pp. 135–148, Inst. of Physics Conf. Series No. 102.

Latham, D. J. (1963) PhD thesis, University of Birmingham.

Lee, P. W. (1972) PhD thesis, Drexel University.

Leech, P. W. (1985) *Met. Trans.*, **16A**, 1900–1903.

Lemaire, J. C. and Backofen, W. A. (1972) *Met. Trans.*, **3**, 477–481.

Lemonds, J. and Needleman, A. (1986a) *Mech. of Matls*, **5**, 339–361.

Lemonds, J. and Needleman, A. (1986b) *Mech. of Matls*, **5**, 363–373.

Le Roy, G., Embury, J. D., Edwards, G. *et al.* (1981) *Acta Metall.*, **29**, 1509–1522.

Le Roy, G. and Ortiz, M. (1989) In *Mechanical Properties of Materials at High Rates of Strain* (ed. Harding, J.) pp. 257–265, Inst. of Physics Conf. Series No. 102.

Lindholm, U. S. (1974) In *Mechanical Properties of Materials at High Rates of Strain* (ed. Harding, J.) pp. 3–21, Inst. of Physics Conf. Series No. 21.

Lindholm, U. S. and Bessey, R. L. (1969) AAFML-TR-69-119.

Lindholm, U. S. and Hargreaves, C. R. (1976) In *2nd Int. Conf. Mech. Beh. of Materials, ICM2*, pp. 1463–1467, Boston.

Lindholm, U. S. and Johnson, G. R. (1983) In *Material Behaviour under High Stress and Ultra-High Loading Rates* (eds Mescall, J. and Weiss, V.) pp. 61–79, 29th Sagamore Army Materials Conf., Plenum Press, New York.

Lindholm, U. S., Nagy, A., Johnson, G. R. *et al.* (1980) *J. Engng. Matls. and Tech.*, **102**, 376–381.

Lipkin, J., Campbell, J. D. and Swearengen, J. C. (1978) *J. Mech. Phys. Solids*, **26**, 251–268.

Litonski, J. (1977) *Bull. L'Academie Polonaise des Sciences*, **25**, 7–14.

Livshshits, B. G. (1956) *Physical Properties of Metals and Alloys*, MASHCHGIZ, Moscow.

Loewen, E. C. and Shaw, M. C. (1954) *Trans. ASME*, **76**, 217–231.

Lu, W. X., Wang, L. L. and Lu, Z. (1986) *Acta Metal.*, **4**, 317.

Lücke, K. and Mecking, H. (1973) *The Inhomogeneity of Plastic Deformation*, pp. 223–250, ASM, OH, USA.

Ludwik, P. (1909) *Elemente der Technologisken Mechanik*, Springer-Verlag, Berlin.

Ludwik, P. (1927) *Z. Vereins deut. Ing.*, **71**, 1532–1538.

MacDonald, R. J., Carlson, R. L. and Lankford, W. T. (1956) *Proc. Am. Soc. Testing of Materials*, **56**, 704–720.

MacGregor, C. W. and Fisher, J. C. (1946) *J. Appl. Mech.*, **13**, A11–A16.

Malvern, L. E. (1951) *J. Appl. Mech.*, **18**, 203–208.

Malvern, L. E. (1969) *Introduction to the Mechanics of a Continuous Medium*, Prentice-Hall, Englewood Cliffs.

Malvern, L. E. (1984) In *Mechanical Properties of Materials at High Rates of Strain*, 1984 (ed. Harding, J.) pp. 1–20, Inst. of Phys. Conf. Series No. 70.

Marchand, A. and Duffy, J. (1987) Brown Univ. Tech. Report. Contract No. N00014-85-K-0597.

Marchand, A. and Duffy, J. (1988) *J. Mech. Phys. Solids*, **36**, 261–283.

Massey, H. R. (1921) *Proc. Manchester Assoc. Engineers*, 21–66.

Me-Bar, Y. and Shechtman, D. (1983) *Mat. Sci. Engng.*, **58**, 181–188.

Merzer, A. M. (1982) *J. Mech. Phys. Solids*, **30**, 323–330.

Mescall, J. and Papirno, R. (1974) *Expt. Mechanics*, **14**, 257–266.

Meyer, L. W. and Manwaring, S. (1986) *Metallurgical Applications of Shock Wave and High-Strain-Rate Phenomena* (eds Murr, L. E., Standhammer, K. P. and Meyers, M. A.) pp. 675–688, Marcel Dekker Inc., New York.

Meyer, L. W. and Pursche, F. (2012) 'Experimental methods'. In *Adiabatic Shear Localization: Frontiers and Advances* (eds. Dodd, B. and Bai, Y.) pp. 21–109, Elsevier, London.

Meyers, M. A. and Pak, H. R. (1986) *Acta Metall.*, **12**, 2493–2499.

Molinari, A. and Clifton, R. J. (1983) *C.R. Acad. Sc. Paris*, **296**, Series II 1–4.

Molinari, A. and Clifton, R. J. (1987) *J. Appl. Mech.*, **54**, 806–812.

Moore, T. D. (1973) *Structural Alloys Handbook*, Mechanical Properties Data Center, Traverse Cig, MI.

Moss, G. L. (1981) *Shock Waves and High-Strain-Rate Phenomena in Metals* (eds Meyers, M. A. and Murr, L. E.) pp. 299–312, Plenum Press, New York.

Moss, G. L. and Pond, R. B. (1975) *Metall. Trans.*, **6A**, 1223–1235.

Murson, D. E. and Lawrence, J. R. (1979) *J. Appl. Phys.*, **50**, 6272–6282.

National Coal Board's Ropeman's Handbook (1982).

Needleman, A. (1989) *J. Appl. Mech.*, **56**, 1–9.

Nicholas, T. (1971) *Expt. Mech.*, **11**, 370–374.

Nicholas, T. and Campbell, J. D. (1972) *Expt. Mech.*, **12**, 441–447.

Nicholas, T. and Lawson, J. E. (1972) *Expt. Mech.*, **20**, 57–64.

Nicolis, G. and Prigogine, I. (1971) *Self-organisation in Non-equilibrium Systems*, John Wiley and Sons, New York.

Nordling, C. and Osterman, J. (1980) *Physics Handbook*, Studentlitterratur, Lund.

Olson, G. B., Mescall, J. F. and Azrin, M. (1981) *Shock Waves and High-Strain-Rate Phenomena in Metals* (eds Meyers, M. A. and Murr, L. E.) pp. 221–247, Plenum Press, New York.

Orowan, E. (1950) BISRA Report MW/F/22/50.

Orowan, E. (1960) *Rock Deformation, Geological Society America Memoir*, **79**, 323.

Osakada, K., Watadani, A. and Sekiguchi, H. (1977) *Bull. J.S.M.E.*, **20**, 1557–1565.

Oxley, P. L. B. (1989) *Mechanics of Machining*, Ellis Horwood, Chichester.

Pak, H. R., Wittman, C. L. and Meyers, M. A. (1986) *Metallurgical Applications of Shock Wave and High-Strain-Rate Phenomena* (eds Murr, L. E., Staudhammer, K. P. and Meyers, M. A.) pp. 749–760, Marcel Dekker Inc., New York.

Pan, J., Saje, M. and Needleman, A. (1983) *Int. J. Fracture*, **21**, 261–278.

Pearson, J. and Finnegan, S. A. (1981) *Shock Waves and High-Strain-Rate Phenomena in Metals* (eds Meyers, M. A. and Murr, L. E.) pp. 113–125, Plenum Press, New York.

Perzyna, P. (1966) *Adv. in Appl. Mech.*, **9**, 243–377.

Pöhlandt, K. (1989) *Materials Testing for the Metal Forming Industry*, Springer-Verlag, Berlin.

Pomey, J. (1966) *Annal.* C1RP, **13**, 93–109.

Recht, R. F. (1964) *J. Appl. Mech.*, **31**, 189–193.

Reichembach, G. S. (1958) *Trans. ASME*, **80**, 525–540.

Rice, J. R. (1977) *Theo. Appl. Mech.* (ed. Koiter, W. T.) pp. 207–220, North Holland, Amsterdam.

Rice, J. R. and Tracey, D. M. (1969) *J. Mech. Phys. Solids*, **17**, 201–217.

Rinehart, J. S. and Pearson, J. (1965) *Behavior of Metals under Impulsive Loads*, Dover.

Rogers, H. C. (1974) *Adiabatic Shearing — A Review*, Drexel Univ. Report for the US Army Research Office.

Rogers, H. C. (1979) *Ann. Rev. Mat. Sci.*, **9**, 283–311.

Rogers, H. C. (1983) In *Material Behavior under High Stress and Ultra-High Loading Rates* (eds. Mescall, J. and Weiss, V.) pp. 101–118, 29th Sagamore Army Materials Conf., Plenum Press, New York.

Rogers, H. C. and Shastry, C. V. (1981) *Shock Waves and High-Strain-Rate Phenomena in Metals* (eds Meyers, M. A. and Murr, L. E.) pp. 285–298, Plenum Press, New York.

Rohde, R. W., Wise, J. L., Byrne, J. G. *et al.* (1984) *Shock Waves in Condensed Matter* (eds Asay, J. R., Graham, R. A. and Staub, G. K.) pp. 407–410, North Holland, New York.

Rosenberg, Z. and Dekel, E. (2012) *Terminal Ballistics*, Springer Verlag, Berlin.

Rosenfcld, A. R. and Hahn, G. T. (1966) *Trans. ASM*, **59**, 962–980.

Ruiz, D. (1991) D Phil thesis, Univ. of Oxford, Dept. of Engng. Science.

Sakui, S., Nakamura, T. and Tsumura, T. (1966) *Materials* (in Japanese), **15**, 247–253.

Samuels, L. E. and Lamborn, I. R. (1978) In *Symp. of Metallography in Failure Analysis* (eds McCall, J. L. and French, P. M.) pp. 167–190, Plenum Press, New York.

Sandusky, H. W., Coffey, C. S. and Liddiard, T. P. (1984) In *Mechanical Properties of Materials at High Rates of Strain* (ed. Harding, J.) pp. 373–380, Inst. Physics Conf. Series 42.

Sargent, P. M. and Ashby, M. F. (1983) Cambridge Univ. Engng. Dept. Report No. CUED/C/MATS/TR.98.

Saxena, A. and Chatfield, D. A. (1976) SAE paper 760209.

Seaman, L. (1983) *Shear Bands*, lectures for presentation at Beijing Inst. Tech.

Seaman, L., Curran, D. R. and Shockey, D. A. (1983) In *Material Behavior under High Stress and Ultra-High Loading Rates* (eds Mescall, J. and Weiss, V.) pp. 295–309, 29th Sagamore Army Materials Conf., Plenum Press, New York.

Seaman, L. and Dein, J. L. (1982) *Non-Linear Deformation Waves.*

Seaman, L. and Shockey, D. A. (1975) Final Report to contract DAAG 46-72-C-0182 AMMRC, Watertown, MA.

Seaman, L., Shockey, D. A., Curran, D. R. *et al.* (1975) Final Report for SRI Contract N00178-74-C-0450.

Seeger, A. (1955) *Phil. Mag.*, **46**, 1194–1217.

Sek, W. (1988) *J. de Physique*, **49**, C3, 371–377.

Semiatin, S. L. and Jonas, J. J. (1984) *Formability and Workability of Metals*, ASM Metals Park, OH.

Semiatin, S. L. and Lahoti, G. D. (1982) *Met. Trans.*, **13A**, 275–282.

Semiatin, S. L. and Lahoti, G. D. (1983) *Met. Trans.*, **14A**, 105–115.

Semiatin, S. L. and Rao, S. B. (1983) *Matl. Sci. and Engng.*, **61**, 185–192.

Semiatin, S. L., Staker, M. R. and Jonas, J. J. (1984) *Acta Metall.*, **32**, 1347–1354.

Senseny, P. E., Duffy, J. and Hawley, R. H. (1978) *J. Appl. Mech.*, **45**, 60–66.

Shaw, M. C. (1984) *Metal Cutting Principles*, Clarendon Press, Oxford.

Shawki, T. G., Clifton, R. J. and Majda, G. (1983) Brown Univ. Tech. Report: ARO DAAG29-81-K-0121/3.

Shirakashi, T. and Usui, E. (1970) *Bull. Japan Soc. of Prec. Engrs.*, **14**, 91.

Shockey, D. A. (1985) Poulter Lab. Tech. Report 004-85, SRI Int. CA. USA.

Shockey, D. A. (1986) In *Metallurgical Applications of Shock Wave and High-Strain-Rate Phenomena* (eds. Murr, L. E., Staudhammer, K. P. and Meyers, M. A.) pp. 633–659, Marcel Dekker Inc., New York.

Shockey, D. A., Curran, D. R. and Decarli, P. S. (1975) *J. Appl. Phys.*, **46**, 3766–3775.

Shockey, D. A. and Erlich, D. C. (1981). *Shock Waves and High-Strain-Rate Phenomena in Metals* (eds. Meyers, M. A. and Murr, L. E.) pp. 249–261, Plenum Press, New York.

Slater, R. A. C. (1965) *Proc. Manchester Assoc. Engrs.*, **5**, 1–45.

Slater, R. A. C. (1977) *Engineering Plasticity, Theory and Application to Metal Forming Processes*, Macmillan Ltd, London.

Sokolovsky, V. V. (1948) *Priskl. Mat. Mekh.*, **12**, 261–281.

Staker, M. R. (1980) *Scripta Metall.*, **14**, 677–680.

Staker, M. R. (1981) *Acta Metall.*, **29**, 683–689.

Stelly, M. and Dormeval, R. (1986) *Metallurgical Applications of Shock Wave and High-Strain-Rate Phenomena* (eds Murr, L. E., Staudhammer, K. P. and Meyers, M. A.) pp. 607–632, Marcel Dekker Inc., New York.

Stelly, M., Legrand, J. and Dormeval, R. (1981) *Shock Waves and High-Strain-Rate Phenomena in Metals* (eds Meyers, M. A. and Murr, L. E.) pp. 113–125, Plenum Press, New York.

Stevenson, M. G. and Campbell, J. D. (1974) Univ. Oxford Report No. 1098/74.

Stevenson, M. G. and Oxley, P. L. B. (1970) *Proc. Inst. Mech. Engrs.*, **185**, 55–71.

Stock, T. A. C. and Thompson, K. R. L. (1970) *Met. Trans.*, **1**, 219–224.

Suzuki, S., Hashizume, S., Yabuki, Y., *et al.* (1968) Report Inst. Ind. Sci., Univ. of Tokyo, 18, No. 3.

Swallowe, G. M. and Field, J. E. (1981) In *Proc. 7th Int. Symp. Deton. ONR*, pp. 24–35.

Swallowe, G. M., Field, J. E. and Horn, L. A. (1986) *J. Matl. Sci.*, **21**, 4089–4096.

Swallowe, G. M., Field, J. E. and Walley, S. M. (1984) In *Mechanical Properties of Materials at High Rates of Strain* (ed. Harding, J.) pp. 443–444, Inst. of Physics Conf. Series No. 70.

Swegle, J. W. and Grady, D. E. (1985) *J. Appl. Phys.*, **58**, 692–701.

Swift, H. W. (1952) *J. Mech. Phys. Solids*, **1**, 1–18.

Tanaka, K., Ogawa, K. and Nojima, T. (1978) *High Velocity Deformation of Solids* (eds Kawata, K. and Shiori, J.) pp. 98–107, Springer-Verlag, Berlin.

Tanimura, S. and Duffy, J. (1984) Brown Univ. Report DAAG 29-81-K-0121/4.

Tanner, R. I. and Johnson, W. (1960) *Int. J. Mech. Sci.*, **1**, 28–44.

Taylor, G. I. (1937) *Proc. Royal Soc.* (London), **A165**, 568–592.

Taylor, G. I. and Quinney, H. (1934) *Proc. Royal Soc.* (London), **A143**, 307–326.

Thomason, P. F. (1969) *Int. J. Mech. Sci.*, **11**, 187–198.

Thornton, P. A. and Heiser, F. A. (1971) *Met. Trans.*, **2**, 1496–1499.

Tian, L. Q. and Bai, Y. L. (1985) Report of the Chinese Institute of Mechanics.

Timothy, S. P. (1987) *Acta Metall.*, **35**, 301–306.

Timothy, S. P. and Hutchings, I. M. (1981) *Proc. 7th Int. Conf. on High Energy Rate Fabrication* (ed. Blazynski, T. Z.).

Timothy, S. P. and Hutchings, I. M. (1984a) *High Energy Rate Fabrication* (eds Berman, I. and Schroeder, I. J. W.) pp. 31–40, ASME, New York.

Timothy, S. P. and Hutchings, I. M. (1984b) In *Mechanical Properties of Materials at High Rates of Strain* (ed. Harding, J.) pp. 397–404, Inst. of Physics Conf. Series 70.

Timothy, S. P. and Hutchings, I. M. (1985) *Acta Metall.*, **33**, 667–676.

Timothy, S. P. and Hutchings, I. M. (1989) In *Mechanical Properties of Materials at High Rates of Strain* (ed. Harding, J.) pp. 127–134, Inst. of Physics Conf. Series 102.

Trent, E. M. (1941) *J. Iron Steel Inst.*, **143**, 401–419.

Trent, E. M. and Wright, P. K. (2000) *Metal Cutting* 4th edition, Butterworth–Heinemann, Woburn.

Tresca, H. (1878) *Proc. Inst. Mech. Engrs.*, **30**, 301–345.

Tsao, M. C. C. and Campbell, J. D. (1973) Oxford Univ. Engng. Lab. Report No. 1055/73.

Tsubouchi, M. and Kudo, H. (1968) *J. Japan Soc. Tech. Plasticity*, **9**, 332–344.

Turley, D. M., Doyle, E. D. and Lamalingham, S. (1982) *Matl. Sci. Engng.*, **55**, 45–48.

Turner, A. (1988a) CEC Report no. EUR 11181EN.

Turner, A. (1988b) *Future Requirements of the Wire Industry*, BSC-BISPA conference.

Turner, A., Betteridge, C. S. and Avery, J. (1984) EUR 8998 en.

Vinh, T., Afzali, M. and Roche, A. (1979) In *3rd Int. Conf. Mech. Behav. of Matls. (ICM3)* (eds Miller, K. J. and Smith, R. F.) pp. 633–642, Cambridge.

von Karman, T. (1911) *Z. Vereins deut. Ing.*, **55**, 1749–1757.

von Turkovich, B. and Steinhurst, W. R. (1986) *SME Manufacturing Tech. Rev.*, **1**, pp. 340–347, 14th N.A. Manuf. Res. Conf., NAMR Inst. of SME, Dearborn, MI, USA.

Wada, M., Nakamura, T. and Kinoshita, N. (1978) *Phil. Mag.*, **38**, 167 185.

Walker, T. J. and Shaw, M. C. (1969) In *Proc. 10th Int. Machine Tool Design Res. Conf.* (eds Tobias, S. A. and Koenigsberger, F.) pp. 241–252, Pergamon Press, Oxford.

Walley, S. M. (2012) 'Strain localization in energetic and granular materials'. In *Adiabatic Shear Localization: Frontiers and Advances* (eds. Dodd, B. and Bai, Y.) pp. 267–310, Elsevier, London.

Walley, S. M., Field, J. E., Pope, P. H. *et al.* (1989a) *Phil. Trans. Roy. Soc. London*, **328**, 1–33.

Walley, S. M., Field, J. E., Pope, P. H. *et al.* (1989b) In *Proc. of the 5th one-day meeting of DYMAT at University of Bordeaux.*

Walley, S. M., Balzer, J. E., Proud, W. G. and Field, J. E. (2000) 'Response of thermites to dynamic high pressure', *Proc. Royal Soc. London, Ser. A*, **456**, 1483–1503.

Walley, S. M., Field, J. E. and Greenaway, M. W. (2006) 'Crystal sensitivities of Energetic Materials', *Mater. Sci. Tech. Ser.*, **22**, 402–413.

Walley, S. M., Siviour, C. R., Drodge, D. R. and Williams, D. M. (2010) 'High-rate mechanical properties of energetic materials', *JOM — J Met*, **62**(1), 31–34.

Wang, L. L., Bao, H. S. and Lu, W. X. (1988) *J. de Physique*, **C3**, 207–214.

Wang, M. J., Hu, R. S. and Liu, P. D. (1990) *Science Bulletin*, **35**, 634–636.

Weiner, J. H. (1955) *Trans. ASME*, **77**, 1331–1341.

Wilkins, M. L. (1980) *Ballistic Materials and Penetration Mechanics* (ed. Laible, R. C.) pp. 225–252, Elsevier Science Publishing Co., Amsterdam.

Wingrove, A. L. (1973a) *Met. Trans.*, **4**, 219–224.

Wingrove, A. L. (1973b) *Met. Trans.*, **4**, 1829–1833.

Wingrove, A. L. and Wulf, G. L. (1973) *J. Aust. Inst. Met.*, **18**, 167–172.

Winter, R. E. (1975) *Phil. Mag.*, **31**, 765–773.

Winter, R. E. and Field, J. E. (1975) *Proc. Royal Soc. London*, **A343**, 399–413.

Winter, R. E. and Hutchings, I. M. (1974) *Wear*, **29**, 181–194.

Woodward, R. L. (1984) *Int. J. Impact. Engng.*, **2**, 121–129.

Woodward, R. L. (1990) *High Velocity Impact Dynamics* (ed. Zukas, J. A.) pp. 65–125, Wiley and Sons Ltd, New York.

Woodward, R. L. and Aghan, R. L. (1978) *Metals Forum*, **1**, 180–184.

Woodward, R. L., Baxter, B. J. and Scarlett, N. V. Y. (1984) In *Mechanical Properties of Materials at High Rates of Strain* (ed. Harding, J.) pp. 525–532. Inst, of Physics Conf. Series No. 70, Bristol.

Work, C. E. and Dolan, T. J. (1953) *Proc. Am. Soc. Testing of Materials*, **53**, 611–626.

Wright, P. K. (1971) PhD thesis, University of Birmingham.

Wright, T. W. (1987) *J. Mech. Phys. Solids*, **35**, 269–282.

Wright, T. W. (2002) *The Physics and Mathematics of Adiabatic Shear Bands*, Cambridge University Press, Cambridge.

Wright, T. W. (2012) 'Theory of adiabatic shear bands'. In *Adiabatic Shear Localization: Frontiers and Advances* (eds. Dodd, B. and Bai, Y.) pp. 215–246, Elsevier, London.

Wright, T. W. and Batra, R. C. (1985a) *J. de Physique*, **46**, Coll. C5 suppl. to No. 8, C5 323–330.

Wright, T. W. and Batra, R. C. (1985b) *Int. J. Plasticity*, **1**, 205–212.

Wright, T. W. and Walter, J. W. (1987) *J. Mech. Phys. Solids*, **32**, 119–132.

Wright, T. W. and Walter, J. W. (1989) In *Mechanical Properties of Materials at High Rates of Strain* (ed. Harding, J.) pp. 119–126, Inst. of Physics Conf. Series No. 102, Bristol.

Wu, F. H. and Freund, L. B. (1984) *J. Mech. Phys, Solids*, **32**, 119–132.

Wulf, G. L. (1979) *Intl. J. Mech. Sci.*, **21**, 713–718.

Xi, Y. and Meyers, M. A. (2012) 'Nanostructural and microstructural aspects of shear localization for materials'. In *Adiabatic Shear Localization: Frontiers and Advances* (eds. Dodd, B. and Bai, Y.) pp. 111–171, Elsevier, London.

Xing, D. (1988) Thesis, Institute of Mechanics, Beijing.

Xu, Y. B., Wang, Z. G., Huang, X. L. *et al.* (1989) *Matls. Sci. and Engng.*, **A114**, 81–87.

Zener, C. (1948) *Fracturing of Metals*, pp. 3–31, ASM, OH.

Zener, C. and Hollomon, J. H. (1944) *J. Appl. Phys.*, **15**, 22–32.

Zerilli, F. J. and Armstrong, R. W. (1987) *J. Appl. Phys.*, **61**, 1816–1825.

Zhang, D. X. (1986) In *Proc. Int. Symp. Intense Dynamic Loading and its Effects*, pp. 1038–1043, Science Press, Beijing.

Index

Printed in the United States
By Bookmasters